Particles and the
Universe

From the Ionian School to the
Higgs Boson and Beyond

Particles and the
Universe

From the Ionian School to the
Higgs Boson and Beyond

Stephan Narison

Laboratoire Univers et Particules de Montpellier, CNRS, France

World Scientific

NEW JERSEY • LONDON • SINGAPORE • BEIJING • SHANGHAI • HONG KONG • TAIPEI • CHENNAI

Published by

World Scientific Publishing Co. Pte. Ltd.

5 Toh Tuck Link, Singapore 596224

USA office: 27 Warren Street, Suite 401-402, Hackensack, NJ 07601

UK office: 57 Shelton Street, Covent Garden, London WC2H 9HE

Library of Congress Cataloging-in-Publication Data
Narison, S., author.
   Particles and the universe : from the Ionian school to the Higgs boson and beyond / Stephan Narison (Laboratoire Univers et Particules de Montpellier, CNRS, France).
      pages cm
   Includes bibliographical references and index.
   ISBN 978-9814644686 (hardcover : alk. paper) -- ISBN 978-9814644693 (softcover : alk. paper)
   1. Particles (Nuclear physics)--History. I. Title.
   QC793.16.N37 2015
   530.1--dc23
                         2015007909

British Library Cataloguing-in-Publication Data
A catalogue record for this book is available from the British Library.

In-house Editor: Ng Kah Fee

Typeset by Stallion Press
Email: enquiries@stallionpress.com

To my family

Foreword

This is a two-in-one book where its main body traces qualitatively the evolution of the concept of particles from the Ionian school era to quantum chromodynamics (QCD) and to the Standard Model of quarks and leptons after the discovery of the Higgs boson. It is not only an enumeration of facts but tries to provide, to a large spectrum of readers, a qualitative understanding of the notion of particles and of the historical (cosmological) evolution of the Universe based on the assumption of the Big Bang for the creation of the Universe.

The discussions are completed by the presentation of some particle physics experiments and of the field of astrophysics/cosmology. Sociological and indirect technology impacts of fundamental researches are discussed and the popularization of the previous fields of research in developing countries by taking the example of Madagascar is emphasized.

In the main body of the book, I have avoided (insofar as possible) technical details. Hence, the book will be attractive to a wide spectrum of audience (public, high-school and university students, teachers and researchers) having a minimum of scientific knowledge and of mathematical notions at the high-school level.

Part VIII of the book presented as Appendices is addressed to advanced readers (Masters degree and PhD students, teachers and professional researchers), who can find the essential technical basis for a better understanding of the field of particle (high-energy) physics and for starting research in this field. These compact notes are useful complements to lectures in quantum field and gauge theories.

My hope is that, after reading this book, unexperienced readers will discover the exciting world of particles and of the Universe, with and in which, we are living. High-school and university students will find it useful for their future direction and their vocation in the field of Physics where they can find in Part VIII, useful complements to their course (Masters degree and PhD) in this field.

• Part I is devoted to a general introduction to particle physics (evolution of physical sciences, basic notions and qualitative description of the evolution of the Universe) and to the theoretical foundations such as quantum mechanics and quantum field theories which serve as a basis for particle physics.

• Part II is a discussion on the modern gauge theories of elementary particles (quarks and leptons) at the subnuclear level: quantum electrodynamics (QED) for the electromagnetic force, quantum chromodynamics (QCD) for the strong nuclear interaction and the electroweak Standard Model (SM) that unifies the weak and electromagnetic interactions. These theories are presented from their birth until the latest progress of researches in the field.

• Part III describes the Higgs boson discovery in July 2012 at CERN which is one of the most fundamental discoveries in particle physics in this century. We also discuss its consequence for Model buildings in the attempts to unify the three microscopic electroweak and strong forces such as Grand Unified Theories (GUTS) and the more ambitious Theory of Everything (TOE) such as string theories which attempt to unify gravitation (macroscopic force) with the previous three microscopic forces.

• Part IV presents some modern research experiments in particle physics for exploring the Universe and the heart of matter. This includes the search for neutrinos from cosmic rays and reactor experiments and a detailed presentation of high-energy e^+e^- colliders such as the large electron–positron, LEP at CERN, Geneva, and the proton–proton colliders such as the large hadron collider, LHC at CERN. The birth of the web www page at CERN is also mentioned.

• Part V discusses, for completeness, the subject of Astrophysics/Cosmology which traces the evolution of the creation of the Universe. Though an exciting field at infinitely large scale, the discussion is, however, willingly not developed as the author is not an expert in these fields.

• Part VI discusses the relation of fundamental research to society (sociological and geopolitical aspects) and the (indirect) technological impacts of research in particle physics. We also present the complex organization of a big experimental group like the LHC. We also discuss the awareness of this field by taking the example of the QCD-conference series in Montpellier started in 1985 and the one in Madagascar where HEP was first introduced in a series of HEPMAD international conferences from 2001, alternately with the QCD–Montpellier conferences. These conferences have been complemented by organizing from 2010 popular conferences in various high-schools, technical superior institutes, universities and french cultural association (Alliance Française) and by creating the iHEPMAD research institute for training Masters and PhD students.

• Part VII is an Epilogue devoted to some concluding comments on the main body of the book.

• Part VIII is an Appendix which presents some useful technical notes for advanced readers (undergraduate, Masters degree and PhD students, teachers and researchers) who wish to have a deeper understanding of some concepts on special and general relativity, quantum mechanics, quantum fields and gauge theories or/and to plan to work in this field. These compact notes are useful complements to lectures in these fields.

• Part IX is an Annex which presents some important dates in particle physics and the Nobel Prize winners.

• Part X corresponds to the Index of the book and to a short and certainly incomplete bibliography which is only given as a guide of references. Classic works are mentioned in the book together with the photos of the corresponding authors. Some selected works of the present author are also listed in the Annex.

• Part XI is a short biography of the author.

Outline

Since the antique era, we have been curious to know the origin and the nature of the Universe. Numerous ancient philosophers and scientists have tried to answer these fundamental questions. It is only now, in the 21st Century, that we can provide a partial answer to these questions, as some significant progresses have been accomplished in particle physics and astrophysics, which are two complementary researches in two opposite scale directions.

On the one hand, this progress is due to our ability to explore the core of matter, with powerful accelerators (where the accelerated particle has the velocity near to that of light), which can reveal their infinitely small deep structures.

On the other hand, big telescopes explore the large structure of the Universe, and may reach the time of its origin.

Presently, these (apparent) two opposite (in scale) researches are found to have a common feature as the conditions required for exploring the smallest structure of matter (quarks and leptons) reproduce the periods which followed the Big Bang (see Fig. 2.4), a model which has been proposed for understanding the origin of the Universe.

Only the most essential aspects of astrophysics/cosmology will be discussed briefly in this book for completeness because the author is not an expert in the field (Part V), while the ones of particle physics or high-energy physics (HEP) which studies the properties of (elementary) infinitely small components of matter and the forces that govern our Universe will be discussed in detail.

Researches in particle physics have made considerable progress during the 20th century, where all observed particles and (with the exception of the macroscopic gravitation force) the three microscopic forces — electromagnetic origin of the light, weak force which mediates the radioactivity and strong force responsible of the nuclear force — can be described successfully within the framework of gauge theories essentially based on symmetries.

Additional symmetries between the electromagnetic and weak forces have been found in 1967, leading S.L. Glashow, A. Salam and S. Weinberg (GSW) in Fig. 7.7 to their unification of an electroweak force in the so-called Standard Model of electroweak interaction, while the nuclear force is described by the theory of quarks and gluons named quantum chromodynamics (QCD). These theories have satisfied all existing high-precision experimental tests. However, if these theories possess symmetry, all observed particles should be massless which are contrary to experimental observations. Therefore, at a certain point, the symmetry of the electroweak theory should be broken for explaining why the photon which mediates the electromagnetic force is massless and why the vector bosons which mediate the weak force are massive.

We already know from the work of J. Goldstone and H. Nambu in 1960–1961 (Fig. 5.8) that some theories of broken symmetry already exist but they introduce a lot of massless particles which we do not observe. However, the question about the way in which the electroweak symmetry is broken remains unanswered. Robert Brout, François Englert, Peter Higgs, Gerald Guralnik, Carl Hagen and Tom Kibble (Fig. 7.5) have proposed that in some theories, the massless Nambu–Goldstone bosons disappear after giving masses to the vector bosons and leaving the photon massless. It may be that some new scalar fields break this electroweak symmetry.

In the GSW theory, one has to introduce four scalar fields where three of them have been eaten by the three heavy gauge bosons $W^{\pm}$ and Z (see Chapter 7) by giving them masses, while the remaining one (the so-called Higgs boson) should be seen in experiments if this way of electroweak symmetry breaking is correct. In this way of breaking, all particles get their masses by interacting with the Higgs boson, while it is only the Higgs boson which does not get its mass from the electroweak symmetry breaking. Experimentalists have searched without success for this strange particle for about 30 years. These continuous long-standing efforts have been recompensed by the recent discovery on 4 July 2012, in the proton–proton collider of the LHC (Large Hadron Collider) at CERN (European Center for Nuclear Research) in Geneva of a scalar particle having a mass $M_H = (125.15 \pm 0.24)$ GeV which looks like the Higgs scalar boson. There was an unprecedented agitation at CERN when the two groups ATLAS and CMS of the LHC decided to present their first results where none of the two groups knew the results of the other. Anyone in and outside the auditorium would have been in haste to know the results. On 4 July 2012, we modified the regular program of the QCD–Montpellier conference to watch directly, by video conference, the presentation of the results at the CERN auditorium. There was large satisfaction among the viewers. The presentation was followed by many applauses when Fabiola Gianotti of ATLAS confirmed the results of Joseph Incandela of CMS (Fig. 8.12), but it was also emotional to see the tears of Peter Higgs after this announcement. It was important before announcing the discovery that the two experiments were independent and both results reached the 5σ statistical significance (the result was correct at 99.9999% confidence level) corresponding to a discovery.

The Higgs-like boson discovery is very important (Englert–Higgs, Nobel Prize of Physics, 2013) (Fig. 7.5) as it has completed the missing puzzle of the Standard Model. It also confirms that our description and understanding of nature within gauge theories and their related symmetry up to the electroweak scale where the electromagnetic and weak forces are unified is correct. We are on the right track. This discovery is a good guide for the future towards which we should orient our research works for building theories beyond the Standard Model which attempt to unify the three microscopic forces (Grand Unified Theory: GUT) and to (eventually) include later on the macroscopic gravitational force (Theory of EveryThing: TOE such as string theories).

Hence, it is now the time to give a retrospective of the evolution of our understanding of matter and the notion of elementary constituents from the philosophy on the notion of particles of the Ionian school, 610 BC until the recent discovery of the Higgs boson on 4 July 2012.

This is the aim of this book which concludes with a short discussion on the Big Bang phenomena assumed to be the origin of the Universe and with a short mention of some speculative theories beyond the Standard Model or the Higgs boson.

As mentioned in the Foreword, this book is a two-in-one book addressed both to non-initiated readers including non-physicists (Part I–VI), while the last part is more technical and contains useful notes for advanced readers (undergraduate, Masters degree, PhD students, teachers and professional researchers).

I sincerely hope that my goal in writing this book, by sharing high-level scientific knowledge without complicated technicalities has been achieved!

Contents

xviii   *Particles and the Universe*

Part I
General Introduction

Chapter 1

Evolution of Physical Sciences

1.1 Greek Philosophers

♣ Ionian School[1]

It is in Miletus, city of Ionia Greek colony in Asia minor, as a long line of philosophers, called Ionians because of their origin, began to think about the very nature of things, on what lies behind their appearances, i.e. on their principle. For Thales (630–570 BC), the original principle is water, for Anaximenes (580–520 BC) air, for Heraclitus (550–480 BC) the fire. In fact, these doctrines of the genesis of the world are other than a secularization of myths cosmogonic and these depersonalized elements take the place of ancestral deities: Zeus will be fire, Hades the air, Poseidon the water and Gaia the Earth. The world emerges from conflicts (chaos) between these antagonistic deities under the arbitration of Zeus. The philosopher Anaximander (610–547 BC) (Fig. 1.1) disciple of Thales, Anaximenes and Pythagoras, founder of the Ionian school, differs from these philosophers. For him, the infinite (no limit) is the primitive chaos containing several elements that are combined and processed to form the Universe, while the

---

[1]Throughout the book, subsections will be numbered succesively by ♣, ◇, ♡, ♠ which are symbols of four playing pinochle cards.

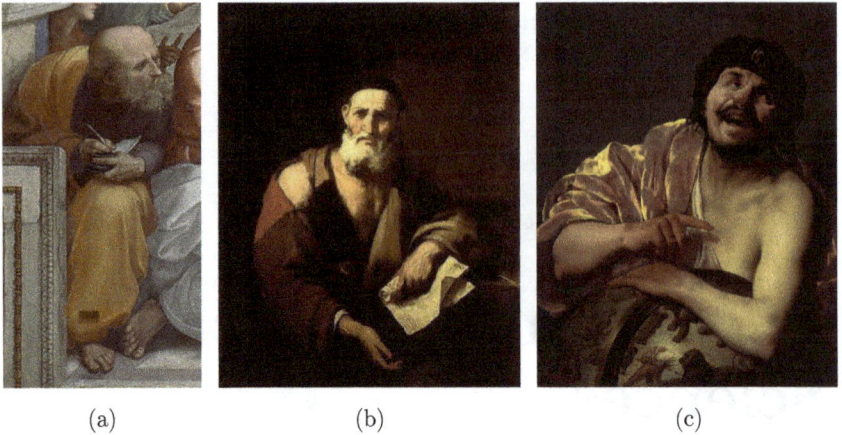

(a)   (b)   (c)

Figure 1.1: (a) Anaximander (from Raphael's *The School of Athens*), (b) Leucippus, (c) Democritus (*Democritus* by Hendrick ter Brugghen).

living beings are the result of the action of the Sun on Earth saturated by humidity and changes in position between the particles of matter. For him, the planets have their own movement and Earth does not fall, because it is at the center of the Universe! All these philosophers accept the uniqueness of the principle which aims to unify the quasi-infinite diversity of phenomena. All refuse the Earth as a principle, unlike the philosopher Xenophanes (560–470 BC) of Ionian descent who emigrated to Sicily and founded the school of Elea. He also considers that it is absurd to represent God in human or in animal form, because God is unique, round like the Universe without a beginning or an end.

◇ The Atomists

One can consider that Leucippus (460–370 BC) of Miletus and his student, Democritus (460–370 BC) of Abdera (Fig. 1.1) are the inventors of *philosophical atomism that is a mechanistic, materialistic theory for designing the world without reference to God and the supernatural.* Atomism of Leucippus was based on the notion of empty and full. It considers that the vacuum is the space of all beings and the full *atoms* (comes from the Greek átomos meaning non-breaking or indivisible) are an infinite number. The Universe would be the consequence of incessant vortical movements of atoms in a vacuum in which collisions will form figures distinguished by their height, weight and speed. Democritus departs from this principle that matter is composed of small particles (atoms) that appears to be continuous. These atoms are full and eternal and move into an empty space. They are not the same and do not have the same form to explain the diversity of nature. The complex bodies are assemblies of simpler body. The four elements (fire, water, air, earth) are composed and result from the vortices created by atoms when they travel in the Universe. Similarly, the soul is formed of atoms round and smooth and is therefore not immortal because it is closely related to the brain.

♡ Aristotelianism

However, these Ionian theories have been refuted by Aristotle (384–322 BC, founder of the "Lycée", the name of his school) (Fig. 1.2). For him, the physicist cannot dissociate the shape and material. Nature is, for human beings, the principle of the more stable movement because it is internal to human beings that he drives. *Time* is continuous according to the anterior and posterior motion, while an *instant* is a limit which determines the before and the after. Also, for him, *the infinite is a circular motion*, Earth is stationary and the Universe is divided in part sublunary (at the bottom of the sphere of the Moon: field of corruption, things which arise and perish subject to the contingency and randomly) and supralunar or heavenly characterized by regularity of movements that occur there! In his treatise

(a)   (b)

Figure 1.2: (a) Aristotle (right) and Plato his master (left) (*The School of Athens*, Raphael), (b) Galileo.

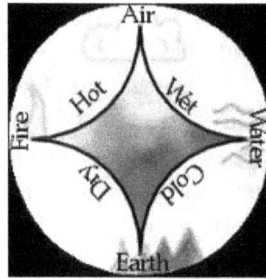

Figure 1.3: Four elements and their phase transition.

of the sky, Aristotle adds to the four elements (air, earth, water and fire: see Fig. 1.3) responsible for changes to the level of the sublunary world by circular generation of elements (cold for earth and water, the wet for water and air, the hot for air and fire), a *fifth element* that he called *first body or ether* and that later became the *quintessence*. This fifth element which does not mix with the other four, allows him to affirm the transcendence of the sky. Aristotle also showed that the heavier body falls faster than a lighter body which was invalidated later by Galileo.

♠ Galilean Revolution

Galileo (1564–1642) (Fig. 1.2) has debunked this Aristotelian doctrine mythical religious of the cosmos (astrology) based on a classification of bodies and human beings (human sciences) and on a stable *geocentrism* where the immobile Earth is the center of the Universe (natural science), proposing the *heliocentric* theory which places the Sun at the center of the Universe and of the solar system (unitary Universe). It thus joins the cosmological system of Copernicus. However, he encountered criticism from the Aristotelian and Roman Catholic Jesuits but, despite this, the Galilean revolution took place. *It is not enough to contemplate the truth (contemplative science). The real must be shown, measured and experienced.* The truth of the Universe is independent of human beings and comes in the form of physical contents of indefinite extension, in which the physical facts are mathematically calculable. In 1603, he experimented from the Bell Tower of a church in Padua, on the fall of bodies to test the hypothesis of Aristotle. It shows that, unlike Aristotle, *two bodies of the same size and same shape drop at the same time regardless of their weight.* It is the air resistance that slows down the fall of a sheet of paper compared to a petanque ball which

(a)   (b)   (c)   (d)

Figure 1.4: (a) R. Descartes, (b) I. Newton, (c) J.-L. Lagrange, (d) L. Euler.

will fall faster. He made other experiences on the straight translation of a
body and their inertia,... and thus laid the foundations of mechanics with
kinematics and dynamics. *The new science involves a new type of scientist,
i.e. a layman of reason and reflection, a man of laboratory, open to collaboration and reflection.* Galileo is considered since 1680 as the founder of the
physics, the first of modern sciences.

♣ Cartesian Thoughts

In 1633, René Descartes (Fig. 1.4) known by his works in mathematics
(Cartesian) and his treatise on the reason (discours de la méthode) postulated that *the Universe* (in which the vacuum does not exist!) *is filled
with animated and swirl particles.* He thus took the idea of the Ionian
school. However, he runs into the mechanics of Galilea where an object
has a straight continuous motion.

1.2 Classical Physics

♣ Rational Mechanics of Newton

Isaac Newton (1643–1727) (Fig. 1.4) continued the work of Galileo on the
mechanics, on the basis of classical or Newtonian mechanics based on three
laws of motion, which is summarized in his 1686 paper "Philosophi Natu-
ralis Principia Mathematica":
• *The 1st law*
It states that in the absence of an external force, any isolated body has
no acceleration, i.e. its speed is constant, or is at rest or has a uniform

rectilinear motion. This law applies in a Galilean or inertial frame which is set experimentally. It reversed the concept of Aristotle which assumes that one must apply a continuous force to keep the body at a constant speed.

• *The 2nd law*

Called fundamental principle of dynamics in translation, it indicates that the acceleration γ of a body of mass m in a Galilean coordinate system is proportional to the external forces F applied to it and is inversely proportional to its mass m: $\gamma = F/m$.

• *The 3rd law*

Called principle of mutual actions, it states that the action is equal to the reaction, i.e. the actions of two bodies, one on the other are equal and opposite directions.

• *Gravitation pull*

To these three laws, can be added the gravitational pull that behaves as the inverse of the square of the distance. In Newtonian mechanics, the energy of a moving body is characterized by its kinetic energy E_c equal to the product of its mass m and the square of its speed v: $E_c = (1/2)mv^2$, and by a potential energy V which is equal to the product of its mass m with the gravitation constant or Newton constant g and the vertical height h where the object falls: $V = mgh$.

◇ Analytical Mechanics of Lagrange

Joseph-Louis Lagrange (1736–1813) and Leonhard Euler (1707–1783) (Fig. 1.4) have improved the Newtonian mechanics giving mathematical formalisms. They introduced *the variational calculus or principle of the least action* (minimum energy to move from point A to point B regardless of the followed road) in 1756 (*Euler–Lagrange equation*: see Appendix D). J.-L. Lagrange published his book on analytical mechanics in 1788 (in this approach, he called *Lagrangian* the difference between the kinetic energy E_c and the potential energy V: $L = E_c - V$. Unlike Newtonian mechanics which treats a point-like object, it considers complex systems and examines the evolution of their degrees of freedom (for instance, in a 1-dimensional space, an object has two degrees of freedom that are the rotation and translation) in an abstract space which has for dimensions the degrees of freedom and which is called configuration space. Thus the state of a system with N degrees of freedom can be represented by a point in a configuration space and the laws of motion are derived from variational calculus.

(a)   (b)   (c)   (d)

Figure 1.5: (a) Joseph Fourier, (b) Sadi Carnot, (c) Christiaan Huygens, (d) Denis Papin.

♡ Classical Thermodynamics

It studies heat and temperature dependent phenomena. It is based on two fundamental principles:

• *The 1st principle*

It concerns the *conservation of energy*. In particular, the energy of an isolated system remains constant.

• *The 2nd principle*

It introduces the *concept of irreversibility* (degradation) of a transformation (the energy of a system passes spontaneously from concentrated to diffuse forms) and the *concept of entropy* (entropy of an isolated system is constant or increasing). Joseph Fourier (1768–1830) (Fig. 1.5) gives a mathematical formulation for heat in 1811. Thermodynamics has been a great success as the science of thermal machines, by the work of Sadi Carnot (1796–1832) in 1824 (Fig. 1.5). The association of mechanics and thermodynamics is the origin of the industrial revolution of the 18–19th centuries, thanks to the invention of the steam engine by Huygens (1629–1695) and Papin (1647–1712) in 1673 (Fig. 1.5).

♠ Classical Electromagnetism and Maxwell's Equations

• *Nature of light*

Since Euclid, one has always asked about the nature of light. In Newton's Corpuscular Theory (1672), light consists of grains; while for Huygens (1678) it is a wave which propagates spherically in a medium that is the ether. Based on his observations of the satellite of Jupiter where the

(a)   (b)   (c)   (d)

Figure 1.6: (a) H. Fizeau, (b) L. Foucault, (c) E. Malus, (d) A. Fersnel.

light emission is not instantaneous, Huygens shows that its speed is greater than $c = 243\,000$ km/s, which will be confirmed in the laboratory from Armand Hippolyte Louis Fizeau (1819–1896) measurement in 1849 and Jean-Bernard Léon Foucault (1819–1868) (Fig. 1.6) in 1862. This value is compatible with the current speed value $c = 299\,792\,458$ m/s. However, by comparing the speed of light in air and in water, Foucault has favoured the *wave nature of light*, which allowed him to explain the diffraction, interference and polarization observed by Augustin Jean Fresnel (1788–1827) in 1818, Etienne Louis Malus (1775–1812) (Fig. 1.6) in the period 1807–1810 and Thomas Young (1773–1829) (Fig. 1.21) in 1801.

• *The electromagnetism theory*

It begins with the observation by Hans Christian Oersted (1777–1851) and André-Marie Ampère (1775–1836) in 1820 and by that of Michael Faraday (1791–1867) (Fig. 1.7) in 1831 that *the electric current induces a magnetic field and vice versa*. Faraday also introduces the concept of the electric field developed from Coulomb's law (Fig. 1.7) (attraction or repulsion of the motionless electrical charges by an electrostatic force proportional to the product of the charges Q and Q' and inversely proportional to the square of the distance separating them r, $F \propto QQ'/r^2$). For several charges, the total force will be the vector sum (a vector has three coordinates in space) of each force. The electric field which is exerted on the charge Q is defined as the total force divided by this charge and is therefore also a vector. Electric and magnetic fields can be viewed by force-oriented lines (Fig. 1.8) that are tangent to the vector field defined at each point of the 2-dimensional plan. We also see in Fig. 1.8, the analogy between the two fields: the negative

Figure 1.7: (a) C. Coulomb, (b) H.C. Oersted, (c) M. Faraday, (d) A.M. Ampere.

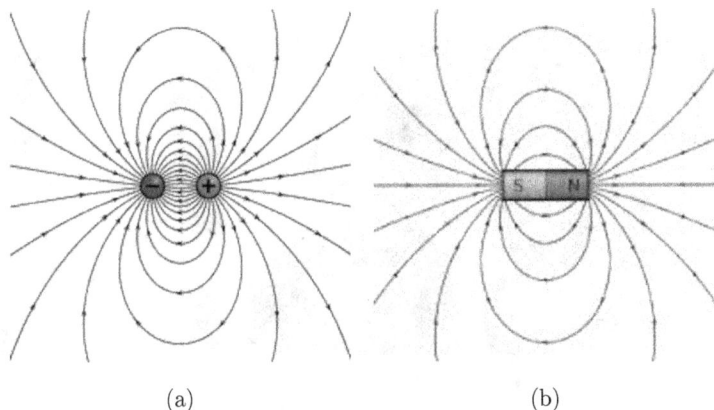

Figure 1.8: Visualization of the tangential electric field to the lines of force between two static charges of opposite signs (dipole, which is attractive) (left) and the magnetic field produced by a magnet (right).

charge corresponds to the south pole of the magnet and the positive charge to the north pole.

• *Maxwell's equations*

The magnetic effects of electricity led James Clerck Maxwell (1831–1879) (Fig. 1.10) to propose in 1860 *a theory unifying electric and magnetic forces* which is formulated by the famous four differential equations of Maxwell (see Appendix D). They describe the behavior of the electromagnetic field and its interaction with matter. These equations predict the existence of a wave associated with oscillations in the electromagnetic field which moves in

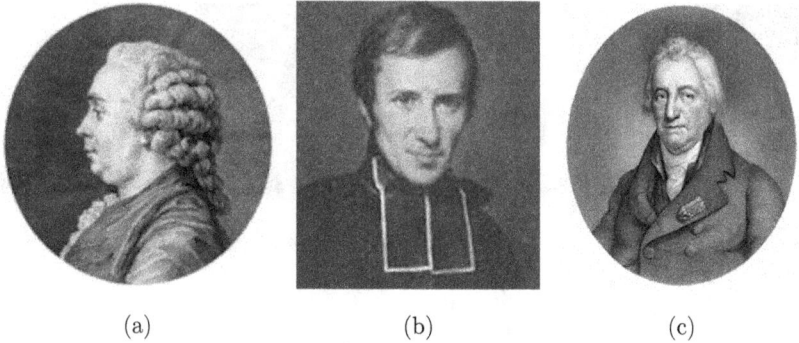

(a)   (b)   (c)

Figure 1.9: (a) P. Trudaine, (b) J.B. Meusnier, (c) C.L. Berthollet.

(a)   (b)   (c)

Figure 1.10: (a) J.C. Maxwell, (b) A.A. Michelson, (c) E.W. Morley.

a vacuum regardless of the oscillating charge. By an experimental measure Maxwell found that the speed of electromagnetic waves is equal to that of light. This led him to the conclusion that *light moves in space as an electromagnetic wave*. However, he thinks that this spread requires the support of waves: the ether which is nothing else than the fifth element of Aristotle. Thus, Albert Abraham Michelson (1852–1931) in 1881, Nobel Prize of 1907, and later Edward Williams Morley (1838–1923) (Fig. 1.10) in 1887, undertook several experiments to seek the ether but without success. This puts in question the existence of ether and the invariance of the speed of light. These issues will be resolved by the Theory of Relativity by Albert Einstein in 1905 (Fig. 1.17).

(a)   (b)   (c)   (d)

Figure 1.11: (a) A.L. Lavoisier, (b) J. Dalton, (c) A. Avogadro, (d) D. Mendeleev.

1.3   Beginning of Modernity

♣ Chemistry or the End of Alchemy

The ideas of Anaximander and Descartes on the concept of particles start to triumph after the first experiences of Antoine Laurent Lavoisier (1743–1794) (Fig. 1.11) in 1789 on the law of conservation of matter (*nothing is lost, nothing is created, everything is transformed*), and on the presence of oxygen during combustion. These results have been strengthened by the experiment of Philibert Trudaine de Montigny (1733–1777) (Fig. 1.9) in 1776 on the analysis of the air which contains 4/5 nitrogen and 1/5 oxygen showing that air is not an element (in the sense given by Aristotle). With Jean-Baptiste Charles Meusnier (1754–1793) and Claude Louis Berthollet (1748–1822) (Fig. 1.9) in 1785 on the synthesis of water, one also shows that the *water is not an element because it is composed of hydrogen and oxygen.*

In 1808, John Dalton (1766–1844) (Fig. 1.11) revived the theory of Democritus and assumed that the atoms are identical in the same element while the atoms of different elements can recombine to form new elements.

In 1811, Amadeo Avogadro (1776–1856) (Fig. 1.11) postulated that at given temperatures and pressures, equal volumes of gases contain the same number of molecules (*Avogadro's law*). In normal conditions: 0°C temperature and a pressure of 1 atmosphere, a volume of 22.4 liter of gas (molar volume) contains $N = 6.023 \times 10^{23}$ molecules.

In 1869 Dimitri Ivanovich Mendeleev (1834–1907) (Fig. 1.11) proposed the periodic table of chemical elements according to their atomic weight (the choice of the unit of atomic weight remains still open: in 1805, John Dalton took for unit weight, the one of hydrogen, in 1961 the international union of chemistry took 1/12 of carbon-12, in 2002 the French CNRS proposed to take the proton but has not yet been validated). At that time there were 69 elements and Mendeleev provided the existence of new elements: scandium, gallium and germanium. We currently have 103 items ranked by increasing number of the atomic number (number of protons in the nucleus of the atom) on this *Periodic Table* (Fig. 1.12) also called *Mendeleev Table*.

◇ Statistical Thermodynamics

• *Thermodynamics*
It has also evolved through the design of molecules and statistical methods. In 1873, Maxwell (Fig. 1.10) and Ludwig Boltzmann (1844–1906) (Fig. 1.13) formalized the mathematical expression of the kinetic theory of gases using the statistical method. *Maxwell thus establishes a link between the elastic collisions of molecules and the 1st law of thermodynamics.*
• *The pressure*
The pressure of a gas on a wall corresponds to the impact exerted by the particles on this wall. A perfect gas has no collisions between particles and the gas molecule size is negligible. If, on the contrary, one considers the size of the molecules, one may access to the *properties of transport* (viscosity, diffusion, bloodshed and thermal conductivity).
• *The temperature*
It is a measure of the hustle and bustle of the particles. It is equal to their average kinetic energy divided by $3k$ where $k = 1.380\ 648\ 8(13) \times 10^{-23}$ Joule/degree Kelvin is the *Boltzmann constant*.
• *Brownian motion*
The trajectory of the molecules inside the gas can be modeled by the Brownian motion (botanist R. Brown (1773–1858) (Fig. 1.13) in 1827. *It is a mathematical description of the random motion of a big particle immersed in a fluid and which interacts only with small molecules of this fluid.* It follows a highly irregular large particle movement as can be seen in Fig. 1.14. The big particle moves in a straight line at constant speed in between collisions with the small particles or the wall, while during the collisions it is accelerated and the direction of movement is changed.

♡ Discovery of the Electron and the Nucleus

In chemistry and statistical thermodynamics, results which involve atoms/molecules have been strengthened by the discoveries of the electron by Joseph Thomson (1856–1940) (Fig. 1.15) in 1897 and of the nucleus

Figure 1.12: Periodic Table of elements in 2013. The atom is labeled thus: above the chemical symbol is the atomic number Z = number of protons; below is the atomic mass $A = Z + N$ where N is the number of neutrons. Isotopes have the same Z but different N. (Courtesy of sciencenotes.org)

(a)   (b)

Figure 1.13: (a) Ludwig Boltzmann, (b) Robert Brown.

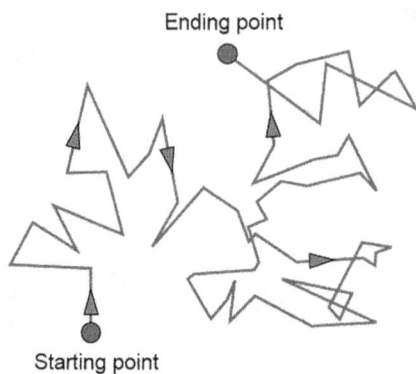

Ending point

Starting point

Figure 1.14: Brownian motion of a particle.

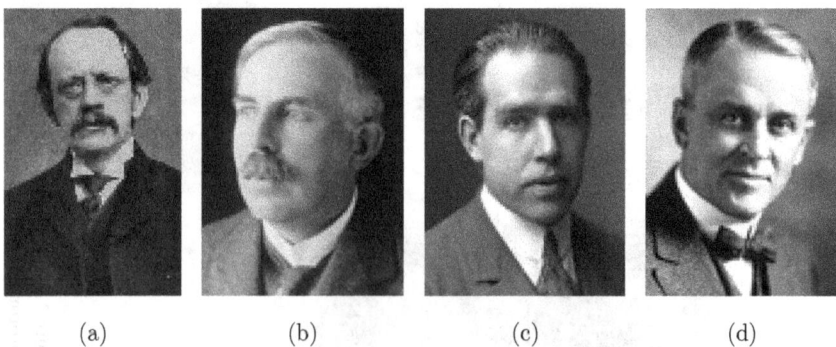

(a)   (b)   (c)   (d)

Figure 1.15: (a) J. Thomson, (b) E. Rutherford, (c) N.H. Bohr, (d) R.A. Millikan.

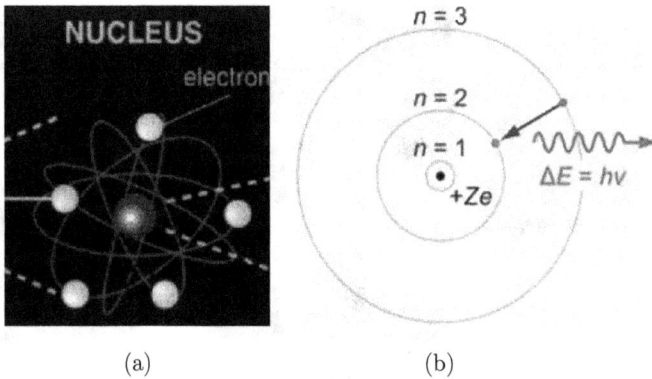

(a)   (b)

Figure 1.16: (a) Atom composed of a nucleus surrounded by electrons. (b) Bohr model showing that electron moves around the nucleus in circular orbits. The jump from an orbit to another leads to an emission of light corresponding to a variation of energy $\Delta E = nh\nu$ where n ($n = 1$ here) is the number of orbits, ν the frequency of light and h the Planck constant.

by Ernest Rutherford (1871–1937) (Fig. 1.15) in 1911 that highlighted the concept of atoms (Fig. 1.16a) (see the atom (Fig. 1.16b) of Niels Henrik David Bohr (1885–1962) (Fig. 1.15)) in 1913 and therefore molecules that are their joints. *For Rutherford, the electrons move around the nucleus like planets around the Sun.* The attractive force between the electron (negative charge) and the nucleus (positive charge) plays the role of the gravitational force for planets, then, the name, *planetary atom model.*

1.4 Quantum Mechanics

While classical mechanics is the theory of matter and energy at the macroscopic level (motion of a body,...), quantum mechanics explains matter and its interactions at the atomic and subatomic scales.

♣ Black Body Experiment and Photoelectric Effect

In the same period of the discovery of electron and nucleus, *Newtonian mechanics cannot explain the emission of electromagnetic radiation from an absorbing black body in thermal equilibrium with its environment.* Its spectrum is only determined by the temperature but not by the body's form or composition. This led Planck (1858–1947) in 1900 (Nobel Prize of 1918, Fig. 1.17) and later, in 1905 A. Einstein [1878–1955, Nobel Prize of

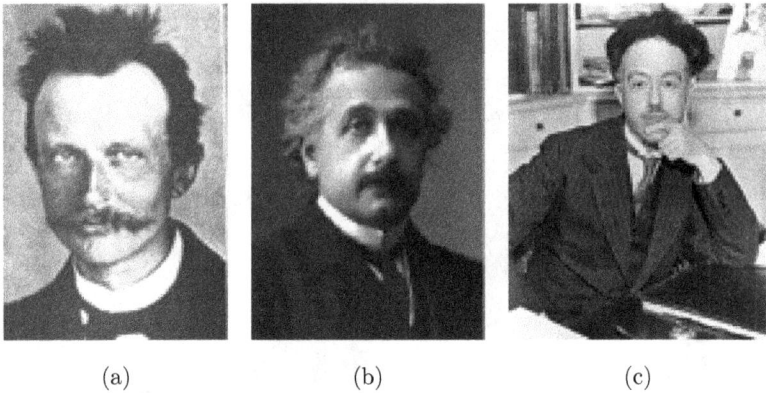

(a)   (b)   (c)

Figure 1.17: (a) Max Planck, (b) Albert Einstein, (c) Louis de Broglie.

1921 (Fig. 1.17) on the *photoelectric effect*[2]] to quantize the energy and therefore to introduce quantum mechanics. In this context, the energy emitted by a blackbody $E = nh\nu$ (*famous Planck formula*) is a non-relativistic (because black body does not move) microscopic phenomenon of the jump of the electron from one orbit to another orbit $n = 1, 2, \ldots$ where $h = 6.626\ 069\ 57(29) \times 10^{-34}$ Joule-second is the *Planck's constant* and ν the frequency of the radiation (light). Passing from one orbit to the other, the electron emits light in the form of quanta or the wave packets which can also be characterized by the wavelength introduced by Louis de Broglie (1892–1987) (Nobel Prize of 1929, Fig. 4.1) in his 1924 thesis: $\lambda = h/p$ where p is the momentum of the state. Robert Millikan (1868–1953, Fig. 1.15) Nobel Prize of 1923, has done experiments to measure the photoelectric effect and Planck's constant. His famous experiment in 1909 with Harvey Fletcher (1884–1981) on the drops of oil charged between two horizontal electrodes of a capacitor plan allowed to measure the electron's charge but also showed that there may not be free particles of fractional charge, which (as we will see later) does not support (at first sight) the existence of elementary quarks inside the nucleons (protons and neutrons).

◇ The Bohr Atom

In 1913, N. Bohr (1885–1962) (Fig. 1.15) provides a model (Fig. 1.16b) that is complementary to that of Rutherford in 1911. He explained that,

---

[2]It is the energy to extract the electron in an atom. It must be greater than its binding energy with the atom. The energy imparted to the extracted electron does not depend on the intensity of the radiation.

in its ground state, the atom is stable and the electrons revolve around the nucleus without emitting radiation (otherwise it overwrites/crashes on the nucleus as predicted by classical mechanics!) in specific quantized orbits. The energy of the electron orbit is well defined through the quantization of its angular momentum. Providing a sufficient amount of energy above a certain threshold, we arrive to the atom's excited states where the electron is passed into higher orbits. To simplify, one always represents the orbits of electrons by circling around the nucleus. However, this is not the reality, because for a quantum theory, the electron has neither a well-defined position nor speed on a circle but must be on an spherical orbital and thus becomes difficult to locate.

♡ Bosons and Fermions: Spin and Statistics

• *Compton effect*
Inspired by the result of Bohr, Einstein (Fig. 1.17) reconsidered Planck's (Fig. 1.17) formula in 1905. He suggested that *an atom in the excited state decays spontaneously but this spontaneous emission of quanta of energy must be accompanied by a stimulated emission due to the presence of quanta of energy equal to that of the quanta to be issued.* This is the Arthur Holly Compton (1892–1962) (Nobel Prize of 1927) (Fig. 1.19) effect, discovered in 1923, in which a quantum of energy, called photon by Gilbert Newton Lewis (1875–1926) (Fig. 1.19) in 1926, collides with an electron and transfers its energy and its amount of movement (momentum). This idea was the origin of *LASER* (Light Amplification by Stimulated Emission of Radiation) rays used industrially in the 1950s. It was in 1923, after having translated the article by S. Bose (1894–1974) (Fig. 1.18), that Einstein has finally understood the quantum property of the photon origin of stimulated emission and the success of the formula of Planck.

• *Photon is a boson*
It is a boson that meets *Bose–Einstein statistics*. Its presence stimulates the emission of new photons in the same quantum state.

• *Electrons are fermions*
They tend to mutually exclude because they obey *Fermi–Dirac statistics*: Fermi (1901–1954) (Nobel Prize of 1938) and Dirac (1902–1984) (Nobel Prize of 1933) (Fig. 1.18). This is commonly the principle of exclusion which Wolfgang Pauli (1900–1958) (Nobel Prize of 1945) (Fig. 1.18) has introduced in 1925.

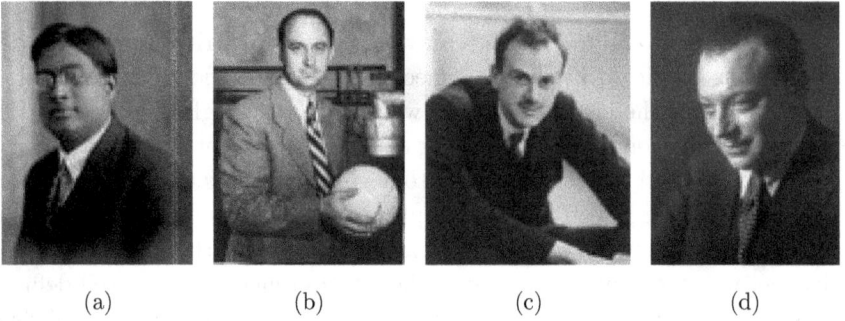

(a)   (b)   (c)   (d)

Figure 1.18: (a) S. Nath Bose, (b) Enrico Fermi, (c) Paul A.M. Dirac, (d) Woflgang Pauli.

(a)   (b)

Figure 1.19: (a) A.H Compton, (b) G.N. Lewis.

• *Spin*

The statistical properties of bosons and fermions are characterized by their intrinsic quantum number called *spin*, s. It is the kinetic or angular momentum of an object rotating on himself that takes account of its mass and shape. This notion of spin puts away even the image of a point particle because a point may not have intrinsic angular momentum. In the case of a fermion, s is a half-integer number $(1/2, 3/2 \ldots)$, and *it is equal to 1/2 for the electron*. Thus projected on an axis, the electron can have spin $+1/2$ or $-1/2$. You can view it in Fig. 1.20a for spin $1/2$ and Fig. 1.20b for spin $-1/2$. Turning on itself the electron creates a magnetic field corresponding to the poles of a magnet South to the North in the case of the spin $1/2$ (direction of arrow) and vice versa in the case of spin $-1/2$. For bosons,

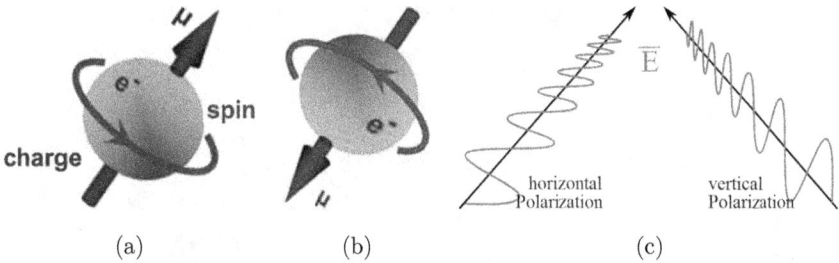

Figure 1.20: (a) electron angular momentum and spin $1/2$ μ turning on himself; (b) electron-spin $-1/2$; (c) horizontal and vertical polarizations of light (photon). E is the electric field.

the spin is an integer ($s = 0, 1, \ldots$). The *Higgs boson* which we shall discuss in Chapter 8 *has a spin 0*. The *photon has spin 1* and is associated to the light. *As it is massless, it can only have 2 degrees of freedom* such that its spin is equal to its helicity. It can be polarized horizontally or vertically (Fig. 1.20c). On the contrary, a *massive spin 1 particle* like a vector boson of the weak interactions or the ρ meson (defined in Chapters 5 and 7) *will have three degrees of freedom*.

♠ Wave–Particle Duality

The ambivalent nature of *the light which is an electromagnetic wave and a quantum of energy (photon)* in quantum mechanics is in agreement with the two (*a priori* divergent) interpretations of Newton and Huygens on its nature. In 1924, Louis de Broglie (1892–1987) extends this duality to matter (Fig. 1.17). Thus, it combines material wave to electron. In general, this *wave–particle duality* is expressed by the relation $E = h\nu$ between energy E of the particle and the frequency ν of the photon. This relation characterizes the spread of the associated wave and/or by the wavelength $\lambda = h/p$ where p is the momentum. Remember that Planck explained the energy emitted by the black body with this formula (previous paragraph).

♣ States and Observables in Quantum Mechanics

Unlike Newtonian classical mechanics, the *body position*, or rather of a particle (microscopic) *is not determined* in quantum mechanics. Also the *energy is quantized*, i.e. that it is no longer continuous like the kinetic energy or potential energy of classical mechanics.

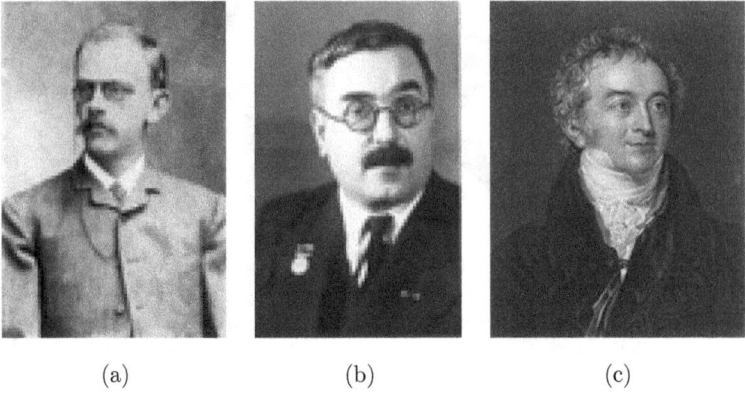

(a)   (b)   (c)

Figure 1.21: (a) David Hilbert, (b) Vladimir Fock, (c) Thomas Young.

• *Wave function*
A *quantum state is characterized by its wave function* denoted by $\psi(x,t)$
where x is the position in space[3] at the time t. In a mathematical language,
the wave function is a vector having three components of the vector space
of David Hilbert (1862–1943) (Fig. 1.21) of complex functions (a generalization of the 3- and 4-dimensions Euclidian space to higher dimensions
proposed in 1900) on which is set a *norm and a scalar product* defined
as $|\psi(x,t)|^2$. The norm of the quantum state corresponds to the modulus
(length) of the wave function and its square is the density of probability of
finding the particle at the point x and at time t. The property of linearity
of the Hilbert space allows the superposition and combination of quantum
states (sum of the vectors that are complex numbers). Vladimir Fock (1898–
1974) (Fig. 1.21) proposed in 1932 a generalization of Hilbert space. In the
Fock space, a state can be formed by n identical particles characterized by
a wave function $\psi_n(x,t)$.

• *Physical observables and operators*
Position x in the space, the momentum, or impulsion or the amount of
motion p, the energy E are examples of *physical observables*. To each observable is associated an *operator* acting on the wave functions of the quantum
state. To the impulsion p is associated an operator $\mathcal{O}$ that characterizes its
variation in the three dimensional space which is its derivative with respect

---

[3]The position x has three components (x_x, x_y, x_z). It is called a vector in this space.

to the space coordinates. These operators also have the property of linearity like vectors but are not necessarily commutative. *Non-commutative means that the $\mathcal{O} \times \mathcal{O}'$ product is not necessarily equal to $\mathcal{O}' \times \mathcal{O}$. This non-commutativity characterizes quantum states.* If the action of an $\mathcal{O}$ operator on a vector $\psi(x,t)$ (for example a wave function) reduces to the product of this vector by a number, it is said that $\psi(x,t)$ *is the eigenvector of the operator $\mathcal{O}$ and has its proper value* λ. In this case one says that the state associated with the eigenvector is invariant under the action of the observable and the measured value of the observable is its proper value which is a real but not a complex number. It is said that the corresponding *operator is self-adjoint.* If, however, the measured state is not in a state associated with an eigenvector, one cannot carryout the measurement with certainty but only in a probabilistic way. This is the case of radioactive β decay or nuclear reaction which is not possible to describe deterministically with equations because one should include observation conditions. *The decay or the reaction will be characterized by a probability amplitude* that can interfere in a constructive or destructive manner because it is a complex number. We shall illustrate it experimentally as follows.

◇ The Experiment of the Young Slits

The experiment of slits made by Thomas Young (1773–1829) (Fig. 1.21) in 1801 uses a monochromatic light source where one stores a plate with two holes between the source and the screen (Fig. 1.22). One then observed on the screen that the two lights interfere to add up (brilliant fringe) or cancel (dark fringe) thus proving their wave nature.

As explained in detail by Richard Philipps Feynman (1918–1988) (Fig. 1.23), Nobel Prize of 1965, in his book Lecture in Physics (1965), interpreted in quantum terms, *Young's slits experiment illustrates the probability amplitude and wave–particle duality.* Replacing the screen by a very sensitive detector, one can register one by one the photon impacts. After a while, these impacts reproduce the interference pattern: accumulation of a number of impacts on the bright fringe and a small number on the dark fringe. This phenomenon cannot be explained by classical mechanics. In quantum mechanics, one can consider that the individual photon is located in a state superimposed once he crossed the slots. From the resulting wave function, one can determine, for each point of the detector, the probability for detection of the quantum. Then, one can demonstrate

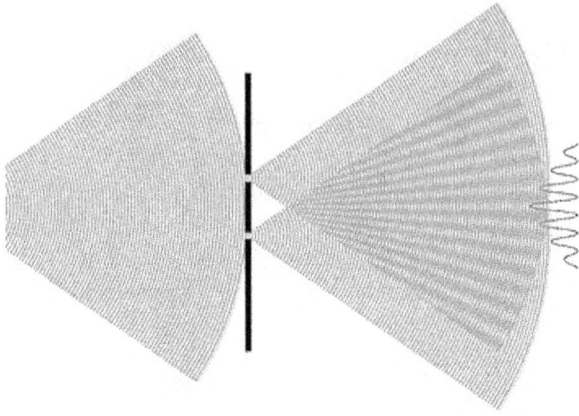

Figure 1.22: The experiment of Young's slits where the bright and dark fringes appear at the interference of two monochromatic lights.

(a)   (b)   (c)

Figure 1.23: (a) Richard Feynman, (b) Erwin Schrödinger, (c) Werner Heisenberg.

that the probability distribution follows from the figure of interference. *This means that the quantum is passed through the two slits and interferes with itself.* However, if one seeks to detect through which slit the photon "really happened", we are at an impasse because the interference disappears. This result calls into question the measurement in quantum mechanics which is a waiver of a deterministic description of individual processes but rather by the statistical method.

♡ Formalism of Quantum Mechanics

• *Heisenberg approach*

In 1925, Werner Karl Heisenberg (1901–1976) (Fig. 1.23) (Nobel Prize of 1932) developed matrix mechanics with Max Born and Pascual Jordan, where he combines the physical properties of the particles as matrices evolving in time and obeying commutation relations. For example, the q *position* and the *amount of movement* (momentum) $p = mv$ of a particle (where $v = dq/dt$ is the speed) are *conjugate variables* obeying the law of non-commutation. This interpretation was later known as the Copenhagen School.

• *Dirac formulation*

For its part, Dirac (Fig. 1.18) offers an alternative formulation of the commutation relations for the defense of his thesis in 1926 by introducing the notations: *bra* $\langle \psi |$ and *ket* $| \phi \rangle$ to emphasize the vector aspect of a quantum state where ψ and ϕ denote the wave-functions which the scalar product is $\langle \psi | \phi \rangle$.

• *Schrödinger approach*

In 1926, Erwin Schrödinger (1887–1961) (Fig. 1.23) (Nobel Prize of 1933) proposes a third approach called mechanical wave based on the wave function which obeys his famous equation. This equation expresses the total energy which is the sum of the kinetic E_c and potential energy V called the *Hamiltonian* $\mathcal{H} = E_c + V$ (compare with the Lagrangian $\mathcal{L} = E_c - V$ which is the energy difference), in terms of the time-variation or derivative operator $(\partial/\partial t)$ of the wave function. Schrödinger shows a little later in 1926 the equivalence between the three approaches.

• *Heisenberg's uncertainty principle*

In 1927, Heisenberg states the principle of uncertainty which expresses that the product of the distance l with energy E or of the quantity of movement q with the position p is a constant: $l \cdot E = q \cdot p \geq \hbar/2$, where $\hbar = h/2\pi$ is the reduced Planck constant, h being the Planck's constant given on page 18. This inequality shows that one cannot simultaneously determine with a good precision the conjugate variables: length l and energy E or the momentum p and position q of a particle. We shall see later that this property is important for building a high-energy physics accelerator. It expresses the fact that *we need large energy if we want to explore the smallest constituents of matter.*

• *Difficulties of quantum mechanics*

However, *quantum mechanics has many conceptual difficulties in conflict with the everyday life* (macroscopic vision ≡ determinism) *like non-locality or non-separability* (entanglement): two particles form a whole plug even

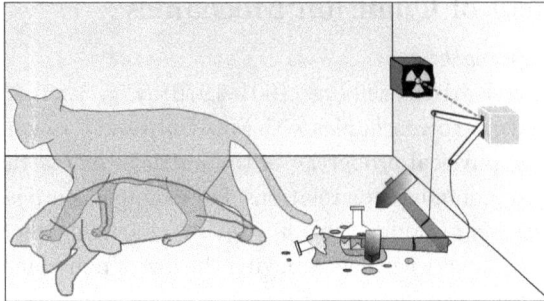

Figure 1.24: Experiment of Schrödinger's cat (courtesy of Dhatfield, wikipedia.org).

(a)   (b)   (c)   (d)

Figure 1.25: (a) Boris Podolsky, (b) Nathan Rosen, (c) Otto Stern, (d) Walter Gerlach.

if they are far apart, like the concept of measure, such as the concept of phenomenon (issue raised and redefined by the Copenhagen school under the direction of Bohr and Heisenberg in 1925–1927), ...

• *The Schrödinger's cat*

To show deficiencies in the course of thoughts of the Copenhagen school, on measurement applied to everday life, Schrödinger in 1935 used the example of the thought experiment of the cat trapped in a box with a vial filled with deadly gases and a radioactive source (Fig. 1.24). If the Geiger counter detects radiation threshold, the bottle will break and the cat dies. *According to the school of Copenhagen, the cat is both dead and alive. However, according to Schrödinger, when you open the box, the cat is either dead or alive.*

• *The Einstein–Podolsky–Rosen (EPR) paradox*

One of the famous questions on the validity of quantum mechanics is that of Einstein, Podolsky and Rosen (1909–1995) (Fig. 1.25) (*EPR paradox*) in 1935 who defended locality (realism) in favor of separability (entanglement). The EPR paradox states that as the position q and impulsion p do not commute, they obey Heisenberg inequality and cannot be determined simultaneously with a good accuracy. However, the difference of position $q_1 - q_2$ commutes with the sum of impulsions or of momenta $p_1 + p_2$, which suggests that one can accurately measure the difference and the sum. This implies that one can measure accurately q_1 or p_1 if q_2 or p_2 is measured. This result contradicts the Heisenberg uncertainty principle.

• *Stern–Gerlach experiment and de Broglie–Böhm theory*

In 1922, Stern–Gerlach [Otto Stern (1888–1969, Nobel Prize of 1943) and Walter Gerlach (1889–1979) (Fig. 1.25)] experiment which consists of sending beams of particles through an homogeneous magnetic field led to the observation of their deflection indicating that electrons and atom particles possess an intrinsic angular momentum analogous to the one of a classically spinning object, but take only certain quantized values. It also shows that only one component of a particle's spin can be measured at one time, meaning that the measurement of the spin along the z-axis destroys information about a particle's spin along the x and y axis. In *de Broglie–Böhm theory* [David Böhm (1917–1992) (Fig. 1.26)], proposed in 1952 where the spin of a particle is represented by operators which do not commute with each other, the results of a spin experiment cannot be analyzed without some knowledge of the experimental setup because the spin is in the wave function of the particle which is in relation to the particular device being used to measure it. De Broglie–Böhm theory is often referred to as a *hidden variable theory, which is causal but not local.*

(a)   (b)   (c)

Figure 1.26: (a) David Böhm (courtesy of wikipedia.org), (b) John Stewart Bell, (c) Alain Aspect (courtesy of CNRS Photothèque/Jerôme Chatin).

• *Bell inequalities and experimental tests*

To explain this paradox, John Stewart Bell (1928–1990) (Fig. 1.26) introduced in 1964, Bell inequalities based on the fact that quantum mechanics is incomplete and that one should bring the notion of *non-locality of hidden variable* or give up the assumption that experiments produce unique results. Bell evokes the principle of *locality* (no influence between 2 distant objects), *causality* (the state of a particle depends only from its initial state) and *realism* (a particle has its own properties conveyed with it). In particular, Bell proved that any local theory with unique results must make empirical predictions satisfying a statistical constraint called *Bell's inequality*. Alain Aspect (1947–) and his group (Fig. 1.26) using EPR-type setup have shown experimentally that Bell inequalities are violated by 42σ which thus confirm the predictions of quantum mechanics without hidden variables. In these *Bell test experiments*, entangled pairs of particles are created; the particles are separated, traveling to remote measuring apparatus. The orientation of the measuring apparatus can be changed while the particles are in flight, demonstrating the apparent non-locality of the effect. The result indicates that a photon can travel as a wave through two places at the same time, even if it can only be observed at one location, as a single particle. The de Broglie–Böhm theory makes the same (empirically correct) predictions for the Bell test experiments as ordinary quantum mechanics because it is manifestly non-local.

1.5   Special Relativity

♣ Lorentz Transformations

Special relativity[4] was developed by Einstein in 1905 in order to describe an object with a large velocity near one of the light c. Under this condition, what we know from classical mechanics should be modified. The kinetic energy of a particle moving with a velocity v is no longer half its mass times v^2. The Maxwell's equations (see Appendix D) which are not invariant by transformations of Galileo or by inertial transformations, are now space-time invariant in the 4-dimensional space of Minkowski space (1864–1909) (Fig. 1.27) under the transformations of Lorentz (1853–1928, Fig. 1.27) (Nobel Prize of 1902) (see more in Appendix A). *Lorentz transformations* preserve the *invariance of the space–time interval*: $s^2 = c^2t^2 - l^2$ where $l^2 = x^2 + y^2 + z^2$ is the square of the vector distance in 3-dimensional

---

[4]General Relativity will be discussed in Chapter 12.

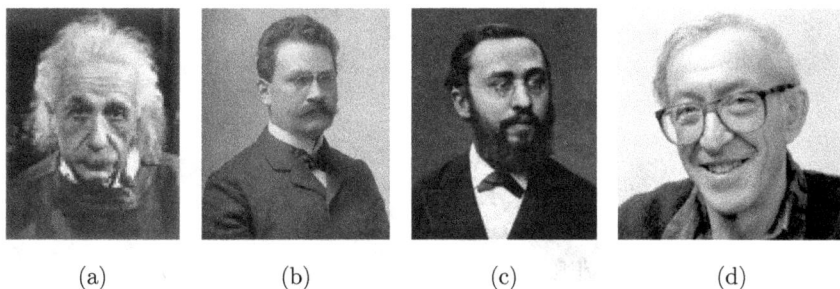

Figure 1.27: (a) A. Einstein, (b) H. Minkowski, (c) H.A Lorentz, (d) I. Shapiro.

space.[5] These transformations also preserve the *invariance of the speed of light* in all inertial frame. However, their significance is not well understood and their consequences on the length dilatation and time contraction remain mysterious.

◇ The Light Speed is the Maximum Speed

To deal with this problem, Einstein accepted the electromagnetic theory which is confirmed experimentally. However, he postulated that *the speed of light c is the maximum speed* (photon almost massless) that one shall not exceed in the Lorentz transformations because if the speed v of the particle is greater than c, it would take the square root of a negative number $\gamma = 1/\sqrt{1 - v^2/c^2}$ called *Lorentz factor* that is not real. In this relativistic case where the velocity of the object is large but remains less than the speed of the light, some other new phenomena occurs.

♡ Innovations of Special Relativity

• *Time contraction and proper time*
We can link the time t in our frame of reference to the time τ in the frame where the clock is at rest by the relationship $d\tau = dt/\gamma$ referring to the invariance of the infinitesimal interval of space–time from a frame change. Looking at the expression of the Lorentz factor γ which is larger than 1 because $v \leq c$, it indicates that the time t is shorter than τ (*time contraction*). The time τ is called *a proper time* of the clock because it is

---

[5]Here, one takes the convention metric $(+, -, -, -)$ for the space–time where $+$ corresponds to the time and the three to the space of coordinates. t is the time and c is the light velocity.

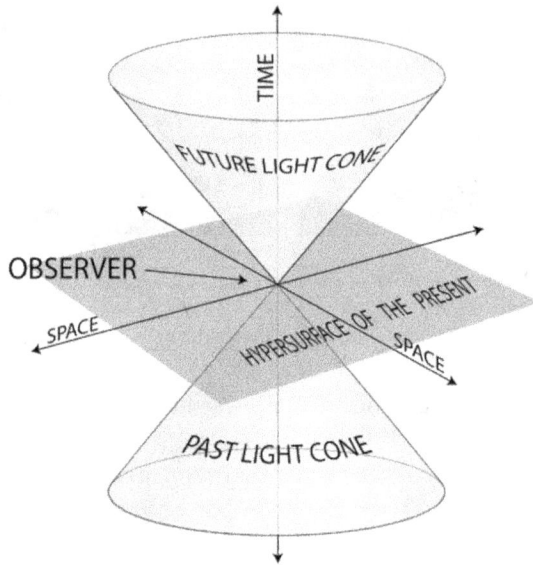

Figure 1.28: Topology of the light cone in 3 spatial dimensions including 2-dimensional space $(x, y) \oplus 1$ time.

connected to the invariant interval of the space–time: $ds^2 = c^2 dt^2 - dl^2 = c^2 \tau^2$ because $dx_\tau{}^2 = dy_\tau{}^2 = dz_\tau{}^2 = 0$ when the clock is at rest. One should note that ds^2 *is not always positive* because its sign depends on the region of space–time (light cone) where the events are measured.

• *Light-cone*

In Fig. 1.28, one shows the *topology of the light cone*. The more one goes towards the past or the future, more the light-cone is large:

— On the surface of the light-cone, the ds^2 space–time interval is zero. One says that the *interval is light-like*. The photon and massless particles reside there.

— The massive particles reside inside the light-cone, where ds^2 is positive because the temporal component dominates over the space component $(c^2 dt^2 \geq dl^2)$. It is said that the *interval is time-like*. In this case, under the causality principle, the photon or a massive particle can go from the past to the future because it passes from a light-cone to another. The temporal component ct (vertical axis) measures the evolution of time at rest. Thus ds^2 enables us to measure the *particle aging*.

— ds^2 is negative outside the light-cone. In this case, the *interval is called space-like*.

• *Energy–momentum and the famous $E = mc^2$ Einstein relation*

The energy E and the three components of momentum p form a four-vector P which is the energy–momentum. In this 4-dimension space–time, its square can be expressed as the difference of the energy square and of its 3-dimension momentum square: $P^2 = E^2/c^2 - p^2 = m^2c^2$. The second equality expresses that its square is invariant (conservation of the energy which is independent of the frame) and can be expressed in terms of the square of the invariant mass m of the particle (more discussions on Lorentz transformations can be found in Appendix A). This relation implies that the energy of a particle moving with a velocity v or having a four-momentum p is: $E = c\sqrt{p^2 + m^2c^2}$. The famous $E = mc^2 \equiv E_0$ formula corresponds to the rest mass of the particle ($p = v = 0$) which indicates that even a particle at rest possesses an energy E_0 (its *proper energy*) which (apparently) is in contradiction with the result of classical mechanics (a particle with zero velocity has zero kinetic energy according to the relation: $E_c = mv^2/2 = p^2/2m$). However, in the limit where the velocity of the particle v is much smaller than the speed of light c ($v \ll c$), one can Taylor expand ($\sqrt{1 + x} \simeq 1 + x/2 + \cdots$ for $x \to 0$) the previous formula in terms of $x \equiv p/mc$ and obtain: $E \approx E_0 + E_c$ where we recover the kinetic energy of classical mechanics: $E_c = p^2/2m = mv^2/2$. This result explicitly demonstrates that classical physics is the low-velocity limit of special relativity.

1.6 Quantum Field Theory

Quantum Field Theory (QFT) combines quantum mechanics with relativity. In this theory, the particles are considered as excited states of quantum fields. In QFT, quantum mechanical interactions between different particles are described by interaction terms between the corresponding fields, which are similar to the ones with charges and electromagnetic fields of Maxwell theories. However, here the fields are not classical but are a superposition of the fields which obey the laws of quantum mechanics.

♣ Harmonic Oscillator and Canonical Quantization

• *Harmonic oscillator in one dimension*

On can illustrate the previous concept using the example of the harmonic oscillator. In a one dimension space–time, an harmonic oscillator corresponds to a system which oscillates vertically at a position x_n due to a return spring and possesses a kinetic energy proportional to the square of its velocity or momentum p_n ($E_c = p_n^2/2m$) and a potential energy proportional to the square of its deviation x_n from its position $V_n(r) = k_n x_n^2/2$; k_n is the spiral spring constant and $\omega_n = \sqrt{k_n/m_n}$ is the oscillator frequency where m_n is the mass of the particle suspended on the spring.

• *Canonical quantization and annihilation and creation operators*

In a canonical quantization, one replaces the momentum p_n and the position x_n (often denoted by q_n) called conjugate variables by operators named *annihilation* a_n and *creation* a_n^+ which respectively decreases and increases by one unit of the Planck quantum energy: $\epsilon_n = h\nu_n = \hbar\omega_n$, where ν_n or ω_n is the frequency of the state n. The annihilation operator a_n annihilates the vacuum (or ground state) $a_n|0\rangle = 0$ and decreases by one unit the quantum number of a state $|n\rangle$: $a_n|n\rangle = \sqrt{n}|n - 1\rangle$. On the contrary, the creation operator a_n^+ increases the quantum number of a state: $a_n^+|n\rangle = \sqrt{n+1}|n + 1\rangle$. Finally, in terms of these operators, the Hamiltonian operator reads: $H = h\nu_n a_n^+ a_n$ where $a_n^+ a_n = (n + 1/2)$, which is in agreement with Planck's result that the energy is quantized (integer $+1/2$ number). It also indicates that the energy of the vacuum $n = 0$, or ground state or fundamental state is not zero (minimum of the potential) but at energy $h\nu/2$ above, which is called *zero-point energy*. It thus implies that the position and momentum of the oscillator in the ground state are not fixed (as they would be in a classical oscillator), but have a small range of variation, in accordance with the Heisenberg uncertainty principle.

• *Generalization to 2-dimensional space*

One can generalize this result to two or more dimensions by considering the example of a cap which oscillates vertically on the surface of the sea where a wave propagates. A field can be considered as a system of infinite harmonic oscillators placed at each point x_n of space and coupled to its neighbours: the water surface would be a model of 2-dimensional field and a quantization of this field would correspond to the quantization of these harmonic oscillators. Working in the momentum space, called Fourier transform of the x-space position, the field can be written as a superposition of $1, \ldots n$ plane-waves characterized by its frequency ω_n and its momentum vector p_n. The coefficients of these plane waves being the annihilation a_n and creation a_n^+ operators. The Hamiltonian of the system is the sum of the Hamiltonians of these independent harmonic oscillators.

(a)   (b)   (c)   (d)

Figure 1.29: (a) O. Klein, (b) H. Weyl, (c) N. Abel, (d) E. Noether.

◇ Klein–Gordon Equation for Scalar Field

In 1926, the Schrödinger equation of quantum mechanics has been generalized by Oskar Klein (1894–1977) and Walter Gordon (1893–1939)
(Fig. 1.29) and also by Vladimir Fock (Fig. 1.21) in the relativistic case. *A
free (without external interaction) scalar particle of spin 0 is described by
the Klein–Gordon equation:* $(p^2 - m^2)\phi = 0$, in the system of natural units
$\hbar = c = 1$, where $\phi(x, t)$ is the scalar field. p^2 and m^2 are respectively the
squares of the four-momentum and of the mass of the associated particle.
Expressing p in terms of its spatial $\vec{p}$ and temporal ϵ (energy) components
$p = (\epsilon, \vec{p})$, the particle energy squared is: $\epsilon^2 = \vec{p}^2 + m^2$.

♡ Second Quantization of Dirac

In taking the square root of the previous expression of energy, one sees that
one can have a positive and a negative energy, while only positive energy
has a physical meaning. However, one cannot reject purely and simply the
negative solution. To remedy this problem, Dirac introduced in 1927 the
second quantization which states that the wave function will be developed
as a sum of annihilation a_n and creation a_n^+ operators of a particle (a_n, a_n^+)
and of its anti-particle (b_n, b_n^+) which is a generalization of the plane wave
discussed before for the harmonic oscillator.

♠ Dirac Equation for Fermions

It is the equation obeyed by a particle of spin 1/2 (fermion) described by
a spinor field $\psi(x, t)$ having four components in 4-dimensional space–time.
Dirac wanted to transform the Schrödinger equation to make it compatible

with the relativity principles such as the invariance by Lorentz transformations. Unlike the Schrödinger equation which describes a scalar field and which corresponds to the square of the momentum and mass, the *Dirac equation* is linear with the mass m and with the momentum p of the fermion particle. It explicitly reads: $(p^\mu \gamma_\mu - m)\psi = 0$ where γ_μ ($\mu = 0, 1, 2, 3$) are 4×4 matrices called Dirac matrices which are a table with 4 lines and 4 columns (their properties are discussed in Appendix E) not to be confused with γ of the Lorentz transformation). The solution of the Dirac equation accurately predicted spectra of the hydrogen atom, while the junk solution of negative energy has led to the concept of anti-particle called positron in the case of electron. The corresponding equation for the positron is $\bar{\psi}(p^\mu \gamma_\mu + m) = 0$ where $\bar{\psi} = \psi^* \gamma_0$ is the conjugate field describing the positron. We will see later (Chapter 7) that in the case of a fermion of zero mass (neutrino), one can also decompose the Dirac spinor field in two *chiral two-component spinors* $\psi_{L,R} = (1 \mp \gamma_5)\psi/2$ called Hermann Weyl (1885–1955) spinors (Fig. 1.29). L and R are respectively the left and right components of the field and γ_5 is a combination of the Dirac matrices, which characterizes chirality and is discussed in Appendix E.

♣ The Quantum Vacuum is not Empty

In reality, the vacuum energy $E_0 = \pm \sum \epsilon$ (additive constant) is infinite because the summation $\sum$ acts on an infinite number of states. In the case of the fermion, and because of the Pauli exclusion principle, it is prohibited to add an electron of negative energy to the vacuum because all possible states are already occupied. However, one can redefine the vacuum as a space having zero particles and zero energy. So everything happens as if the energy states do not exist but are only virtual. Assuming that a negative energy electron is missing in the vacuum, this hole of negative energy represents the positron e^+. If the electron falls into this hole, there is *annihilation of a pair* $e^+ e^-$. If e^+ is ejected from the hole by any interaction and thus acquires a positive energy, there is *creation of a pair* $e^+ e^-$. This immediately emitted pair is reabsorbed by the photon. However, the mass of the emitted pair violates the law of conservation of energy, but this is possible in quantum physics through the Heisenberg uncertainty principle. The pair exists for a very short time Δt and can therefore have great energy because $\Delta E \sim 1/\Delta t$. This phenomenon is called the *vacuum polarization* in analogy to electrostatics where an insulation (which is not in a vacuum) polarizes when placed in an electric field. This heuristic model of Dirac, which seems to apply only to fermions also works for bosons.

Chapter 2

Anatomy of the Universe

2.1 Across the Universe: The Two Infinities

Before addressing our presentations on particle physics, it is important to locate the different scales of our Universe and the corresponding research areas. In Fig. 2.1, we choose as the origin of position 10^0 meter $= 1$ meter, where we can make our observations with the naked eye.

♣ Infinitely Small

When we move to the left to go to the infinitely small:

• *At the scale of thousandth of a meter* (10^{-3}) *and micron* (10^{-6}), we work in *biology and biochemistry* and we study *macromolecules and viruses.*

• *At the scale* of $10^{-(9\sim12)}$ m, we study *chemistry and atomic physics*, and we study respectively the *molecules and atoms.*

For the previous two types of researches, we use ordinary or/and electronic microscopes.

• *At the scale of* $10^{-(12\sim15)}$ m, we study the *nucleus and nucleons*, and *nuclear physics*. The cyclotron is the machine used for this research.

• To research in *particle physics or high energy physics*, let us explore the matter at *a smaller scale of the order of* $10^{-(15\sim18)}$ m. For this purpose, we use linear or/and circular *particle accelerators* and big cameras called

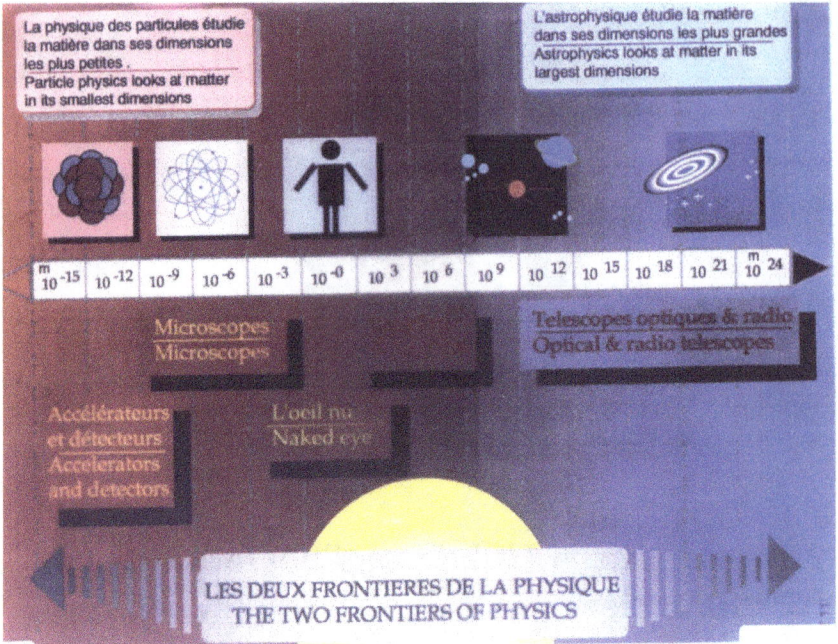

La physique des particules étudie la matière dans ses dimensions les plus petites.
Particle physics looks at matter in its smallest dimensions

L'astrophysique étudie la matière dans ses dimensions les plus grandes
Astrophysics looks at matter in its largest dimensions

| m 10^{-15} | 10^{-12} | 10^{-9} | 10^{-6} | 10^{-3} | 10^{-0} | 10^{3} | 10^{6} | 10^{9} | 10^{12} | 10^{15} | 10^{18} | 10^{21} | m 10^{24} |

Microscopes
Microscopes

Telescopes optiques & radio
Optical & radio telescopes

Accélérateurs et détecteurs
Accelerators and detectors

L'oeil nu
Naked eye

LES DEUX FRONTIERES DE LA PHYSIQUE
THE TWO FRONTIERS OF PHYSICS

Figure 2.1: The different scales of the Universe (courtesy of CERN).

detectors. We shall discuss in detail these points in the next chapters (Part II to IV).

• The earlier areas of research are summarized in Fig. 2.2 where one also shows the *equivalence between the distance and energy* (see Table 3.1) as well as the different devices to perform the corresponding researches.

◇ **Infinitely Large**

Now, we move to the right towards the infinitely large *at a distance of the order of* $10^{21\sim23}$ m where we study *Astrophysics*. For this purpose, one uses telescopes or/and satellites to observe and analyze the *stars, planets and cosmic rays*. We discuss this topic in Part V.

2.2 The Four Forces of Nature

The four forces that govern nature and which we see around us are: *strong, electromagnetic, weak and gravitational forces*. They are classified according to their intensity and range in Fig. 2.3.

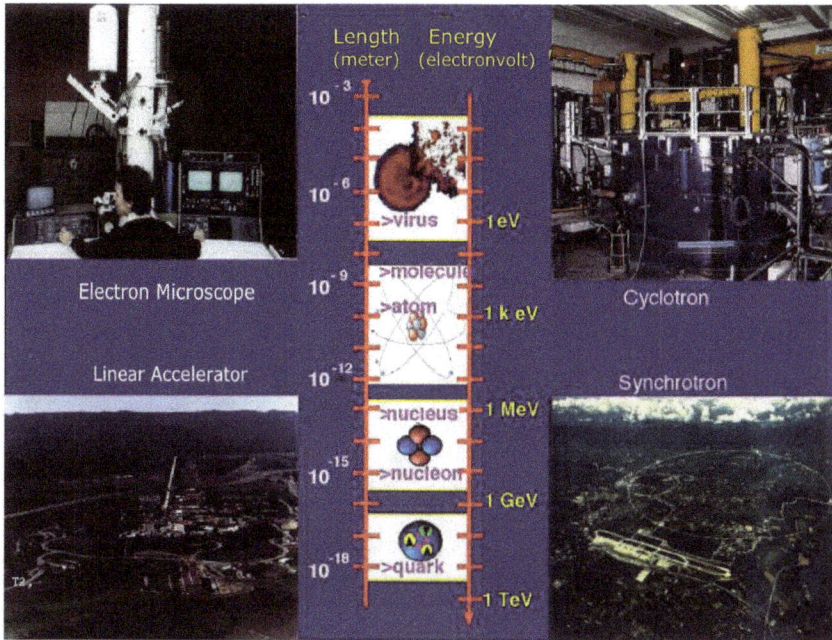

Figure 2.2: Equivalence between distance and energy according to the Heisenberg uncertainty principle. Measuring devices.

♣ Nuclear Force or Strong Force

In admitting that the nucleus is composed of protons of positive charge and neutrons of zero charge, it is clear according to Heisenberg (1933) that the force which stabilizes the nucleus is not the electromagnetic force because the charged protons must repel among themselves, while this force has no effect on neutrons. *A new force very intense and of short range with the size of the nucleus (1 Fermi = 10^{-15} m) must exist.* This force will be called nuclear force or *strong force* and is at *the origin of the binding of the nucleus.* It is 1000 times stronger than the electromagnetic force. According to Yukawa in 1930, this force is mediated by the π meson. We will see later, in the context of *Quantum Chromodynamics* (QCD) (color theory: chromo in Greek) which is the *modern theory of nuclear physics* at the scale of *quarks* (elementary constituents of the proton and the neutron), that this strong interaction is mediated by massless particles called *gluons* which glue quarks together (hence its name) to form the proton and the neutron. *These massless gluons which are bosons of spin 1 (vector bosons) are colored with eight types of colors* .

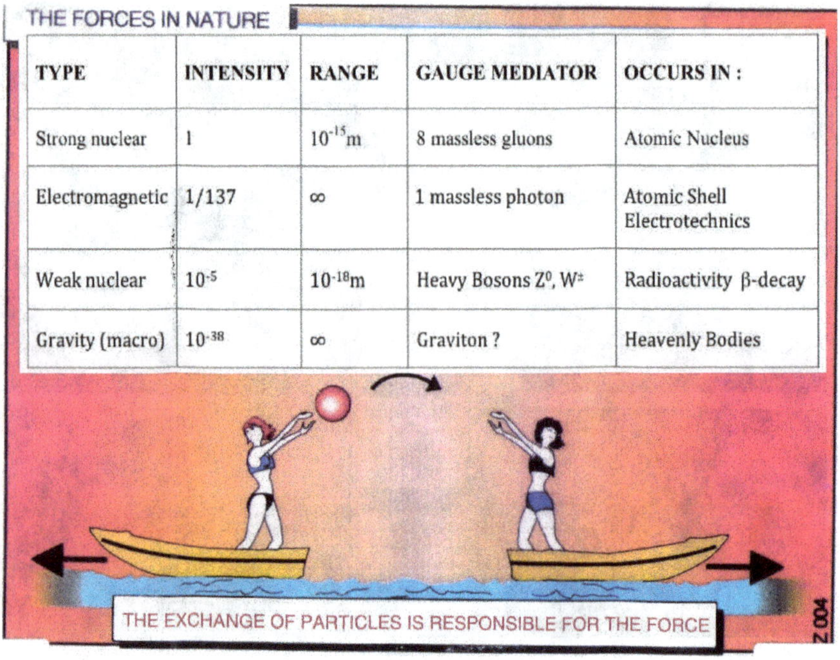

TYPE	INTENSITY	RANGE	GAUGE MEDIATOR	OCCURS IN :
Strong nuclear	1	10^{-15}m	8 massless gluons	Atomic Nucleus
Electromagnetic	1/137	∞	1 massless photon	Atomic Shell Electrotechnics
Weak nuclear	10^{-5}	10^{-18}m	Heavy Bosons Z^0, $W^{\pm}$	Radioactivity β-decay
Gravity (macro)	10^{-38}	∞	Graviton ?	Heavenly Bodies

THE FORCES IN NATURE

THE EXCHANGE OF PARTICLES IS RESPONSIBLE FOR THE FORCE

Figure 2.3: The four forces of nature.

◇ Electromagnetic Force

It is described by Maxwell's equations (see Appendix D) and is mediated by the photon having an almost null mass ($m_{\gamma} \leq 10^{-18}$ eV). *Electromagnetic force is responsible for the light* where one can notice its ambivalent nature: *light is a particle because of the photon, while it is a wave as it propagates.* This force is thousand times weaker than the nuclear force intensity and has an infinite range because the mass of the photon is almost zero. The *photon is the gauge field of Quantum Electrodynamics* (QED) which is a field theory to describe this interaction (see Chapter 4).

♡ Weak Force

Almost at the same time as the nuclear force, another type of interaction is discovered. *It is responsible for β radioactivity from the nucleus.* This force is said to be weak because the decay probability in β radioactivity is small. A time for a neutron to decay by β radioactivity to lose a half of his total is

15 min i.e. at almost the macroscopic level. This force is 100 times weaker than the electromagnetic force. Fermi (Fig. 1.18) has formulated the theory of this interaction in 1933. In the gauge theory of the *Standard Model* at the level of quarks and leptons, *this force is mediated by massive vector bosons* (spin 1) $W^{\pm}$ and Z^0 which are the analogues of the photon but are very heavy having respective masses of 80 and 92 GeV (80 times and 92 times the mass of the proton).

♠ Gravitational Force

This force is known since Galileo and Newton and is responsible for the apple that falls on our head or the fact that the satellite that revolves around the Earth does not deviate from its trajectory. It is supposed to be mediated by the *graviton (spin 2 particle)* which one has not yet seen. Unlike the other three previous microscopic forces, its intensity is very weak, of the order of 10^{-38} times weaker than the strong force. It is said that it is a *macroscopic force* as it acts on objects that we see with the naked eye.

♣ Duality Between Force and Exchange of Particles

One may also notice in Fig. 2.3 these charming ladies who, by exchanging a balloon, split the boat into two parts. This illustrates the fact that the *exchange of particles* (ball) *produces forces and vice versa at the microscopic level.* This is the duality between force and exchange of particles in high-energy physics.

2.3 Understanding the Origin of the Universe

Our main goal is to understand the origin of the Universe and the infinitely small constituents of matter.

♣ For the Believers

In the beginning God/Allah/Zanahary (the Malagasy believe in God) created the Heavens and the Earth (Genesis 1:1, Bible for Christians).

The beginning lies in the thought (the word), and the thought is in God, and thought is God. It is through thought that everything was created (Saint John 1:1-3, Bible for Christians).

So, by faith, you can stop here because it is God who created everything.

◇ What Said Science?

The Universe began (origin of a reference time[1]) by the brutal explosion of a singular point with an infinitely large energy, it was about 13.8 billion years ago. This is the phenomenon of the *Big Bang*, the name is derived from the irony of the physicist Fred Hoyle during a broadcast on the BBC on "The nature of things". Just after the explosion, elementary (infinitely small) particles were produced which then recombined gradually with time to form successively the nucleons (protons and neutrons), the nucleus, atoms, molecules, stars and matters as illustrated in Fig. 2.4. Detailed comments on this figure will be given in Part V.

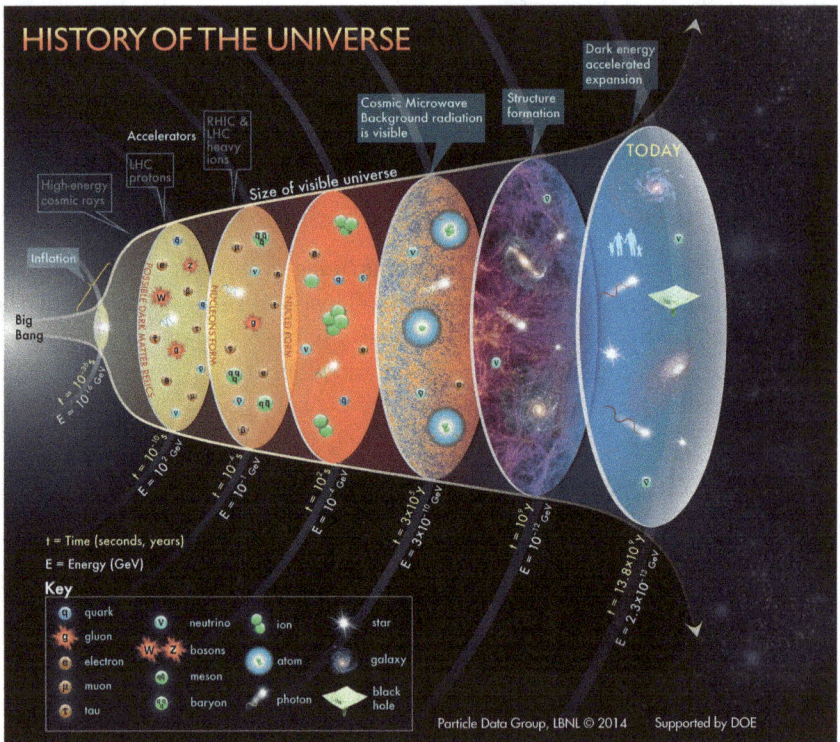

Figure 2.4: The Big Bang hypothesis and the cosmological evolution of the Universe.

---

[1]To avoid a drift metaphysical and many frame discussions, we choose a Cartesian coordinate system with an origin time $t = 0$. When the Pope John Paul II met S. Hawkings, he said: before the Big Bang is ours, after is yours!

High-energy physics (HEP) or particle physics research is to test the Big Bang hypothesis at the level of the laboratory by exploring the tiniest constituent parts of matter using accelerators (mini Big Bang!), reactors or/and cosmic rays. In the HEP accelerators, one succeeds to create the medium just after the Big Bang explosion with deliberated particles.

Astrophysics or cosmology studies the Big Bang phenomena, the large scale of the Universe and the epoch of its creation *via cosmic rays messengers using large telescopes and satellites.*

These two aspects of research though apparently opposite are complementary. They will be discussed in the subsequent chapters.

Chapter 3

Basics of Particle Physics

3.1 Conservation Laws and Symmetries

♣ Noether Current and Charge Conservation

The notion of current and electric charge can be generalized to quantum fields. The Klein–Gordon and Dirac equations for a free field lead to the conservation of the current $J(x,t)$ expressed in terms of this field (see Appendix D) and which is called Noether current after Emmy Noether (1882–1935) (Fig. 1.29). *The current conservation means that its infinitesimal variation in the space–time is zero:* $(\partial_\mu J^\mu = 0)$. To these currents, one can also associate a *charge* Q defined as the integral (sum) over the space of the temporal component J_0 of the current: $Q = \int d^3x J_0(x)$ (the electric charge is associated to the electromagnetic current). In terms of particle numbers, the charge operator has, as an eigenvalue, the difference between the number of particles N and antiparticles $\bar{N}$: $Q = \sum_n (N_n - \bar{N}_n)$ where n characterizes the states (momentum and spin) of a particle.

◇ Difference between Boson and Fermion

It may be noted that the eigenvalue of the charge is the same for the boson and for the fermion. In addition to the spin that we discussed previously, it is only through the Hamiltonian operator $\mathcal{H}$ that one can differentiate them (see more in Appendix B).

♡ Invariances

Remember that observable physical phenomena must not depend on a frame and must have an absolute sense. This is, for example, the case of gravity or Coulomb force between two opposite charges. *Invariance is important in physics because it is related to conservation laws.* These invariances are:

• *Time translation invariance*

It implies the *conservation of energy.*

• *Rotation invariance*

It corresponds to the *conservation of the angular momentum.*

• *Gauge invariance*

It corresponds to the *charge conservation but also requires that the gauge boson mass is zero.* In the case of electromagnetism, the electric charge for an isolated system remains constant as for example the case of a single perfect capacitor. In this case, it is called: *1st species global gauge invariance.* Consider now the case of an isolated object with a constant total charge and assume that the charge varies in each point of the object. This variation of charge must be compensated by an opposite and simultaneous variation in another space. However, an instantaneous transmission of information between two points in space is not possible in special relativity because it cannot exceed the speed of light. In the case of electric charge, this feature, called *2nd species local gauge invariance*, can be explained by Maxwell's equations through the interchangeability of the electric field E and the magnetic field B according to the frame where the observer in special relativity stands: the field E produced by a charge will be perceived as a field B by an observer in uniform motion. For the observer, the charge is moving and therefore corresponds to an electric current which generates a magnetic field. This local invariance will be widespread in gauge theories. In this case, a gauge transformation corresponds to multiplying the gauge field by a phase $e^{i\theta}$ (rotation of angle θ). *The transformation is global if θ does not depend on the position but it is local otherwise.*

♠ Symmetries

They are as important as the invariance in physics. *There are global symmetries which do not change with the position and the local symmetries that vary from one point to another* as for example the change of phase of a particle field. We give below some examples of global symmetries which play an important role in particle physics:

• *The symmetry of space–time*

The symmetry of space–time indicates that *one can swap the positions of observers without modifying the observed phenomena.* In this case, it is said that the space–time symmetries form the *Poincaré group.*

• *The P parity symmetry*

Also called mirror symmetry, *it replaces the spatial coordinates of a vector by their opposite:* $X_i \rightarrow -X_i$ for $i = 1, 2, 3$. One can illustrate it as the image of the object in a mirror. Strictly speaking, one has yet to turn in the same plane as the mirror but this last operation can be ignored, because the laws of physics are invariant by this operation (isotropy of space). The amount of motion is invariant under P but this is not the case of the spin (analogous to the angular momentum of a spinning top) that changes sign. We will see that in opposition with strong and electromagnetic interactions, weak interactions violate parity.

• *CP Symmetry*

A theory has the CP symmetry if it is *both invariant under simultaneous transformation of charge conjugation C* (change of the particle by its antiparticle) *and by parity P* (space inversion). One can illustrate it in Fig. 3.1 where an electron of spin $1/2$ is replaced by a positron of spin $-1/2$. *CP violation explains the matter–antimatter asymmetry in our Universe and is therefore responsible for its stability* (see Chapter 4).

• *Symmetry and CPT theorem*

It is a simultaneous conversion by charge conjugation, parity and time

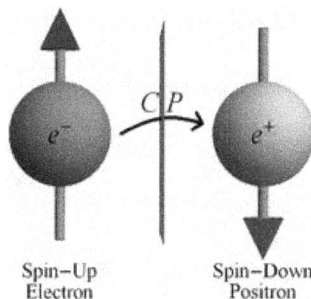

Figure 3.1: Transformation of an electron in positron CP symmetry.

reversal. The CPT theorem shows that any quantum field theory invariant under the Lorentz transformations with an Hermitian conjugate Hamiltonian must have CPT symmetry. This symmetry suggests that a mirror image of our Universe where matter is replaced by antimatter (C symmetry) where objects have moments inverse kinetics (P symmetry) and where time is reversed (T symmetry), would be exactly the same as our Universe. Till now, one has never observed this symmetry violation.

• *The Abelian symmetry group* $U(1)$

Let us consider the set of points on a circle with center O of radius OM_0 and operations that pass a point M_{n-1} to an another M_n of the circle which will be characterized by the angle of rotation θ_n, see Fig. 3.2. One believes that the ensemble forms *the unitary rotation group called* $U(1)$ which conserves the length (radius OM_0) on the circle because it satisfies the following properties of a group: (1) it has a neutral element which is rotation zero: $\theta = 0$; (2) there is a reverse operation of angle $-\theta$ which brings point M_1 back to point M_0; (3) the sum of two rotations is an element of the group: the angle θ_1 to go from M_0 to M_1 and θ_2 from M_1 to M_2 is equal to the rotation angle: $\theta_1 + \theta_2$ of M_0 to M_2 and thus belongs to the group; (4) the operation is associative: to go from M_0 to M_3, one can either first go from M_0 to M_1 (θ_1), then go from M_1 to M_3 ($\theta_2 + \theta_3$); or one can go from M_0 to M_2 ($\theta_1 + \theta_2$), then from M_2 to M_3 (θ_3): $\theta_1 + (\theta_2 + \theta_3) = (\theta_1 + \theta_2) + \theta_3$. One says that the group is Abelian [the name of the mathematician Niels Henrik Abel (1802–1829) (Fig. 1.29)] because the sum of two rotations does not depend on what order the two rotations are made: $\theta_1 + \theta_2 = \theta_2 + \theta_1$.

• *The non-Abelian symmetry group* $SU(N)$

We can also generalize these rotations on a sphere in $N+1$ dimensions. In this case, the operations are no more commutative: $\theta_1 + \theta_2 \neq \theta_2 + \theta_1$. Then, one said that the symmetry group $SU(N)$ (S as special and U as unitary)

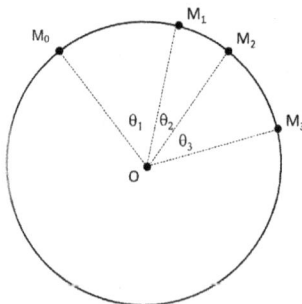

Figure 3.2: Circle of radius OM_i with different angles of rotation θ_i.

is non-abelian. We will see later that these $U(1)$ and $SU(N)$ groups play an important role in the theory of particle physics (for more discussions on the properties of these groups, see Appendix C).

3.2 Observables in Particle Physics

♣ Scattering Matrix or S-Matrix

In particle physics, one is always led either to measure collisions between particles, or to study their decays. Knowing the initial state of a system (a set of free particles), one wants to find the probabilities of the various final states (another set of free particles). If the initial state is represented by $|i\rangle$ and the final state by $\langle f|$ (ket and bra notation of Dirac), this probability can be represented by the sum $\sum_f |f\rangle\langle f|S_{fi}|i\rangle$ on all final states. $S_{fi} \equiv \langle f|S|i\rangle$ is called scattering matrix or S-matrix. Its modulus squared gives the probability of transition from initial $|i\rangle$ towards a determined final state $|f\rangle$. It is unitary: $\sum_n |S_{ni}|^2 = 1$ which indicates that the sum of the transition probabilities of a given initial state to all final states should be equal to 1. We can still explain this relationship by expressing that the state of the system in the absence of interactions between particles remains unchanged so the S-matrix is equal to 1. By explicitly showing the momentum conservation (sum of impulsions P_f of the final states = sum of impulsions P_i of the initial states), one can relate S_{fi} to the *scattering amplitude* T_{fi} defined in Eq. (B.52) of Appendix B. It is also usual to introduce the *"reduced" invariant amplitude* $\mathcal{M}_{fi}$ via the relation in Eq. (B.53). Then the goal is first to evaluate the invariant amplitude $\mathcal{M}$ using Feynman diagram techniques which we shall discuss later on.

♦ Decay Rate or Width of a Particle

It is the probability of a fission of a particle $|i\rangle$ to an arbitrary number of particles $\langle f|$. It is connected to the square modulus $|\mathcal{M}|^2$ of the invariant amplitude as given explicitly in Eq. (B.57). The width is characterized by the half-lifetime of a particle which is the inverse of its decay rate. Like for the cross-section, we also have total and differential decay rates.

♡ Cross-section

It is the apparent surface that must be given to the target particle such that the incident particle beam scatters to give the expected reactions.

Therefore, it is the probability of transition of particles $|i\rangle$ to particles $\langle f|$ and is also proportional to the square modulus $|\mathcal{M}|^2$ of the invariant amplitude (see Eq. (B.58)). Having the dimension of a surface, it is expressed in *barn* (1 barn $= 10^{-24}$ cm^2). The efficiency of an apparatus is measured as the inverse of cross-section per unit of time and is called *luminosity* which is the number of events produced per unit of cross-section and per unit of time. A *total cross-section* corresponds to the sum of all possible partial cross-sections for a given reaction. A *differential cross-section* corresponds to the probability of reactions inside an infinitesimal variation of kinematical variables like energy, angle,...

♠ Polarization of a Particle

It measures the average value of the projection of the spin (intrinsic internal quantum number) projected on a certain axis. This is referred to as *transverse or longitudinal polarization* depending on the type of projection (see Fig. 1.20 in the case of the photon).

3.3   Particle Physics Units of Measurement

♣ The Electronvolt as a Unit of Measurement

When measuring a physical phenomenon, one must always set the units of measurement. We will see that instead of using three units of measurement (electron-volt, kilogram, meter and second), the relationship between these three quantities allow us to use just a single unit. *It is customary to use the electronvolt (eV) in particle or high energy physics.* 1 eV corresponds to the energy that must be provided to carry 1 free electron through a potential difference of 1 volt: **1 eV $= 1.602\,2 \times 10^{-19}$ Joule.**

◇ Equivalence between Mass and Energy

Using the well-known formula $E = mc^2$ and taking as unit $c = 1$ (so-called *natural units*), we obtain the equivalence between mass and energy where: **1 eV $= 1.782 \times 10^{-36}$ kg,** whose multiples are: 1 keV $= 1000 = 10^3$ eV, 1 MeV $= 1\,000\,000 = 10^6$ eV, 1 GeV $= 10^9$ eV and 1 TeV $= 10^{12}$ eV.

♡ Equivalence between Distance and Energy

Quantum mechanics gives us also the *uncertainty principle of Heisenberg* (see Fig. 1.23), which expresses that *the product of the distance with energy is a constant*: $l \times E = \hbar/2$ where: $\hbar = h/2\pi$ is the reduced Planck constant. This expression also expresses that if we want to explore the matter in its infinitely small structure, we will need to provide a lot of energy. As mentioned earlier (Chapter 1), this feature is important for building particle accelerators. Always taking as unit $c = 1$ but also $\hbar = 1$ (so-called natural units), we can deduce the equivalence between the distance and the inverse of energy: 2×10^{-16} **meter** $= 1/\textbf{GeV}$.

♠ Equivalence between Energy and Length with Time

The two previous relations enable us to work with *only eV units instead of kg and meter*, which will simplify our measurement works. Using the natural units $\hbar = 1$, one can also relate the inverse of time to the energy: $1/s \equiv 1\ s^{-1} = \textbf{6.582} \times \textbf{10}^{-22}\ \textbf{MeV}$, while using $c = 1$, one can also relate s^{-1} to the length in meter: $1\ s^{-1} = \textbf{299 792 458 m}$ which means that 1 m is the length of a path traveled by the light in the vacuum during a time interval $1/299\ 792\ 458$ s.

♣ Units and Physical Constants

A complete list of conversion units and physical constants is given in PDG [5]. Among these, and for the purpose of the next sections, we present some of them in Tables 3.1 and 3.2.

Table 3.1 High-energy physics conversion constants and units.

Name	Value
Speed of light c	299 792 458 m s^{-1}
Reduced Planck constant $\hbar \equiv h/2\pi$	1.054 572 66(63)$\times 10^{-34}$ J s = 6.582 122 0(20)$\times 10^{-23}$ MeV s
Units where $\hbar = c = 1$	
Energy	1 eV = 1.602 177 33(49) $\times 10^{-19}$ J
	1 GeV = 10^3 MeV = 10^6 keV = 10^9 eV
Mass	1 eV/c^2 = 1.782 662 70(54) $\times 10^{-36}$ kg
Length	1 GeV^{-1} = 0.197 327 053 fm = 0.197 327... $\times 10^{-13}$ cm
Lifetime	1 GeV^{-1} = 65 821 220 $\times 10^{-25}$ s
Decay rate	1 GeV = (1/65 821 220) $\times 10^{25}$ s^{-1}
Cross-section	1 GeV^{-2} = 0.389 379 66(23) $\times 10^6$ barn

Table 3.2 Some high-energy physical constants.

Parameter	Symbol	Value
Proton mass	m_p	938.272 046(21) MeV/c^2
Neutron mass	m_n	1.293 332 2(4) MeV/c^2
$\pi^\pm$ meson mass	$m_{\pi^\pm}$	139.570 18(35) MeV/c^2
π^0 meson mass	m_{π^0}	4.593 6(5) MeV/c^2
Electron mass	m_e	0.510 999 06(15) MeV/c^2
Electron neutrino mass	m_{ν_e}	≤ 2 eV/c^2
Muon mass	m_μ	105.658 357(5) MeV/c^2
Muon neutrino mass	m_{ν_μ}	≤ 0.19 MeV/c^2
Electron charge	e	1.602 177 33(49)$\times 10^{-19}$ C
Electron radius	$r_e = e^2/4\pi\epsilon_0 m_e c^2$	2.817 940 92(38)$\times 10^{-15}$ m
Electron anomaly	$a_e \equiv \frac{1}{2}(g_e - 2)$	115 965 218 076(27) $\times 10^{-13}$
Muon anomaly	$a_\mu \equiv \frac{1}{2}(g_\mu - 2)$	116 592 091(63) $\times 10^{-11}$
Strong coupling constant	$\alpha_s = g^2/4\pi$	0.325(8) at M_τ
		0.118(3) at M_Z
Fine structure constant	$\alpha = e^2/4\pi\epsilon_0\hbar c$	1/137.035 999 58(52) at m_e
		1/128 at M_Z
Fermi coupling constant	$G_F/(\hbar c)^2$	1.166 39(2)$\times 10^{-5}$ GeV^{-2}
Weak mixing angle	$\sin^2\theta_W$ at M_Z	0.231 5(4)
$W^\pm$ boson mass	M_W	80.33(15) GeV/c^2
Z^0 boson mass	M_Z	91.187(7) GeV/c^2
Higgs boson mass	M_H	126.0(6) GeV/c^2

3.4 Birth of Particle Physics

♣ Nucleus

We have seen that it took 25 centuries for the atomic concept to become a scientific theory, while a few dozen years were enough to discover three subatomic levels of elementarity including the atomic nucleus, nucleus components which are protons and neutrons (nucleons) and later the quarks. These discoveries gave rise to the *particle physics that studies the smallest constituents of matter and the basic forces that govern it*. While the proton was discovered by Rutherford in 1911 by observing that the nucleus is the whole of the mass of the atom, and it must have a positive charge to compensate the negative charge of the electron that revolves around him, the neutron did highlight only in 1932 by James Chadwick (1891–1974) (Nobel Prize of 1935) (Fig. 3.3). The proton and the neutron which have very similar masses of 938.27 MeV and 939.57 MeV are considered to be two states of a particle called *nucleon*.

(a) (b) (c) (d)

Figure 3.3: (a) J. Chadwick, (b) Frédéric and Irène Joliot-Curie, (c) C.D. Anderson, (d) S.H. Neddermeyer.

◇ Neutrino (ν)

Chadwick also noticed in 1914 in his experience on the radioactivity β decay [(emission of an electron or a positron: anti-electron) that the produced particle does not have a well-defined energy as it does not correspond to the difference of mass of the initial and final nucleus]. In 1930, Pauli proposes that a new particle neutral (he called it neutron) of very small mass emitted at the same time as the electron would take the missing energy. After the discovery of the true neutron, Fermi has renamed (with the agreement of Pauli) this particle *neutrino ν (small neutron in Italian)*. The anti-electron neutrino has been discovered many years later in 1956 by Clyde Cowan and Frederick Reines (1918–1998) (Nobel Prize of 1995) (Fig. 11.1) in the inverse β-decay reaction: $\bar{\nu}_e$+proton$\rightarrow e^+$+neutron (electron antineutrino scattering off a proton into a positron and a neutron).[1]

♡ Positron (e^+)

In bombarding nuclei [Aluminium with atomic number $Z = 13$ (number of protons) and atomic mass $A = 26$ (number of protons + neutrons)] with a beam of α particles (consisting of $^4_2$He with 2 protons and 2 neutrons), in 1934, Frédéric (1900–1958) and Irène Joliot-Curie (1897–1956) (Nobel Prize in Chemistry of 1935) (Fig. 3.3), provocate an artificial nuclear transmutation accompanied by radiation which is that of *β inverse radioactivity. It is a proton which transforms into neutron plus a positive electron.* Explicitly,

---

[1]More discussions on the experimental searches and discoveries of different neutrino species can be found in Chapter 11.

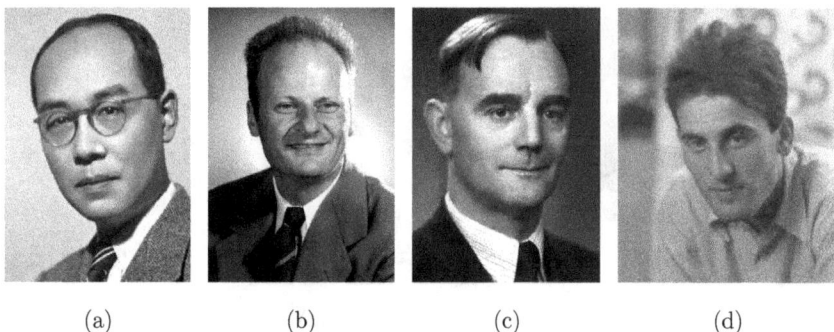

(a)   (b)   (c)   (d)

Figure 3.4: (a) H. Yukawa, (b) H. Bethe, (c) C.F. Powell, (d) G. Occhialini.

the chain reaction is: $^{26}_{13}\text{Al} +^4_2\text{He} \rightarrow \ ^{30}_{15}\text{P} \rightarrow \ ^{30}_{14}\text{Si} + e^+\nu_e + n$. This particle which is the positron, predicted by Dirac by his equation called anti-electron was discovered in 1932 by Carl David Anderson (1905–1991) (Nobel Prize of 1936) (Fig. 3.3). An antineutrino which accompanies the positron is also produced in this β inverse radioactivity. The fact that the proton transforms into a neutron plus a positron or *vice versa* a neutron to a proton plus an electron indicates the symmetric role of the proton and neutron inside the nucleus independently of their charge. This feature signals the presence of new nuclear forces which is not the electromagnetic one.

♠ Pion (π Meson) and Muon (μ)

By analogy with the photon, Hideki Yukawa (1907–1981) (Nobel Prize of 1949) (Fig. 3.4), postulated in 1935 that *there must be a particle (meson) having a mass of the order of 100–200 MeV intermediate between the mass of the electron and proton (origin of the name meson: mesos in Greek which means middle)* responsible for the short-range nuclear force. In 1936, C.D. Anderson and his student S.H. Neddermeyer (1907–1988) (Fig. 3.3) discovered in cosmic radiation a particle that has 200 times the mass of the electron that was falsely identified to the pion of Yukawa. This result has been confirmed by J.C. Street (1906–1989) and E.C. Stevenson in a cloud chamber in 1937. However, its unexpected property of being able to pass through a great thickness of material without interactions, indicates that it does not participate in the strong interactions. At that time, the muon seemed to be unwanted for particle physics such that I.I. Rabi (1898–1988, Nobel Prize of 1944) asked the famous question: *who ordered that?* Hans Bethe (1906–2005) (Nobel Prize of 1967) (Fig. 3.4) and, in 1947, Robert

Marshak (1916–1992) (Fig 7.3), suggested that this particle, called *muon* *(μ), with a mass 105.66 MeV is the result of the decay of the researched meson* needed in the Yukawa theory. This meson baptized *π meson or pion of mass 139.57 MeV has been discovered in cosmic rays* in 1947, by Cecil Frank Powell (1903–1969) and G. Occhialini (1907–1993) (Fig. 3.4) (Nobel Prize of 1950), thus confirming the hypothesis of Bethe and Marshak (Fig. 3.4).

♣ Anti-particles

• *Difference from particles*
The antiparticle is the particle seen in a mirror. This leads to change in its spin (1/2 integer for a fermion and integer or zero for a boson), its helicity (projection of the spin on its speed) and its charge. Positron (positive electron) is the first discovered antiparticle but it is not the only one. The consistency of the theory of fields (see Sec. 1.6) suggests that each particle is associated with an antiparticle. Actually to the muon (μ^-) is associated the anti-muon (μ^+). The pion exists in three states (π^-, π^0, π^+). Neutral particles such as the photon and the π^0 are their own antiparticles. However, this is not true in the case of the neutron because of its composition in quarks. As it is the assembly of 2 quarks of charge $-1/3$ called d (down) and a quark of charge $+2/3$ called u (up) (see Chapter 5 and Table 6.1), the antineutron will consist of 2 antiquarks d of charge $+1/3$ and an antiquark u of charge $-2/3$ and therefore differs from the neutron. The discovery in 1955, by Emilio Gino Segré (1905–1989) and Owen Chamberlain (1920–2006) (Nobel Prize of 1959) (Fig. 3.5), of the antiproton at the laboratory

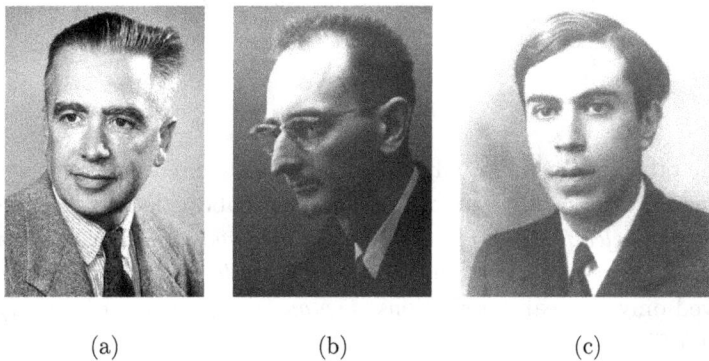

(a)　　　　　(b)　　　　　(c)

Figure 3.5: (a) Emilio Segré, (b) Owen Chamberlain, (c) Ettore Majorana.

level has confirmed the existence of antimatter (though not numerous in our Universe). *When a particle meets an antiparticle, it annihilates to form other matters like the photon* that will then decay into an electron–positron pair or some other pairs. *Antimatter has not yet been discovered till now (2015) in the natural state, which is good news for the stability of our Universe.*

• *Fermions classification*

We can classify the fermions in two categories:

— *Dirac particles* like the electron and the muon, which are massive and which differ from their antiparticles.
— *Particles of Ettore Majorana* (1906–1938?) which is the same as its antiparticle. Doubt still remains on the case of the neutrino (but also on the death or disappearance of Majorana in 1938!) because, at the moment, we saw the neutrino in its left helicity and the antineutrino in its right helicity. This problem remains topical in supersymmetric theories beyond the standard model (see Part III).

◇ Hadrons

The family of particles that can *interact via the strong nuclear interaction* are called hadrons (from the Greek word *hadros* which means strong). These hadrons are divided into two groups: *mesons*, the lightest being the pion, which are bosons (integer spin) and the *baryons* (comes from the Greek word *barys* meaning heavy), the lightest being the proton, which are fermions (spin 1/2 integer) (see Chapter 5). Notice that *the lightness of the proton ensures the stability of our Universe* because it means the proton cannot decay into some other particles.

♡ Leptons

On the other hand, the family of particles that *cannot interact via strong interactions* are known as leptons. They are subdivided into *charged leptons like the electron and muon* subfamily involved in both electromagnetic and weak interactions and subfamily of *neutral leptons* known as neutrinos involved only in weak interactions. *Leptons are fermions of spin 1/2* (see Chapter 7).

3.5 Concluding Remarks

In the subsequent chapters, we shall discuss in details the modern developments of particle physics both in theory and experiments where the number of discovered particles has exploded during the last 50 years as can be seen in the recent Particle Data Group (PDG) compilation [5]. Part II will be devoted to the description of modern theory of forces described by gauge theories. These theories are respectively: Quantum Electrodynamics (QED), Quantum Chromodynamics (QCD) and the Standard Model of electroweak interactions. Models beyond these standard models will be shortly reviewed in Part III. Some experimental aspects of particle physics and in particular the LHC at CERN where the Higgs boson has been recently discovered in 2012 will be discussed in Parts III and IV. In Part VI, we also present the (indirect) social and technological impacts of fundamental researches by taking the example of the LHC.

Part II
Modern Theories
of Forces

Chapter 4

Quantum Electrodynamics (QED)

4.1 QED is a $U(1)$ Gauge Theory

• *Prototype gauge theory*

Quantum electrodynamics (QED)[1] is the *theory of the electron and photon quantum fields*. It is the prototype of the quantum field and gauge theories and is a quantum and relativistic version of classical electromagnetism.

• *Abelian $U(1)$ group and gauge invariance*

QED is described by the Abelian $U(1)$ rotation group on a circle.

— It remains invariant by multiplying the electron field $\psi(x, t)$ by a phase $\exp[i\theta(x)]$ where $\theta(x)$ is the rotation angle as in Fig. 3.2 but depends on the position x of the particle.

— It is also invariant by changing the electromagnetic field $A_\mu(x)$ (called gauge field) by a new field: $A_\mu(x) = A_\mu(x) - \partial_\mu\theta(x)$ where $\partial_\mu \equiv \partial/\partial x_\mu$ is the derivative (small variation) of $\theta(x)$ in the space–time.

— These transformations are called gauge transformations.

---

[1]More developed and technical discussions on QED can be found in Appendix D.

Figure 4.1: (a) P. de Fermat, (b) R.P. Feynman, (c) F. Villars.

4.2 Path Integral and Feynman Diagrams

♣ Principle of Least Action

In 1948, Feynman (Fig. 4.1) pointed out that if light is a wave and hence a field, principle of Fermat (1603/1608–1665) (Fig. 4.1) (the light path is the minimum optical path) can be deduced from the more general principle of Huygens (Fig. 1.5), which stipulates that the amplitude of a field is the sum of the contributions of all the field waves that depart from a source S and arrive at the observer $\mathcal{O}$ after reflection on the mirror. Waves will then browse multiple paths, the sum allows light to choose one that minimizes its time of travel. *In the case of very low wavelengths (geometrical optics), this minimum path is that of the so-called distance of Fermat.* By applying this summation to the motion of the particles considered as quanta of the field, Feynman gets, in the classical limit (the reduced Planck constant $\hbar = 0$), the principle of least action corresponding to Fermat's principle in geometrical optics. He thus managed to unify the wave and the particle.

◇ Amplitudes (S-Matrix) and Path Integral

In the approach of least action, to calculate the transition amplitude $\mathcal{A}$ (or S-matrix) to go from initial point A at the time t_i to a final point B at the time t_f, must be considered the sum of all paths that meet the initial and final conditions. Each path is characterized by its weight that is the exponential of the classical action S with respect to the variation of the Lagrangian L (difference between kinetic E_c and potential energy V) in 3-dimensional space multiplied by $(i/\hbar)$. In terms of an equation, it means that: $S = (i/\hbar) \int dt \, L$ (see more in Appendix D). *The complete path or functional integral corresponds to the summation over an infinite number of complex weights* (infinite number of variables of integration). Then, *it permits to calculate the transition amplitude $\mathcal{A}$.*

♡ Perturbation Theory

However, the summation for getting the transition amplitude $\mathcal{A}$ is, in principle, infinite and, then, *it is impossible in practice to extract the transition amplitude. In order to circumvent this problem, one proceeds by approximation series in terms of a small parameter expansion.* In the case of QED, this small parameter is the square of the electric charge e (in natural units: $\hbar = c = 1$):

$$a \equiv \frac{\alpha}{\pi} = \frac{e^2}{4\pi^2},$$

where $\alpha = 1/137$ is the fine structure constant. It enters via the tree level (lowest order) interaction Lagrangian $\mathcal{L}_I^0(x) = -e : \bar{\psi}\gamma_\mu\psi : A^\mu$ between the electron ψ and photon A^μ fields (see more in Appendix D), where the index zero indicates that the interaction is at the lowest order (tree level) of the perturbation theory. Here we have introduced the normal ordered product notation : : which means that the annihilation operators are to the right of the creation operators (see Appendix B). Therefore, in natural units, the amplitude $\mathcal{A}$ can be written as a sum of amplitudes evaluated at each order of the perturbation theory: $\mathcal{A} = \sum_n a^n \mathcal{A}_n = \mathcal{T} \exp\left(i \int d^4x \, \mathcal{L}_I^{(0)}(x)\right)$, when expressed in terms of the interaction Lagrangian $\mathcal{L}_I^0$. $\mathcal{T}$ is called *chronological ordering or time ordered product* which means that: $\mathcal{T}\mathcal{L}_I^0(t_1)\mathcal{L}_I^0(t_2)\ldots\mathcal{L}_I^0(t_n) = \sum_i \theta(t_{i_1} > t_{i_2} > \cdots > t_{i_n})\mathcal{L}_I^0(t_{i_1})\mathcal{L}_I^0(t_{i_2})\ldots\mathcal{L}_I^0(t_{i_n})$, i.e. for a time interval decreasing from left to right [see Eq. (B.36)]. The previous discussions just indicate that the transition amplitude $\mathcal{A}$ can be evaluated order by order of perturbation theory in series of a.

♠ Feynman Diagrams

To simplify and systematically implement the program of perturbative calculations, *Feynman illustrated the action of QED by introducing his famous diagrams that have precise rules to calculate the coefficients of these perturbative developments* (see more in Appendix E). These *Feynman diagrams*

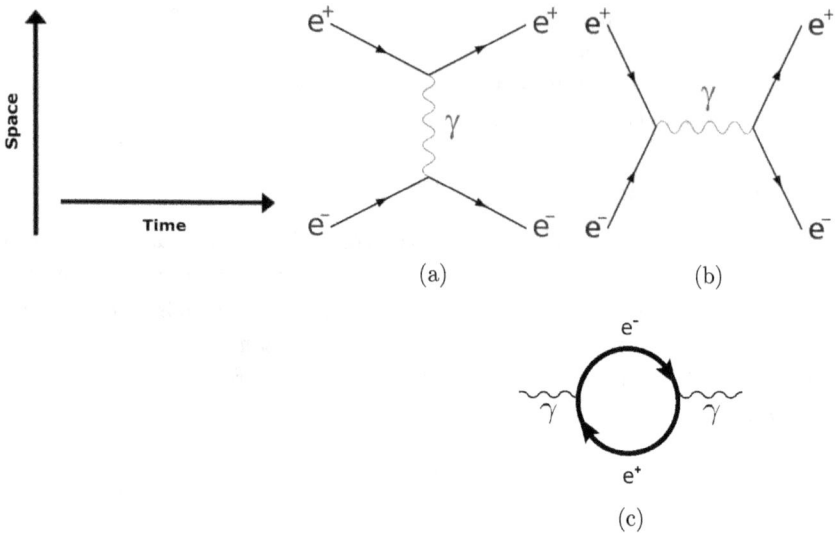

Figure 4.2: Space and time axes indicate the evolution and interaction distance: (a) e^+e^- interaction, (b) e^+e^- annihilation, (c) vacuum polarization.

are the most useful and powerful tools of contemporary particle physicists for evaluating physical processes. In this framework, *the electron line and propagator are represented by an oriented line, while the photon is represented by a wavy line. Each point of interaction called vertex has a definite weight proportional to the electric charge* and deduced from the QED Lagrangian $\mathcal{L}_I^0(x)$. For instance, to lowest order of the expansion in $\alpha = e^2/4\pi$ where e is the electric charge, the interaction between an electron (e^-) and a positron (e^+) is represented by the diagram of Fig. 4.2a, while the one between e^- and e^+ annihilation is given by Fig. 4.2b. The vacuum polarization, where a pair e^+e^- is created and then disappears spontaneously, is represented by Fig. 4.2c. These different Feynman rules are summarized in Appendix E for different Standard Model gauge theories (QED, electroweak and strong interactions).

4.3 Renormalization Program in QED

However, making these perturbative calculations, we meet infinities that have no physical meaning. To resolve these problems, methods (not very elegant in the mathematical sense) have been introduced in 1930 by Ernst C.G. Stuekelberg (1905–1984) (Fig. 4.4) and in 1940 by Feynman (Fig. 4.1), Freeman Dyson (1923–), Julian Schwinger (1918–1994) and Shin'ichiro

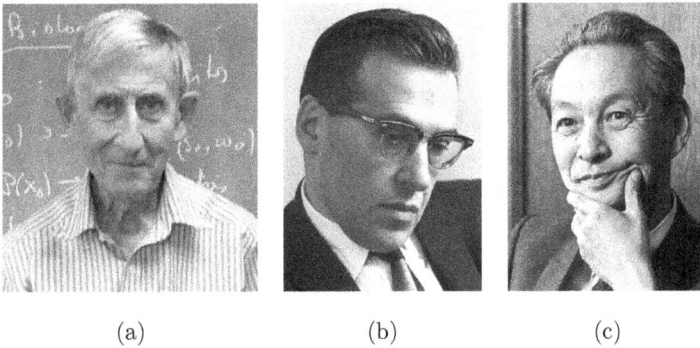

Figure 4.3: (a) F. Dyson (courtesy of Monroem, wikipedia.org), (b) J. Schwinger, (c) S. Tomonaga.

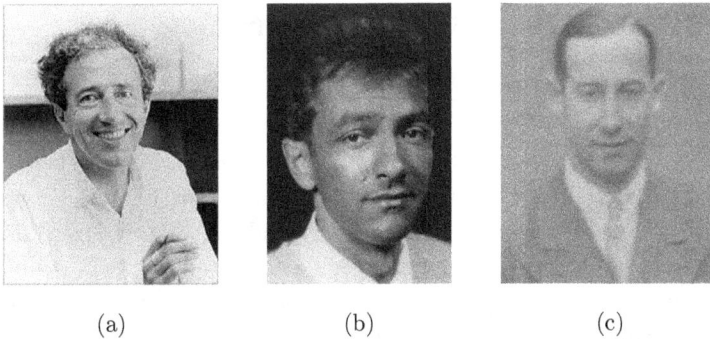

(a)   (b)   (c)

Figure 4.4: (a) F. Low (photo by Kurt Gottfried, courtesy of AIP Emilio Segrè Visual Archives, Gottfried collection), (b) A. Petermann (courtesy of CERN), (c) E.C.G. Stueckelberg (courtesy of GFHund, wikipedia.org).

Tomonaga (1906–1979) (Nobel Prize of 1965, except Dyson) (Fig. 4.3). They consist to absorb these infinities obtained by calculations in the high energy region by introducing an ultraviolet cut-off of infinite energy (*Pauli–Villars regularization*) of W. Pauli (Fig. 1.18) and Felix Villars (1921–2002) (Fig. 4.1) and by a redefinition of the observable (*renormalization*). Thus, an observable is separated into two categories: the naked observable and the renormalized or dressed observable. The *naked observable* is that it would have been obtained if one had no interaction. Infinity appears when calculating with the naked observable which is not measurable experimentally because measuring it one has to interact with it. The *renormalized observable* which is measurable is finite. These operations are only possible in the case of a *renormalizable theory* which is the case of QED. QED is renormalizable because it has no free parameters and depends only on the charge of the electron and its mass. QED processes become finite once one renormalizes these two quantities. However these quantities depend on the *energy of renormalization ν* (or *subtraction point*) but this dependence is not arbitrary because it is governed by differential equations called *renormalization group equation* introduced first by A. Petermann (1922–2011), and E.C.G. Stueckelberg in 1953 (Fig. 4.4), and in 1954 by M. Gell-Mann (Fig. 5.1) and F. Low (1921–2007) (Fig. 4.4) expressing that a *physical observable $\mathcal{O}$ should not depend on the renormalization point ν* which means that its total derivative with respect to ν is zero: $d\mathcal{O}/d\nu=0$ (see more in Appendix E). *In this sense, the theory is not a fundamental theory but is an effective theory* which depends on two parameters: the mass and charge of the electron which are not adjustable parameters because they are physical.

4.4 Beyond the Dirac Theory

Willis Eugene Lamb (1913–2008) and Polykarp Kusch (1911–1993), Nobel Prize, 1955 (Fig. 4.5), shortly employed at the Columbia University in New York have collaborated and pursued the work of I.I Rabi (1898–1988) (Fig. 4.5) (Nobel Prize of 1944) on his resonance method. In this method the spectra of the atoms are studied by radio waves and the details of the spectra can thereby be investigated much more accurately than before. Kusch and Lamb participated during the war in the extensive work on radar technique which was then being performed. Because of the great progress in this field the resonance method could be much improved.

(a)   (b)   (c)

Figure 4.5: (a) I.I. Rabi, (b) W.E. Lamb, (c) P. Kusch.

♣ The Lamb Shift

According to the theory of Dirac in 1928, the electron's energy levels in the hydrogen atom should depend only on the value of the principal quantum number n ($n = 1$ for a ground state and $= 2$ for an excited state) which is called fine structure. But physicists were puzzled when some perturbative calculations, including quantum corrections, suggested that these energies might differ by tiny amounts, while the experimental evidence was unclear. In 1947, Willis Eugene Lamb (1913–2008) (Fig. 4.5) (Nobel Prize of 1955) and his student Robert Curtis Retherford (1912–1981) measured the energy shift when the electron in a hydrogen atom is in the second energy level $n = 2$. They found that the $2P$ (orbital quantum number $L = 1$) level is at a slightly lower energy than the $2S$ (orbital quantum number $L = 0$) level. In a standard symbolic notation: $^{2S+1}L_J$ where S, L, J are respectively the total spin, orbital and total angular quantum numbers, this corresponds to the energy levels between a $^2P_{1/2}$ and $^2S_{1/2}$ state of the hydrogen atom which is about 1000 MHz. As this energy shift behaves like α^5, it has been used to extract the value of the hyperfine constant $\alpha = 137.0368(7)$.[2]

◇ The Anomalous Magnetic Moment of the Electron

The discovery of Kusch and Henry Michael Foley (1917–1982) refers to the magnetic moment of the electron.

It had been known since long that the electron is a small magnet. By interacting with the photon, the electron represented by l in Fig. 4.6

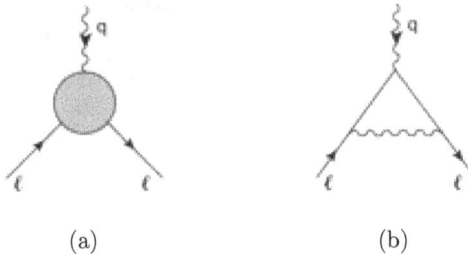

(a)　　　　　　　　(b)

Figure 4.6: (a) Magnetic moment, (b) Lowest order correction.

---

[2]Some other systems like positronium, hyperfine splitting, neutron Compton wavelength, atom recoil-measurements,...have been also used for extracting α.

acquires a magnetic moment, defined by $\mu = (g/2)(e\hbar/2m)$ where g is strictly equal to 2 in the classical case as predicted by the Dirac equation. To take account of the quantum corrections, represented by the black box of Fig. 4.6a, one defines the lepton anomalous magnetic moment $a_l = (g-2)/2$. The first quantum correction to the magnetic moment resulting from the exchange of a virtual photon is given by Fig. 4.6b.

In 1948, Kusch and Foley found that $a_e \equiv (g-2)/2$ is equal to 0.001 19(5) which requires the inclusion of quantum corrections to the classical Dirac equation. Indeed, the inclusion of the lowest order quantum correction induced by the diagram in Fig. 4.6b obtained by Schwinger in 1949 (Fig. 4.3) gives a value $a_e^{(2)} = \alpha/2\pi = 0.001\ 16$ in good agreement (within the errors) with the experimental value.

4.5   Some High-Precision Tests of QED

♣ The Electron Anomaly and Structure Constant α

Since the first result of Schwinger in 1949, Karplus and Kroll have included in 1950 the fourth order contribution which comes from the Feynman diagrams depicted in Fig. 4.7 which involve two virtual (internal) photon lines and the vacuum polarization mentioned in Fig. 4.2. The analytical result, which reads numerically $-0.328\ 48(\alpha/\pi)^2 = -0.000\ 017\ 7$, is the starting of high-precision measurements and higher order calculations for testing QED within its renormalization program. At present, the higher QED calculations are known up to an order of α^5 as shown by the T. Kinoshita Group (Fig. 4.8). Including (tiny) hadronic and weak interaction corrections, the theoretical prediction is reviewed by E. de Rafael (Fig. 4.8) in [6] which

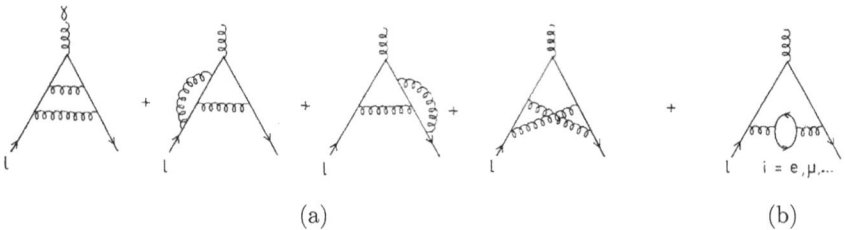

(a)   (b)

Figure 4.7: (a) Two internal photon lines, (b) Vacuum polarization.

(a)   (b)

Figure 4.8: (a) T. Kinoshita, (b) E. de Rafael.

is: $a_e^{th} = 159\ 652\ 181.82(78.35) \times 10^{-12}$, while the present measurement leads to a value [6]: $a_e^{exp} = 159\ 652\ 180.73(0.28) \times 10^{-12}$, which is 0.25 part in a hundred billion. In order to illustrate the fineness of these numbers, *Feynman wrote that, if we measure the distance between New York and Los Angeles, the precision would be the thickness of a human hair.*

Conversely, one can retrieve the value of the constant hyperfine (comparing a_e^{exp} and a_e^{th}. In this way, one can deduce [see e.g. E. de Rafael (Fig. 4.8) in [6]]: $1/\alpha = 137.035\ 999\ 1736(342)$, which is in perfect agreement with the experimental measurement [2010 CODATA recommended values: http://physics.nist.gov/cuu/Constants/]: $1/\alpha = 137.035\ 999\ 0740(440)$.

◇ The Muon Anomaly

To test the lepton nature of the muon, J. Bailey *et al.* (Fig. 4.9) has proposed in 1959 to measure the muon anomalous magnetic moment at the CERN muon storage ring where small deviation from $g = 2$ should be expected. In 1961, the first measurements of the muon g−2 reach a precision of 2%. In 1974, the third experiment (see Fig. 4.9) has been installed in the south–east of the SPS. The first result in 1979 confirmed the theory to a precision of 0.0007%. An almost similar precision of the order of 10^{-11} is obtained for the muon anomalous magnetic moment (Fig. 4.10): $a_\mu^{exp} = 11\ 659\ 2091(54)(33) \times 10^{-11}$, which differs from the Standard Model predictions by: $\Delta a_\mu = a_\mu^{exp} - a_\mu^{SM} = 288(63)(49) \times 10^{-11}$, which is about 3.6σ discrepancy when the total errors are added quadratically.

One should notice that the main theoretical uncertainty comes from the estimate of the hadronic contributions affecting the photon propagator (hadronic vacuum polarization). Recent estimates based on the new

Figure 4.9: Members of the 3rd g−2 experiment (S97 - CERN Muon Storage Ring Collaboration): *Front*: J. Bailey, H. Drumm, G. Petrucci, G. Lebee. *Back*: W. Von Rueden, G. Fremont, E. Picasso, F.J.M. Farley, J.H. Field, W. Flegel, F. Krienen, K. Muehlemann.

measurements of the $e^+e^- \rightarrow$ hadrons total cross-section by BaBar and BESIII groups [8] support previous results from the τ hadronic width by ALEPH [9] but disagree with previous $e^+e^- \rightarrow$ hadrons data from the KLOE group of Frascati [10], which reduce the theoretical and experimental difference to 2.4σ: $\Delta a_\mu = a_\mu^{exp} - a_\mu^{SM} = 198(84) \times 10^{-11}$. At this 2.4σ level, one cannot firmly claim a discrepancy between the Standard Model predictions and the experiment. However, one should, first, clarify the experimental situation on $e^+e^- \rightarrow$ hadrons data before a definite conclusion can be drawn.

The measurements of the anomalous magnetic moments and of the fine structure constant are the most precise tests done in particle physics. As already mentioned by Feynman, these precisions would correspond to the thickness of a human hair if we measure the distance between New York and Los Angeles.

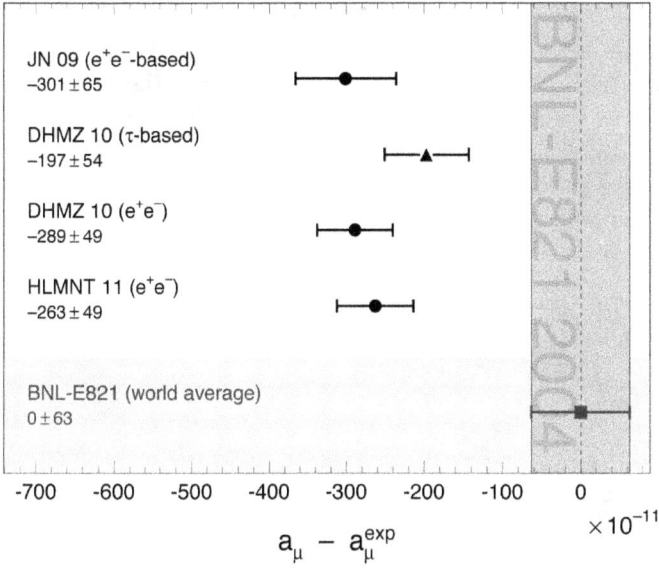

Figure 4.10: Theoretical predictions and experimental measurement (blue) of the $(g-2)/2$ of the muon in 2013 (see [6, 5]).

4.6 Success of QED

The successes of the theoretical QED predictions for the electron anomalous magnetic moment and for the fine structure constant at this high-level accuracy indicate the unexpected validity of the QED renormalization program and of perturbation theory though a dirty game (renormalization program) has been used for eliminating infinities (renormalization) from the theory.

However, at this level of precision and as mentioned earlier, there is a significant disagreement between the experience (coloured blue region on Fig. 4.10) and different theoretical predictions of the muon anomalous magnetic moment a_μ. This result may indicate the existence of a new phenomenon not yet considered until now in the so-called Standard Model of electroweak interaction which we shall discuss later in Chapter 7. However, first, it is mandatory to recheck the theoretical predictions and the experimental measurement of the hadronic contributions before a definite claim for the presence of new phenomena beyond the Standard Model of electroweak interactions.

This result for the muon anomaly also indicates the complementary role of the precise low energy experiments and of the high-energy accelerator like the LHC which we shall discuss in Part IV for new physics searches. The former effect enters through quantum loop corrections, while the latter is a direct effect related to the observation of new particles beyond the Standard Model theories.

Chapter 5

Pre-QCD Era

5.1 $SU(3)$ Classification of Hadrons

During 1950–1960, we have seen a proliferation of hadrons (mesons and baryons) that motivated their classification. In so doing, one has introduced new internal quantum numbers in addition to the spin which differentiates a fermion (spin $1/2$, $3/2$,...) from a boson (spin 0, 1, 2,...).

♣ Isospin Symmetry $SU(2)$ and Baryon Number

The proton and the neutron belong to the family of baryons. They have been considered as two different nucleon states by assigning a new intrinsic quantum number: *isospin* described by the global symmetry group $SU(2)$ of the strong interaction because the proton and the neutron, although with different charge, react the same way with the strong interaction (independence of charge). This symmetry is analogous to the spin of quantum mechanics. *The proton is the state of isospin $+1/2$ and the neutron is the state of isospin $-1/2$.* In general, the hadrons form isospin multiplets I. Each member of the multiplet is characterized by the third component I_3 on the z-axis of the isospin and takes $2I+1$ values. For the baryons, I takes $1/2$ integer values, while it takes integer values for mesons. Thus *for the pion, there are three states of isospin $-1, 0, +1$ corresponding to the states*

71

π^-, π^0, π^+. However, as the proton and the neutron have not quite the same masses, one can say that the isospin symmetry is an approximate symmetry. As the charge and component I_3 of the isospin are preserved by the strong interaction, one can also introduce the *baryon number $\mathcal{B}$ to differentiate baryons ($\mathcal{B} = 1$) from mesons ($\mathcal{B} = 0$) by the relationship $Q = I_3 + \mathcal{B}/2$,* where Q is the electric charge. One can easily check that the proton, with $I_3 = +1/2$ isospin and baryon number $\mathcal{B} = 1$, has a charge $Q = +1$, the π^- which has $I_3 = -1$ and $\mathcal{B} = 0$ has a charge $Q = -1\ldots$

♢ Strangeness S and Hypercharge Y

The discovery of K mesons which are 3.57 times heavier than the pions *was strange at that time because they form an isospin doublet ($I = 1/2$) while their baryon number $\mathcal{B}$ is zero.* On the other hand, the hyperon Λ is a baryon $\mathcal{B} = 1$, while it is in a state of isospin singlet ($I = 0$). This led in 1953, Kazuhiko Nishijima (1926–2009) and in 1956 Murray Gell-Mann (1929– , Nobel Prize of 1969) (Fig. 5.1) to extend the symmetry group $SU(2)$ of isospin to a group $SU(3)$ by introducing two new intrinsic quantum numbers: *the strangeness S and the hypercharge Y defined by $Y = \mathcal{B} + S$.* $S = 0$ for non-strange hadrons such as the nucleon and pion. It is equal to -1 for the K^- and the K^0 and $+1$ for the anti-kaons $\bar{K}^+$ and $\bar{K}^0$ which are strange particles. The charge becomes: $Q = I_3 + (Y = \mathcal{B} + S)/2$. In a mathematical language, the internal symmetry of the strong interaction group is therefore the product of the isospin symmetry $SU(2)$ by the one $U(1)$ of the hypercharge.

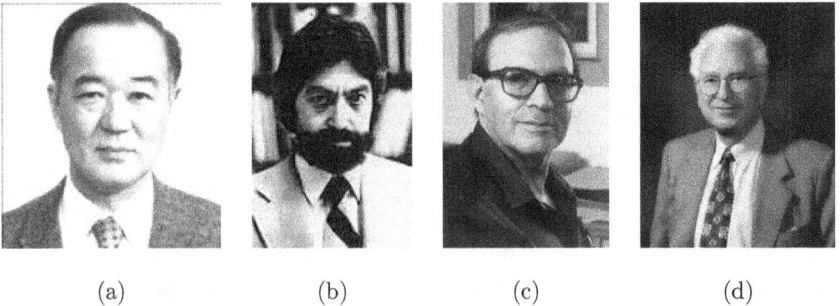

| (a) | (b) | (c) | (d) |

Figure 5.1: (a) K. Nishijima, (b) G. Zweig, (c) Y. Neeman, (d) M. Gell-Mann.

♡ $SU(3)$ and $U(3)$ for Hadrons

Looking closer at the experimental results, it can be seen that one can classify the previous hadrons to multiplets of the symmetry group $SU(3)$. This classification is given in Fig. 5.2 for pseudoscalar mesons of spin 0 (octet) and baryons of spin 1/2 (octet) and 3/2 called hyperons (decuplet). Octet classification has been called *eightfold way* by Gell-Mann. A similar classification exists for scalar mesons of spin 0 and vector mesons of spin 1 (ρ, ω, ϕ and K^*). However, one finds that the masses of vector mesons of spin 1 are always heavier than those of pseudoscalar mesons of spin 0. The ρ is 600 MeV heavier than the π. The K^* is twice heavier than the K ... Explanations for these observations will be given below in the theory of Regge and in QCD later (see Part II Chapter 6). The masses of all known hadrons can currently be found in *Particle Data Group* (PDG) [5] site: http://pdg.lbl.gov/.

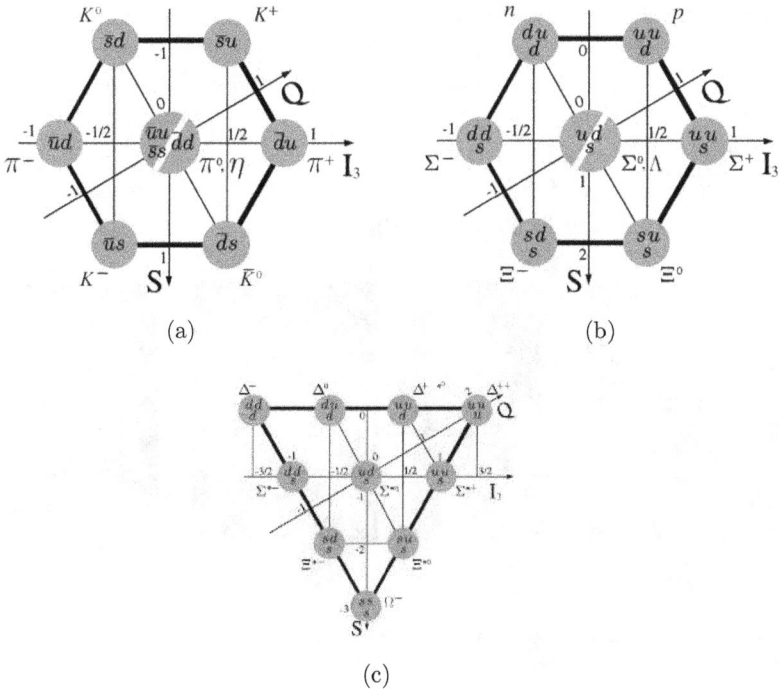

(a)

(b)

(c)

Figure 5.2: (a) Pseudoscalar Mesons octet; (b) Baryons octet; (c) Baryons decuplet: I_3 axis of isospin, S strangeness, Q charge.

5.2 Regge Theory for Hadrons

Regge theory[1] is based on the approach of the S-matrix (also called transition matrix) and the principle of bootstrap developed by Tulio Regge (1931–, Fig. 5.3), Geoffrey Chew, Steven Frautschi, Stanley Mandelstam, Vladimir Gribov to understand the strong interaction by using the properties of analyticity of the S-matrix without assuming the locality of interaction of the theory of fields (gauge theory like QED) to calculate a diffusion or scattering process.

♣ Regge Theory

In 1959, T. Regge noted that by solving the non-relativistic scattering potential Schrödinger equation, it is useful to consider the angular momentum l as a complex number with which he showed that, for a wide class of potentials, only singularities of the scattering amplitude $\mathcal{A}(s)$ in the complex plane l are called *Regge poles*. If these poles are obtained for integer and positive l values, they correspond to bound states or resonances. To illustrate this approach, consider the interaction: $\pi^+\pi^- \to \rho \to \pi^+\pi^-$ where the vector meson ρ is unstable and can be exchanged in the t (*annihilation*) or s (*diffusion*) *channel* (Fig. 5.4). s and t are the square of energy transfer in each channel: $s = (p_1 + p_2)^2$ and $t = (p_1 - p_2)^2$ if p_1 and p_2 are the incoming momenta of 2 incoming pions. Mesons are represented by two lines of opposite directions (in fact this is like a u quark and an antiquark

(a)	(b)	(c)	(d)

Figure 5.3: (a) T. Regge, (b) G. Veneziano, (c) M. Froissart, (d) A. Martin (courtesy of Maximilien Brice, CERN).

[1]For a comprehensive review, see e.g. the book of V. De Alfaro, S. Fubini, G. Furlan and C. Rossetti [4].

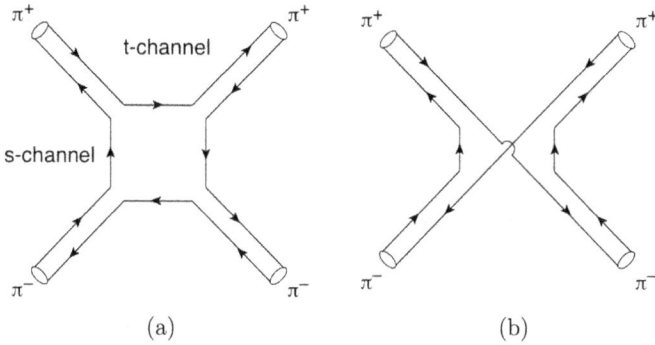

Figure 5.4: Diagrams representing the diffusion $\pi^+\pi^- \rightarrow \rho \rightarrow \pi^+\pi^-$: (a) planar; (b) non-planar.

d of the quark model). Fig. 5.4a is a *planar diagram* and theory shows that it is dominant compared to Fig. 5.4b called *non-planar* (because the meson lines intersect). If one explains this reaction in terms of potential, the vector meson ρ of spin 1 is described in the t-channel by the Coulomb potential that is the inverse of the radius $1/r$, while in the s-channel (t and s are also called *Mandelstam variables*), it is described by a potential $l(l+1)/r^2$, which plays the role of potential barrier, i.e. the total potential is less effective for large values of l, thus explaining why the spin 1 vector meson (the ρ) is 600 MeV heavier than a meson of spin 0 (the π). At high energies, it is also shown that the amplitude of scattering $\mathcal{A}(s,t)$ behaves like $s^{\alpha(t)}$ where $\alpha(t) = l$ is called the intercept or slope of Regge trajectory.

◇ Pomeron and Odderon

This theory of Regge has also introduced the concept of *pomeron* in 1961 [2] and of *odderon* (its odd partner under P parity and charge conjugation C like the photon) in 1973 by L. Lukaszuk and B. Nicolescu [11], *a family of particles with increasing spin on the trajectory of Regge* to explain the slow growth of the cross-sections in hadron collisions. This growth is bounded by the famous *Froissart–Martin bound* (Fig. 5.3), in 1962, expressing that *the collision cross-section may not grow faster than* $\log^2 s$. One consequence of the pomeron hypothesis is that the pp and $\bar{p}p$ differential cross-section $d\sigma/dt$ with the respect to the transfer energy square t are expected to be equal at high energies, while the odderon is responsible for the difference

---

[2] A recent comprehensive review can be found in the book by S. Donnachie *et al.* [4].

between pp and $\bar{p}p$ differential cross-section $d\sigma/dt$ at small values of t but large s.

5.3 Veneziano Amplitude

♣ Veneziano Amplitude for Hadrons

In 1968, Gabriele Veneziano (1942–, Fig. 5.3) shows that *the (planar) diagrams in the s and t channels* (Fig. 5.5) *are equivalent by crossing symmetry* (the sum of the resonances exchanged in the s channel is the same as the one in the t channel). The amplitude can be expressed in terms of the mathematical Euler Gamma (Γ) or Beta (B) function: $\mathcal{A}(s,t) = \Gamma[-\alpha(t)]\Gamma[-\alpha(s)]/\Gamma(-\alpha(t) - \alpha(s)) \equiv B[-\alpha(t), -\alpha(s)]$ which is the Euler beta function. This relation expresses that there are infinite numbers of resonances which appear as poles on the real axis $\alpha(t) = l = \alpha(s)$ with positive integer l and they lie on Regge trajectory with an universal slope $\alpha(t) = l$. The experimental results show that observed hadrons are mostly on Regge trajectories characterized by their slope and which are almost parallel (same slope). For example: $\alpha_\rho(t) \simeq 0.5 + 0.9t$ for mesons: ρ, ω, A_2, f, g, while for nucleons: $N(939), N(1688), N(2220)$, one has $\alpha_N(t) \simeq -0.3 + 0.9t$.

◇ Veneziano Amplitude: Mother of String Theories

Regge theory and the Veneziano model were very fashionable in the 1960–1970 but has been abandoned once Quantum Chromodynamics (QCD) based on a local gauge theory arrived. However, *the Veneziano amplitude is at the origin of string theory* currently used for attempts to unify the three

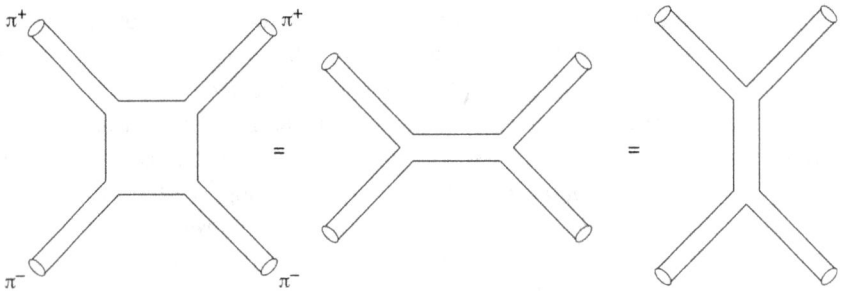

Figure 5.5: Dual-resonance (planar) diagrams of the diffusion: $\pi^+\pi^- \to \rho \to \pi^+\pi^-$, the 2nd and 3rd diagrams are respectively the s and t channels.

microscopic forces with macroscopic gravitational force. We shall discuss this topics in Chapter 9.

5.4 Quark Model

♣ Birth of the Quark Model

• *The Quark Model*

The classification of hadrons by the $SU(3)$ group paved the way for the Quark Model initiated in 1964 by Gell-Mann and George Zweig (born 1937) (Fig. 5.1). The contributions of K. Nishijima (1926–2009) and Neeman (1925–2006) (Fig. 5.1) in 1961 have also played an important role. *Gell-Mann assumes that hadrons are composed of more elementary particles called quarks.* The name quarks, meaning white cheese in German, come from a passage from a poem by James Joice in the novel Finnegans Wake: "Three quarks for Muster Mark... that tells the life of Mr Mark or Mr Finn with his three children." Gell-Mann alludes to the three quarks in the proton. This model suggests that the proton and the neutron and in general hadrons are not elementary but are the assembly of more elementary particles called quarks. As summarized in Fig. 5.6, we start from matter (hand) and after successive blows of hammer (energy supply) to abolish it. Thus, we passed successively from the molecule to the atom where we discovered that it contains a nucleus surrounded by a cloud of electrons (we will see later that the electron is part of the family of leptons). Peering into the nucleus, we find that it contains the nucleons (protons and neutrons). Providing even more energy, we come to find that *the nucleons are composed of three quarks (spin 1/2 fermions)* that, so far, are the smallest constituents of matter. In the case of $SU(3)$ (called symmetry flavor group), it is necessary to assemble the three light quarks u, d and s (flavors of quarks) to form hadrons (which comes from the Greek meaning hadros = strong). In this model:

— The u *(up)* quark has a charge of $+ 2/3$ and an isospin $I_3 = +1/2$. The d *(down)*, s *(strange)* quarks have charge $-1/3$ and respectively an isospin $I_3 = -1/2$ and 0 in unit e of the electron charge.

— Quarks have *three colors*: *blue, green and red* which are new quantum numbers. They assemble to form non colored (white) hadrons. The s quark has a strangeness $S = -1$ while for u and d, $S = 0$. The quarks compositions in the proton, neutron and pion are given on Fig. 5.7.

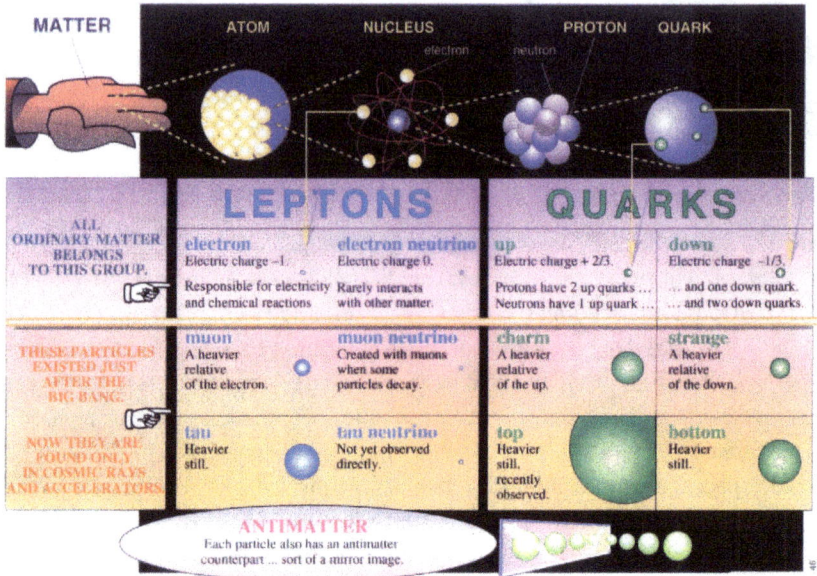

Figure 5.6: Quarks and leptons (fermions): leptons e and μ with their associated neutrinos and light quarks u, d, s discovered before 1962. Other heavier particles: τ and its neutrino, quarks c, b and t have been discovered later (see Chapters 6 and 7). (Courtesy of CERN & scientific information service)

— These quarks are bound together by the nuclear force (wavy line in Fig. 5.7) which were later known as gluons in (QCD) (analogue of the photon of QED).

• *Hadrons classification*

More generally, we show in Fig. 5.2 different quark compositions of mesons of spin 0 octet and baryons of spin 1/2 (octet) and 3/2 (decuplet) of the flavor symmetry group $SU(3)$. Similar figures are obtained for the nonet of vector $(\rho, \omega, \phi, K^*)$, its axial-vector partners and the more puzzling scalar mesons. *Mesons are bound states of a quark and an antiquark while baryons are assemblies of three quarks.* One can see in Fig. 5.7 that the neutron which is formed of udd has a charge $2/3 - 1/3 - 1/3 = 0$ and the positively charged pion formed by $\bar{d}u$ has a charge $-(1/3) + 2/3 = +1$. The isospin of the proton is: $1/2 + 1/2 - 1/2 = +1/2$ and that of the neutron is: $1/2 - 1/2 - 1/2 = -1/2$. The one of the positively charged pion is: $-(-1/2) + 1/2 = +1$.

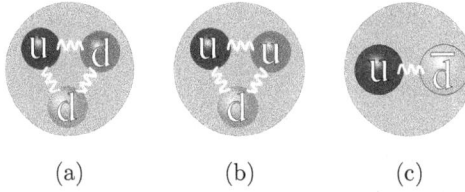

(a)   (b)   (c)

Figure 5.7: (a) Neutron, (b) Proton, (c) Pion. (Courtesy of Arpad Horvath, wikipedia.org)

◇ $SU(3)$ Symmetry Breaking and Discovery of Ω^-

• *Light quark constituent masses*

The discovery of baryons in pion–nucleon and kaon–nucleon scattering experiments has convinced on the viability of the Quark Model. For example, in the case of the octet (spin 1/2), the discovery in 1963 of $\Sigma^+(uus)$ of mass 1189 MeV in the reaction $\pi^- + p \to K^- + \Sigma^+$ and then $\Sigma^+ \to p + \pi^0$ compared to the proton $p(uud)$ of mass $M_p = 938$ MeV allowed to deduce that the difference in mass of the d and s quarks is: $m_s - m_d = 250$ MeV, while that between the proton and neutron $n(udd)$ is 1.3 MeV showing that $m_u - m_d$ is almost 0. Thus, it can be deduced that $m_u \simeq m_d \simeq M_p/3 = 313$ MeV which implies $m_s \simeq 560$ MeV. Mass differences, between baryons containing s quarks with those who do not, indicate $SU(3)$ *flavor symmetry breaking that is quantified by the difference in mass between the strange s and u, d quarks.* But, as these quarks are not observed because they are not released from the baryons at this low energy of the order of the mass of baryons (1–1.5) GeV (*phenomenon of confinement*), these masses are called *effective or constituent masses* because they include in their values the binding energy inside the nucleus. *These masses are in contrast with the perturbative masses called current masses used in QCD perturbation theory.* With these results it is possible to deduce the masses of baryons with 2 or 3 s quarks respectively having a strangeness $S = -2$ and $S = -3$ once we know the masses of the baryons containing 0 or 1 s quark (strangeness $S = -1$). A baryon with strangeness $S = -2$ decays cascading of strangeness to produce the proton of strangeness $S = 0$ at the end. Thus, the $\Xi^-(dss)$ will decay to $\Lambda^0(uds) + \pi^-(\bar{u}d)$ and then the $\Lambda^0(uds)$ is going to disintegrate in turn to $p(udd) + \pi^-(\bar{u}d)$. This decay process looks amazing, as the direct decay: $\Xi^- \to p + \pi$ is prohibited by the strong interaction because the 2 final states contain no s quark. *This process is of weak interactions type* where the s quark will turn into u by the exchange of a W^- boson (see Chapter 7) and W^- will give in turn a pair $\bar{u}d$ that will form the π^-.

• *The $\Omega(sss)$ baryon*

One of the important consequences of the Quark Model is the existence of $\Omega^-(sss)$ of spin 3/2 with a strangeness $S = -3$. If one uses the m_s value $\simeq 560$ MeV obtained previously, one predicted that the $\Omega^-(sss)$ mass should be around $560 \times 3 = 1680$ MeV. However, this value is below the threshold of production of the pair $\Xi^0(uss) + K^-(\bar{u}s)$, of which the sum of mass is $1315 + 494 = 1809$ MeV. Thus, the Ω^- must be stable with respect to the strong interaction and can only disintegrate by the weak interaction. The observation, in 1965 by the group from Brookhaven National Laboratory (BNL) of the Ω^- baryon at 1672 MeV predicted by the Quark Model at 1680 MeV as well as other predicted baryons and mesons have convinced the viability of the Quark Model. This success also indicates that we are on the right track for the understanding the structure of known nuclear matter.

◇ Algebra of Currents and Chiral Symmetry

Encouraged by the success of his model of quarks, Gell-Mann suggests that the vector V, electric current [associated with the electric charge Q (Noether current)] and its axial analogue A (its chiral partner) have a symmetry relative to the product of group $SU(3)_L \otimes SU(3)_R$ called chiral symmetry. L and R means the Left $(V - A)$ and Right $(V + A)$ components of V and A. In this theory, we can better understand the properties of pseudoscalar mesons (π and K as well as the axial partner A_1 of the vector meson ρ where A_1 is the meson associated to the axial current, while the π is associated to its derivative). *The symmetry of the group is partially broken explicitly by the masses of π and K or (equivalently) by the masses of the quarks u, d, s that are small but not zero. π and K are called pseudo-particles of Yoichiro Nambu in 1960 (1921–2015, Nobel Prize of 2008), G. Goldstone in 1961 (1933–) (Fig. 5.8). The spontaneous breaking of the symmetry group is made dynamically by the quark condensate* which is the average value in the vacuum of the quantity $\langle 0|\bar{\psi}\psi|0\rangle$ where ψ represents the field of quarks u, d, s. By analogy with electromagnetism, are then derived several theorems based on the (almost) conservation of axial current. The best-known is that of *Gell-Mann, Oakes and Renner (PCAC relation in 1968 : partially conserved current)* which is written: $m_\pi^2 f_\pi^2 = -(m_u + m_d)\langle 0|\bar{u}u + \bar{d}d|0\rangle$ where

(a)   (b)   (c)   (d)

Figure 5.8: (a) Y. Nambu, (b) G. Goldstone, (c) S. Weinberg, (d) J. Bjorken.

$f_\pi = 130$ MeV called pion decay constant controls the decay $\pi \to \mu + \nu_\mu$ and m_q is the perturbative current (not the constituent) quark mass.

◇ Weinberg Sum Rules

Using the fact that the *chiral symmetry is also realized asymptotically at high-energy* (the difference of the integrated two-point spectral functions associated to the vector and axial-vector current vanishes), S. Weinberg (1933–, Nobel Prize of 1979, Fig. 5.8) and also T. Das, V. Mathur and S. Okubo (DMO) in 1967 derived the famous *superconvergent Weinberg sum rules that allowed for example to connect the mass of the vector meson ρ to the one of its axial partner A_1* : $M_{A_1} = \sqrt{2}M_\rho = 1100$ MeV which is (within the approximation that it is derived) in good agreement with the experimental value: $M_{A_1} = 1230$ MeV. Some other relations were also obtained such as the one of *Goldberger–Treiman* [Marvin L. Goldberger (1922–2014), Sam B. Treiman (1925-1999) (Fig. 5.9)] in 1958 *relating the coupling f_π of the pion to the nucleon–nucleon pair coupling $g_{\pi NN}$*, the one relating the nucleon mass and coupling to the axial current g_A : $g_{\pi NN}f_\pi = M_N g_A$ or the one that connects the ρ coupling to the $\pi^+\pi^-$ pair: $g_{\rho\pi\pi} = M_\rho/(2f_\pi)$. These results are in good agreement with the experimental data and mark the triumph of the approach based on chiral symmetry and current algebras for understanding the hadron phenomena before the advent of QCD.[3]

---

[3]For a comprehensive review, see e.g the book by De Alfaro *et al.* in [4].

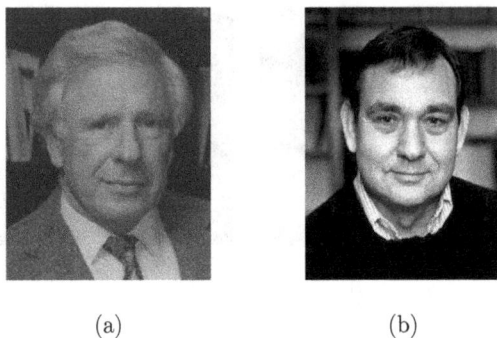

(a)   (b)

Figure 5.9: (a) M.L. Goldberger, (b) S.B. Treiman.

5.5 Parton Model and Bjorken Sum Rules

This model was proposed by Feynman in 1969 to describe the structure of hadrons in order to see their interactions at high energy. It assumes that *hadrons consist of pointlike and independent sub-particles called partons.* One can test this model in the deep inelastic (with very high-energy) scattering experiment of an unpolarized electron bombing a target of protons as shown in (Fig. 5.10) where the (virtual) photon encounters a parton which brings to a fraction $x(0 < x < 1)$ of the total proton momentum p (Fig. 5.10). This variable $x = -q^2/p \cdot q$ called *Bjorken variable* in 1969 [James Bjorken (1934–) (Fig. 5.8)] , where q is the photon momentum, can be expressed in terms of the mass M_p of the proton, the square of the momentum q of the photon, the difference in energy between the initial

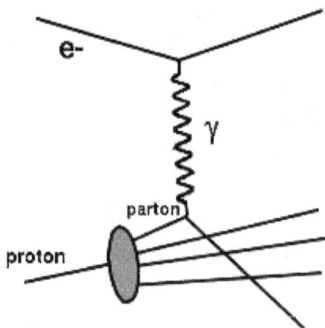

Figure 5.10: Deep inelastic scattering: $e^-p \to e^-H$.

and final electron $\nu \equiv E_i E_f = p \cdot q / M_p$. As we are going to measure the final electron without taking account of final hadrons denoted H, we then say that the *reaction is inclusive*. The *deep inelastic region* corresponds to the limit where $-q^2 \equiv Q^2$ and $W^2 = (p+q)^2$ are very large compared to the square of the mass of the proton M_P^2. The parton model (*Bjorken sum rules*) assumes that the total cross-section of the scattering is the product of the cross-section of the active elementary parton with a function $F_2(x)$ called *proton structure function* which depends only on the value of x but not on p and q (*scale invariance*). *The structure function is the probability of finding the parton inside the proton* and parametrizes the

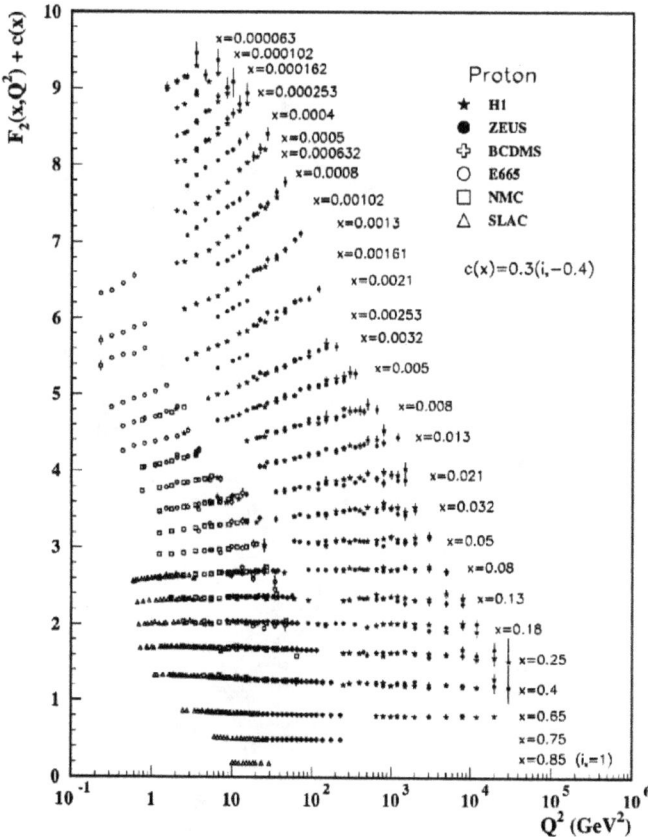

Figure 5.11: Measurement of the so-called $F_2(x)$ proton structure function in $e^- p$ and νp experiments and agreement with parton model and QCD predictions for different values of x as function of $-q^2 = Q^2 > 0$.

square of the hadronic amplitude called *hadronic tensor* and is the proton matrix element of the time-ordered product of the electromagnetic current $(J_\mu(x) =: \bar{\psi}\gamma_\mu\psi :$ where $\psi(x)$ is the electron field) which is formulated as: $\langle p|\mathcal{T}J_\mu(x)J_\nu(0)|p\rangle$. Considering that the partons of the proton are quarks of spin $1/2$, the parton model was validated early in 1969 at SLAC (Stanford Linear Accelerator) by the electron on a proton experiment which was followed later by some other experimental results. These recent results are summarized in Fig. 5.11 where we see that for $0.08 < x < 0.85$, we do have a scale invariance (independence in Q^2).

Chapter 6

Quantum Chromodynamics (QCD)

6.1 QCD Color Gauge Theory

As we reported previously, Quantum Chromodynamics (QCD).[1] (which comes from the greek word *chromo = color*) or color theory is *a modern version of nuclear physics that applies now at the level of quarks and gluons* but not only to that of the nucleon (proton and neutron). *This theory is also an improved version of the Quark Model* originally proposed by Gell-Mann where quarks and gluons are glued together (confinement of quarks and gluons) to form hadrons which are colorless particles.

♣ Color of Quarks and the Puzzle of Δ^{++} Baryon

The $\Delta^{++}(uuu)$ of mass 1232 MeV is composed of three quarks u and has a spin 3/2 (see Fig. 5.2c). As the quarks have a spin $\pm 1/2$, they must all have the same $+1/2$ spin direction to form the Δ^{++}. However, since quarks are fermions, they obey the Dirac statistics and cannot be in the same quantum state (Pauli exclusion principle). To remedy this problem, W. Greenberg (1932–), J. Han (1934–) and Y. Nambu (1921–2015) (Fig. 6.1) proposed a new type of principle called para-statistical exclusion of rank 3.

---

[1] For more detailed discussions on QCD, see e.g. [33].

(a)   (b)   (c)

Figure 6.1: (a) H. Fritzsch and M. Gell-Mann (courtesy of H. Fritzsch), (b) Y. Nambu and O.W. Greenberg, (d) M.-Y. Han.

In 1971, Gell-Mann and Harald Fritzsch (1943–) (Fig. 6.1) retook this idea but proposing that *quarks should have three colors: red, blue and green* in order to satisfy the Pauli exclusion principle or Fermi–Dirac statistics. In this case, the Δ^{++} and in general the baryons would be in an antisymmetric state with respect to the quantum number of the color. *They can then be described by an $SU(3)_c$ color group* which should not be confused with the group $SU(3)$ of flavor for the classification of the hadrons of the previous Quark Model.

◇ QCD is an $SU(3)_c$ Gauge of Color

Starting from the group $SU(3)_c$ of color, one can use a QED analogy that is described by the group $U(1)$. *Color plays the role of the electrical charge*, but instead of having a single charge as the electron, one must have three charges associated with the three colors of the quark. *Gluons play the role of the gauge field* but instead of having one field like the photon, one has eight gauge fields for gluons.[2] However, when the photon interacts with the electron, it does not change its charge, while *the gluon changes the color of the quark*.

When two electrons repel because they have the same charges, *3 u quarks that form the Δ^{++} glue together because they are confined*. To account for these different facts, Chen-Ning Yang (1922–) and Robert Mills (1927–1999) (Fig. 6.2) have generalized the results of the $U(1)$ abelian group

---

[2]In general, for a group $SU(N)$, we have N colors of quarks and $N^2 - 1$ colored gluons. Mathematically, one says that quarks are in the fundamental representation of the group and the gluons are in its adjoint representation.

Figure 6.2: (a) C.N. Yang, R. Mills, (b) M. Veltman, (c) G. 't Hooft.

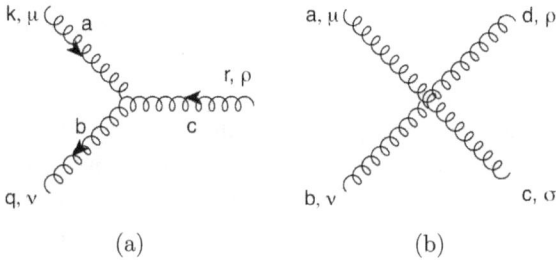

Figure 6.3: 3 and 4 gluons couplings in QCD.

of QED to the $SU(3)_c$ *non-abelian QCD group for building a gauge theory* with a local invariance where the phase of the transformation depends on the position x in the space–time of the particle. In this group, the *gluons have eight colors* because they are in the *adjoint representation* of the group while the *quarks have three colors* because they are in the *fundamental representation*. The QCD Lagrangian is very similar to that of QED (see more in Part VIII) but other new features such as the *self-interactions among gluon fields* (absent in QED) appear due to the non-abelian nature of the $SU(3)_c$ group. This is due to the fact that the photon is electrically neutral and can only interact with the electron or positron, while gluons are colored (color charge) and can self-interact or change the color of the quark. Then, one has new possible vertices (interaction) with 3 or 4 gluons (Fig. 6.3).

Gerald 't Hooft (1946–) and Martinus Veltman (1931–) (Nobel Prize of 1999) (Fig. 6.2) showed that the theory is *renormalizable*, i.e. finite, as one can absorb the infinities. Therefore, all the artillery of regularization and renormalization in QED to extract finite results using perturbation

(a)   (b)

Figure 6.4: (a) C.G. Bollini, (b) J.J. Giambiagi.

theory can be applied. However, instead of using (as in QED) an ultra-violet (UV) energy cut-off to remove infinities [called *Pauli–Villars regularization* (Figs. 1.18 and 4.1)], 't Hooft and Veltman and independently, Carlos Guido Bollini (1926–) and Juan José Giambiagi (1924–1996) (Fig. 6.4) have introduced the technique of *dimensional regularization* (one leaves the dimension n of a space–time arbitrary during calculations and takes, once the calculations are completed, the limit $n = 4 - \epsilon$ where, $\epsilon \to 0$). In this approach, the *infinity appears as poles* (singularities in $1/\epsilon$) which is more convenient because the method respects the gauge invariance of the theory (for a review, see e.g. [32]).

6.2 QCD Asymptotically Free Theory

♣ The QCD Coupling α_s

The success of the parton model at high energy also shows that *quarks are free states at high energy* and that the strong force governed by the coupling constant g of the $SU(3)_c$ group (the analogue of the electric charge e for QED), becomes smaller at high energy. One then says that QCD is an *asymptotically free theory* because the square of its coupling $\alpha_s = g^2/4\pi$ decreases for increasing energy. This is contrary to *the case of QED where the square of the effective charge increases with energy*. For QED, it passes from the value $\alpha(0) = e^2/4\pi = 1/137$ to $\alpha(92) = 1/128$, when the energy varies from 0 to 92 GeV at the Z boson mass. G. 't Hooft (Fig. 6.2) pointed out, in his famous remark at the Aix-en Provence conference in 1972, that *the coupling of non-abelian Yang–Mills gauge theory decreases with*

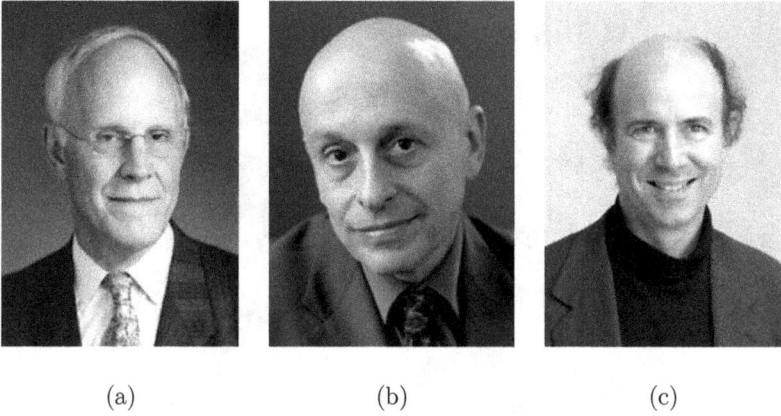

(a)   (b)   (c)

Figure 6.5: (a) D. Gross, (b) D. Politzer, (c) F. Wilczek.

*increasing energy because the corresponding (so-called) β-function is nega-
tive.* However, he did not make any connection of this remark with strong
interactions. It is only in 1973 that David Gross (1941–), David Politzer
(1949–) and Franck Wilczek (1951–) (Fig. 6.5) demonstrated *asymptotic
freedom for non-abelian gauge theory due to the negative sign of β-function*:
$\beta_1 = -(1/6)(11N_c - 2n)$ (to lowest order of α_s) which controls the renormalization of the couplings ($N_c = 3$ for three colors and $n \leq 11N_c/2$
is the number of flavours of quarks: $u, d, s, \ldots$) and showed that strong
interactions can be described with this non-abelian gauge theory with an
unbroken $SU(3)$ group. One can see that β_1 is negative for QCD ($N_c = 3$)
if the number of flavours of quarks does not exceed 16 while it is equal
to $n/3$ and is positive for QED ($N_c = 0$ and $n = 3$ for three families of
leptons). This idea was strongly supported by Harald Fritzsch, Gell-Mann
(Fig. 6.1) and Heinrich Leutwyler (1938–, Fig. 6.6) in 1973, and has lead
to the presently known theory of QCD. By solving the renormalization
group equations, the coupling, written as a function of the energy, reads:
$\alpha_s(Q) = \alpha_s(\nu)/[1 - (\alpha_s(\nu)/\pi)\beta_1 \log(Q/\nu)]$ where ν is called a subtraction
point or subtraction energy at which the renormalized coupling constant is
defined as an initial solution of the renormalization group differential equation. Log is the natural logarithm function. It may be noted that for $Q = \nu$
one has: $\alpha_s(Q) = \alpha_s(\nu)$. This equation explicitly shows that for negative β_1
(QCD case), the coupling constant decreases with energy (while it increases
in the case of QED). At the energy Q, one can also rewrite this equation

(a)   (b)

Figure 6.6: (a) H. Leutwyler, (b) S. Bethke.

in a more convenient form as:

$$\frac{\alpha_s}{\pi}(Q) = \frac{1}{-\beta_1 log(Q/\Lambda)}, \quad \text{where } \Lambda = \nu \, \exp\left[\frac{\pi}{\alpha_s(\nu)\beta_1}\right]$$

is a characteristic scale of the QCD theory. Its value, of the order of $\Lambda = (200-350)$ MeV depending on the number of the excited quark flavors, is obtained from the experimental results in Fig. 6.7 from the compilation of Siegfried Bethke (1954-) (Fig. 6.6) in PDG [5]. *The value of Λ also indicates the lowest extreme limit of validity of perturbation theory*, i.e. for energy below this value, the non-perturbative part of QCD dominates and the perturbative calculation is no more valid. One can notice that, in the case of QED, Λ has a huge value because the β_1 function is positive and the exponential expression increases. Therefore, it is expected that the per-turbation theory of *QED works well in the region of low energy* accessible experimentally. This explains the high-precision success of the QED cal-culations like the anomalous magnetic moment a_l of the lepton (Chapter 4). Returning to QCD, it is therefore expected that the coupling must decrease as the inverse of a logarithm when the energy Q increases (*asymp-totic freedom or asymptotically free theory*). This prediction is verified with very good experimental accuracy from the mass of the heavy lepton τ of

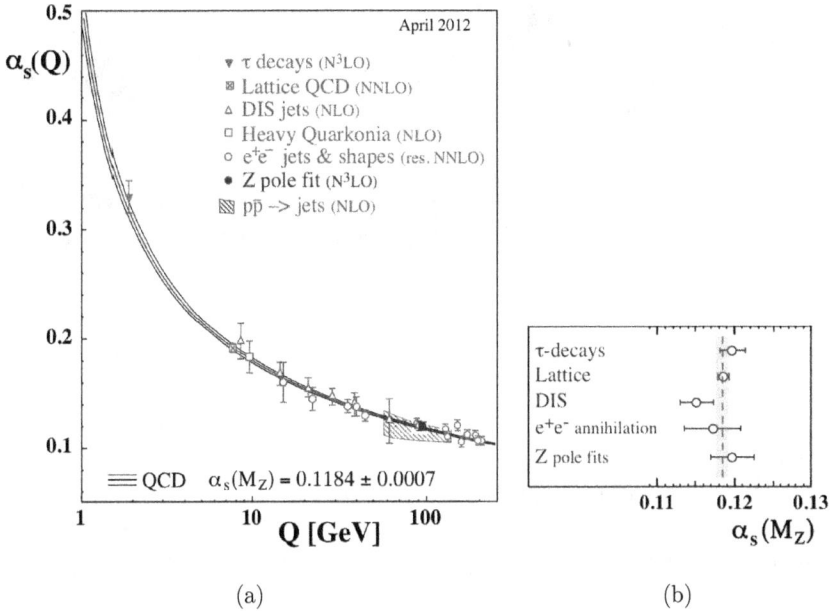

Figure 6.7: (a) Value of $\alpha_s(Q)$ as a function of Q; (b) value of $\alpha_s(Q)$ for $Q = M_Z$.

1.8 GeV [3] at which the value of $\alpha_s(M_\tau) = 0.325(8)$ (as first obtained by Eric Braaten (1953–), the author and Antonio Pich Zardoya (1957–) (Fig. 6.6) and improved more recently in [9,35]) until the mass of the boson Z of 92 GeV (Fig. 6.7a) at which its mean value from different sources is: $\alpha_s(MZ) = 0.1183(7)$ (Fig. 6.7b) as compiled by Bethke [13] in PDG [5].

◇ Testing Scale Invariance of the Parton Model

Despite the success of the Quark Model, one never could highlight direct particles of fractional charge $2/3$ or $-1/3$ [Robert Andrews Millikan and Harvey Fletche experiment in 1909 (Fig. 1.15) on the drops of charged oil between two horizontal electrodes of a capacitor plan] because these quarks are confined into the hadrons. One then turned to accelerators to search for quarks in a free state. By identifying the partons of the proton with spin $1/2$

---

[3] The heavy lepton τ having a mass 1.8 GeV was discovered by Martin Perl (1927–, Nobel Prize of 1995) (Fig. 6.8) in e^+e^- experiments by identifying the pair $e^+\mu^-$ originated from the decay of the pair $\tau^+\tau^-$ produced that is cleaner than the e^+e^- or $\mu^+\mu^-$ pair which may come from hadronic decays.

(a)   (b)   (c)

Figure 6.8: (a) M. Perl, (b) E. Braaten, (c) the author, A. Pich.

quarks, one can measure their structure functions in the electron–proton, neutrino–nucleon, proton–proton, antiproton–proton or electron–positron collisions. As mentioned previously (Chapter 5), the measurement of the *structure function F_2 of the proton* (Fig. 5.11) in electron–proton experiments helped to confirm the parton model composed of quarks of spin 1/2. Indeed, *for relatively large x ($x > 0.1$), F_2 does not depend on Q^2 as* predicted by the model of the partons (scale invariance). However, *when x becomes smaller, we see that F_2 increases with Q^2*. This variation indicates a violation of scale invariance.

• *DGLAP equation and scaling invariance violation*
Violation of the scale invariance is due to the quantum corrections [commonly known as radiative corrections in $\alpha_s(Q)$] or/and to non-perturbative effects to the parton model. These corrections are obtained through the Gribov–Lipatov (1972), Altarelli–Parisi (1977) and Dokshitzer (1977) (DGLAP)[4] (Fig. 6.9) evolution equation issued from the renormalization group equation. One can interpret this equation of DGALP by a fractal model of the partons: in the case of the electron–proton scattering with an exchange of a virtual photon (Fig. 6.10), a parton from a parent proton will first emit a gluon before interacting with the photon. Thus the quark that will be probed by the detector is the son of parton parent and so on. The DGLAP equation allows to calculate the probability (depending on Q) to find the parton of the generation n inside the $n-1$ generation. The agreement between measurements of the F_2 structure function at small x and the predictions of DGLAP (in continuous lines) shown in Fig. 5.11 confirms the DGLAP equation of evolution.

---

[4]V. Gribov (1930–1997), L. Lipatov (1940–), G. Altarelli (1941–2015), G. Parisi (1948–), Y. Dokshitzer (1951–).

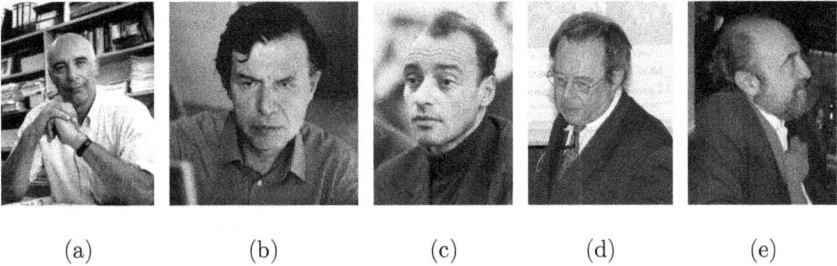

Figure 6.9: (a) G. Altarelli, (b) G. Parisi, (c) V. Gribov, (d) L. Lipatov, (e) Y. Dokshitzer.

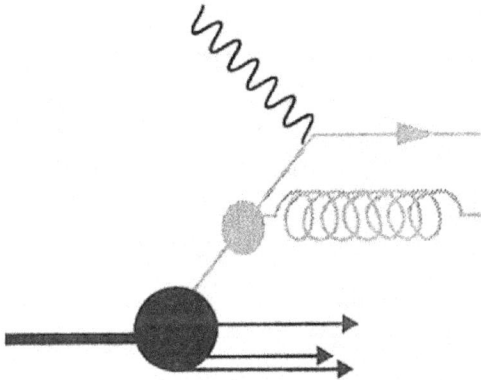

Figure 6.10: Parton parent (black ball) and son (brown ball) with emission of a gluon (spring) in ep scattering. The wavy line is the virtual photon exchange.

6.3 Deep Inelastic Neutrino Scattering

Some other sum rules of the type of the *Bjorken sum rules* for the proton structure function $F_2(x, Q^2)$ shown in Fig. 5.11 have been proposed for exploring the internal structure of the proton. Instead of using the electron, one considers the (anti) neutrino off-proton scattering inclusive processes.

♣ Adler Sum Rule

This sum rule presented by Stephen Adler (Fig. 6.19) in 1966 relates the proton structure function $F_2(x, Q^2)$ associated to the DIS inclusive process: $\nu(\bar{\nu}) + p \rightarrow l^{\pm} +$ hadrons, where here there is an *exchange of the W-boson instead of the photon* in the ep scattering process. The sum rule states

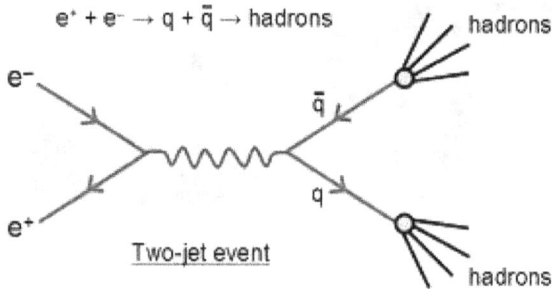

Figure 6.11: Schematic view of a jet of quark–antiquark in the e^+e^- collision.

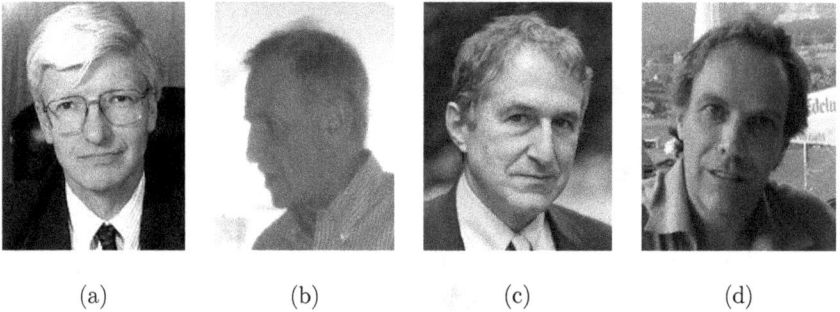

| (a) | (b) | (c) | (d) |

Figure 6.12: (a) C. Llewellyn Smith, (b) A. Kataev, (c) R.L. Jaffe, (d) G.M. Shore.

that the integral of their difference over the x fraction of parton inside the proton is constant: $\int_0^1 (dx/x) \left(F_2^{\bar{\nu}p} - F_2^{\nu p} \right) = 2$, and is not affected by the QCD quantum corrections.

◇ Gross–Llewellyn Smith Sum Rule

The Gross–Llewellyn Smith (GLS) sum rule presented by David Gross (Fig. 6.5) and Chris Llewellyn Smith (1942–) (Fig. 6.12) in 1969 is also applied to the $\nu(\bar{\nu})$ off-proton DIS process but *involves the parity violating part (called $F_3(x, Q^2)$) of the hadronic tensor*. It measures the sum of the quark and anti-quark distributions inside the proton and reads: $(1/2) \int_0^1 dx (F_3^{\bar{\nu}p} + F_3^{\nu p}) = 3(1 + \text{QCD corrections})$, where the QCD corrections are: $-\alpha_s/\pi + \dots$.. This sum rule has been used in the literature [see e.g. Andrei Kataev (Fig. 6.12) and collaborator papers] for extracting from the data the value of the QCD coupling α_s and power corrections which enter through the QCD corrections.

6.4 Polarized DIS Process and the Proton Spin

This is the extension of the *ep* scattering process to the one of *μp* where the *proton target is polarized* (spin oriented in a given direction). This process has been used for testing the *John Ellis (1946–, Fig. 6.15) and Robert Jaffe (1943–, Fig. 6.12) sum rule:* $\Gamma_1(Q^2) = \int_0^1 dx\, g_1(x, Q^2)$ *where g_1 is the singlet part of the parity violating structure function of the proton.* This quantity *is expected to measure the spin content of the proton.* In the naïve quark constituent model picture of the proton $(p = uud)$, the Ellis–Jaffe sum rule predicts that, for three quarks flavors u, d, s, $6\Gamma_1$ possesses a singlet $G_A^{(0)} \equiv \Delta\Sigma \equiv \Delta u + \Delta d + \Delta s$, a triplet $G_A^{(3)} \sim \Delta u - \Delta d$ and an octet $G_A^{(8)} \sim \Delta u + \Delta d - \Delta s$ pieces, where Δq *is the polarized quark distribution function* which is the difference between the quark spin oriented up with the one of the quark spin oriented down $(\Delta q = q^\uparrow - q^\downarrow)$ distributions inside the proton. Using the quark model the predicted values of $G_A^{(3)}$ and $G_A^{(8)}$ are well reproduced by experiments from hyperon and β-decays. In the Ellis–Jaffe sum rule, the quark model predicts: $\Delta\Sigma \simeq (0.6 \pm 0.2)$, while the measurement from European Muon Collaboration (EMC) experiment in 1988 indicates that this quantity has a much smaller value $\Delta\Sigma \simeq (0.19 \pm 0.17)$ which has been confirmed and improved later on by the SLAC and by some other recent experiments. However, an agreement is found with the unpolarized Bjorken sum rule within the experimental error for both experiments. This disagreement with the Ellis–Jaffe sum rule prediction for $\Delta\Sigma$ can be interpreted as an evidence that *the spin of the proton is not carried only by the quarks and where the gluon or/and sea quarks play an important role.* An avalanche of papers appeared after this experimental result. In particular, in a series of papers, Graham Shore (1953–) (Fig. 6.12) and G. Veneziano (Fig. 5.3) attribute this feature to the presence of the gluon inside the derivative of the quark–antiquark singlet current which is sandwiched between two proton states (proton matrix elements). It evokes that the renormalized proton coupling to the quark pseudoscalar current $\bar{u}\gamma_5 u$ is proportional to the square root of the q^2-derivative $\chi'(0)$ evaluated at $q^2 = 0$ of the non-perturbative two-point function called *topological susceptibility:* $\chi(q^2) = i \int e^{iqx}\langle 0|\mathcal{T}Q_R(x)Q_R(0)|0\rangle$ where $Q_R(x)$ is the renormalized of the bare gluon density $Q_B(x) = (\alpha_s/8\pi)Tr(G^{\mu\nu}\tilde{G}_{\mu\nu})$; Tr means trace and $\tilde{G}_{\mu\nu} = (1/2)\epsilon_{\mu\nu\rho\sigma}G^{\rho\sigma}$ is the parity violating piece of the gluon field strength tensor $G_{\mu\nu}$. As $Q_R(x)$ mixes with the quark current $\bar{u}\gamma_5 u$

under renormalization, one should make a careful treatment of these matrix elements for consistency. In a companion paper, Shore, Veneziano and the author [39] have performed some phenomenological applications of these ideas by extracting the slope of the topological susceptibility using QCD spectral sum rules for explaining the EMC data. They have also proposed to test these ideas in the photon–photon scattering process where the gluons are replaced by photons.

6.5 Some Experimental Tests of QCD

♣ Observing the Jets of Quarks and Gluons

• *What are the jets of quarks?*
They are illustrated in Fig. 6.11 in the e^+e^- collision to give a pair of quark and antiquark $\bar{q}q$. The quark and the antiquark move away but the strong interaction that binds them with a string prevented them to separate because this string tension increases with the distance between $\bar{q}$ and q (it is also called phenomenon of *anti-screening of gluons*). With the energy gained, one can then create new $\bar{q}q$ pairs which will give birth to new hadrons (phenomenon of *hadronization*) which are issued in the same directions as the initial $\bar{q}q$ pair. They will then gather around directions which remember the ones of the initial $\bar{q}q$ pair. These sets of *hadrons that have large transverse momenta are called the jets.*
• *Jets in e^+e^- collisions*
The 1st quark jets of the type in Fig. 6.13 has been observed by Gail Hanson (1947–, Fig. 6.14) *in the MARKII experiment of SPEAR*, the SLAC, Stanford e^+e^- collider and confirms the prediction of the Feynman Parton Model. In 1976, John Ellis, Marie-Katharine Gaillard (1937–) and Graham Ross (1929–) (Fig. 6.15) show that in any hard collisions (high energy), produced quarks will also emit gluons which will, in turn, emit *jets of gluons* in addition to quark–antiquark jets. In 1977, George Sterman (1946–, Fig. 6.14) and S. Weinberg (Fig. 5.8) show that one can define, rigorously in QCD, the jet cross-sections after a careful treatment of infrared and collinear singularities. As a consequence, John Ellis, Alvaro De Rújula (1944–), Emmanuel Floratos (1947–) and Mary K. Gaillard (Fig. 6.15) showed that antenna patterns of *gluon radiation* can be calculated in QCD from which one can show an evidence for gluon radiation though individual three-jet events could not be distinguished. *Gluon jets were seen for the first time in e^+e^- experiments at DESY Hamburg in 1979* (Fig. 6.13b) by S.L.

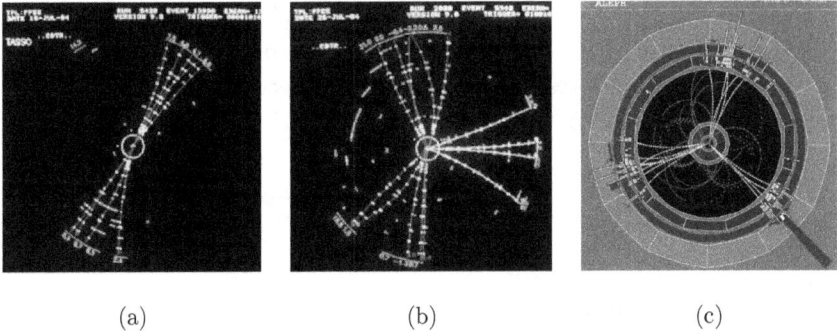

Figure 6.13: Observation of jets in the collision e^+e^- of DESY–Hamburg: (a) jet of quark and antiquark; (b) quark–antiquark jet and gluon. (c) quark–antiquark jet and gluon observed at LEP by ALEPH.

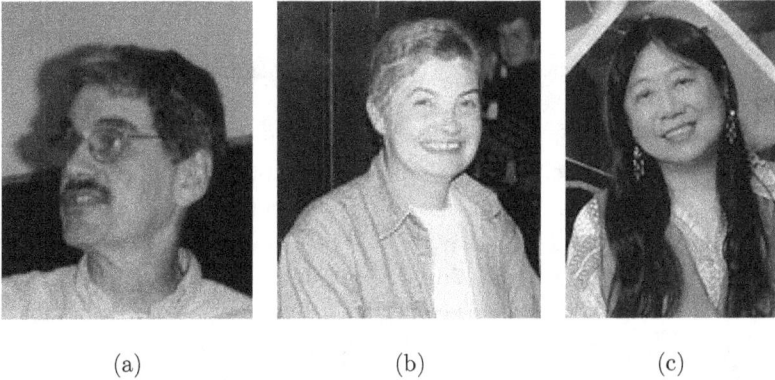

Figure 6.14: (a) G. Sterman, (b) G. Hanson, (c) S. Wu.

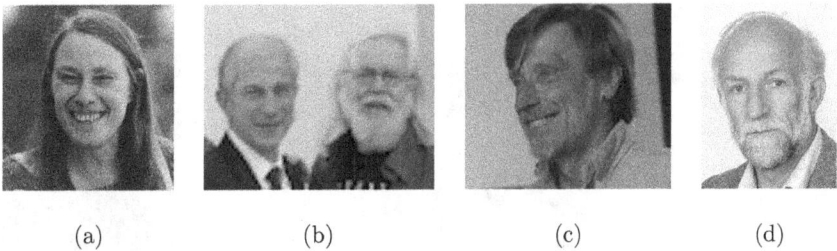

Figure 6.15: (a) M.K. Gaillard, (b) E.G. Floratos, J. Ellis, (c) A. De Rujula, (d) G. Ross.

Wu (1960–, Fig. 6.14) and confirmed at CERN by the ALEPH group in the e^+e^- collisions in LEP (Large Electron Positron collider) (Fig. 6.13c). After a such discovery, some tests have confirmed the vector spin one nature of the gluons as expected from $SU(3)_c$ gauge theory. Three-jets have been used at LEP for measuring with high-precision the value of the QCD coupling α_s while four-jets have been used for testing the three-gluon coupling shown in Fig. 6.3 which is a characteristic of the non-abelian gauge theory nature of QCD.

- *Jets in pp and $\bar{p}p$ colliders and Charpak Detector*

The pp (proton–proton) ISR (Intersecting Storage Ring) collider of 1500 GeV energy was built at CERN (see Fig. 10.6). It allowed to detect the possibility of a hard collision (large transverse momentum) between an extracted quark of a proton with that from the other proton. These quarks are then going to give rise to the jets of particles. This experimental observation was made by placing detectors to 90° over the beams, which necessitated the use of a new type of detector with multiple-wire proportional chambers (commonly called *chamber multiwire detector or Charpak detector*) invented in 1968 by Georges Charpak (1924–2010, Fig. 6.16), Nobel Prize of 1992. This detector has replaced that of the *bubble chambers* used before and has the advantage of allowing observing ionized particles whose data can be processed by computer. In the same way as the ISR, the SPS (Super Proton Synchrotron), 7 km in circumference, where it circulates in reverse an antiproton, was also built with an energy of 270 GeV for each proton (see Fig. 10.6). The SPS has helped the UA1 and UA2 (Underground Area 1 and 2) groups to detect jets of quarks and gluons issued at a wide angle and with large transverse momenta. These QCD jets will play an important role in the discoveries of gauge bosons W and Z of weak interactions that we shall discuss later.

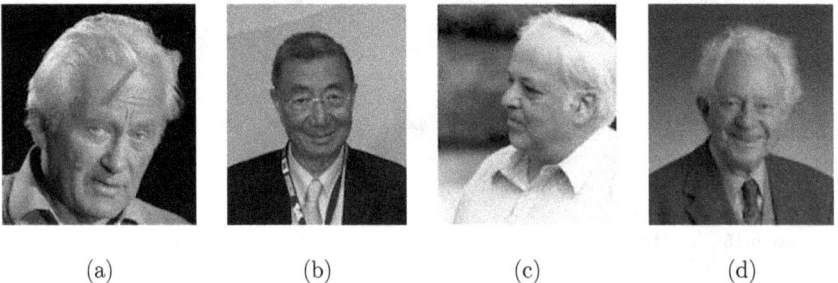

| (a) | (b) | (c) | (d) |

Figure 6.16: (a) G. Charpak, (b) S. Ting, (c) B. Richter, (d) L. Lederman.

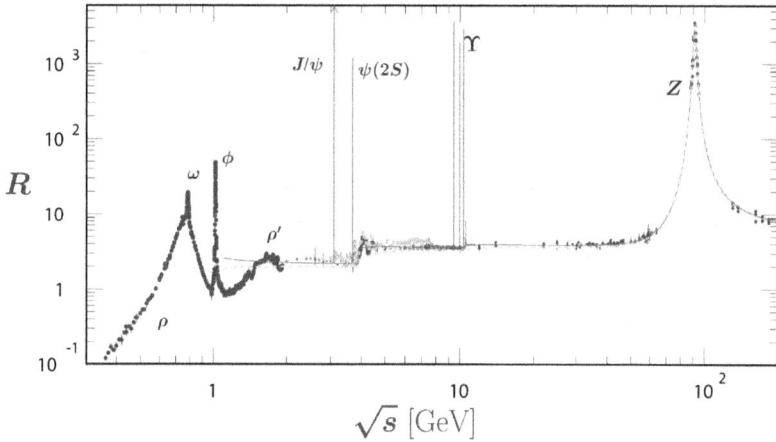

Figure 6.17: Ratio of cross-section for the production of hadrons and $\mu^+\mu^-$ in e^+e^- collisions.

◇ Searching the Heavy Flavours of Quarks

We have already discussed the flavors of the light quarks u, d, s in the Quark Model. Other flavours of heavier quarks have been discovered in new experiences:

• *The c charm quark*
In 1974, the c charm quark was discovered independently by Samuel Ting (1936–) and Burton Richter (1931–) (Fig. 6.16) (Nobel Prize of 1976, awarded for the observation of the J/ψ ($\bar{c}c$) meson in e^+e^- experiments) (Fig. 6.17). It will be discussed in the Standard Model that the discovery of the charm quark of electric charge $Q = +2/3$ was a milestone for the understanding of the theory. To characterize it, one has assigned an intrinsic quantum number $C = +1$ called the *charm* similar to the strangeness $S = -1$ for the s quark.

• *The b beauty or bottom quark*
In 1977, the bottom or beautiful quark b was discovered in proton–proton collisions having 400 GeV energy by the group at Fermilab (Chicago) of Léon Lederman (1922–, Fig. 6.16), Nobel Prize of 1988, by an observation of the meson Υ bound state of $\bar{b}b$. This result was confirmed in e^+e^- experiments (Fig. 6.17). The b-quark has a charge $Q = -1/3$ and its new intrinsic quantum number is $B = -1$ called the *beauty* (not to be confused with the baryon number $\mathcal{B}$!).

• *The t top quark*

It was also discovered in 1994–1995 by the CDF and D⊘ groups of Fermilab in the Tevatron experiment of 1.8 TeV energy proton–antiproton collision. Unlike other quarks, its mass is very heavy: $M_t = (173.5 \pm 1.0)$ GeV which is equivalent to the mass of an atom of tungsten or gold (see Table 6.1). It has a very short lifetime of 10^{-25} s, which does not allow it to hadronize or to form bound states so it is not confined unlike other quarks. It has a charge $Q = +2/3$ and its associated quantum number is $T = +1$ called *top* or *truth*.

• *Summary of flavours of quarks*

We saw that we have discovered six types of quarks that are shown in Table 6.1. In addition to their respective charge, to each is assigned a new intrinsic quantum number. The formula connecting the charge to these quantum numbers is always: $Q = I_3 + Y/2$, *where I_3 is the third component of isospin and Y hypercharge but here* $Y = B + S + C + B + T$ *with B, the baryon number and $S, ..., T$ the quantum numbers of quark flavors.* Note that the baryon number of a quark is $+ 1/3$, while the antiquark has an opposite charge and baryonic number. We summarize the quantum numbers and the masses of each quark on Table 6.1. Note that the masses that appear there are not effective or constituent Quark Model masses. These are the perturbative masses that appear in the QCD Lagrangian that are solutions of the renormalization group equation and their values are function of energy (decreases when the energy increases). *The masses of the light quarks u, d, s are evaluated at 2 GeV energy* which is a typical scale of the corresponding hadronic masses. *Masses of heavy c, b and t quarks are evaluated at the value of the corresponding mass. Unlike leptons, quarks are not observable as free states (except the top quark) because of confinement.* Therefore, we

Table 6.1 Flavor of quarks and their intrinsic quantum numbers. $B = 1/3$ is their baryon number. $Q = I_3 + Y/2$ with I_3 the third component of isospin and $Y = B + S + C + B + T$ is the hypercharge. The last column is the so-called quark running masses. Quarks are fermions of spin 1/2.

Flavor	I_3	S	C	B	T	Q	Mass [MeV]
u	$+1/2$	0	0	0	0	$+2/3$	2.8 ± 0.2
d	$-1/2$	0	0	0	0	$-1/3$	5.1 ± 0.4
s	0	-1	0	0	0	$-1/3$	96.1 ± 4.8
c	0	0	$+1$	0	0	$+2/3$	1261 ± 12
b	0	0	0	-1	0	$-1/3$	4177 ± 11
t	0	0	0	0	$+1$	$+2/3$	$(173.2 \pm 0.9) \times 10^3$

need theoretical approaches to estimate their values. These approaches are, for instance, explained in [30–32] and PDG [5].

♡ Counting Colors

Colors counting are also essential to confirm the theory. We saw this example in the case of the Δ^{++} baryon. We give here a few examples:

• *Neutral pion decaying into two photons*
In the language of Feynman diagram and the hypothesis where the pion is a bound state of quark and antiquark d and u: $|\pi^0\rangle = (1/2)|\bar{u}u - \bar{d}d\rangle$ (Fig. 5.2), the process $\pi^0 \to \gamma\gamma$ is described by the diagrams in Fig. 6.18. In calculating these diagrams, Jacques Steinberger (1921–, Fig. 6.19) (Nobel Prize of 1988) has shown in 1949 that its amplitude is proportional to $N_c(Q_d^2 - Q_u^2)$ where N_c is the number of colors of quarks (here $N_c = 3$), $Q_d = -1/3$ and $Q_u = +2/3$ are the charges of the quarks d and u, in units of the electric charge e:

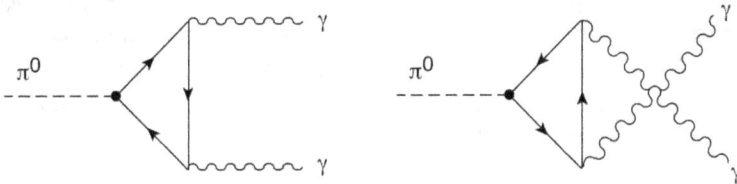

Figure 6.18: Decay of $\pi^0 \to \gamma\gamma$ in the Quark Model where the quarks u and d are circulating in the arrowed triangle. The second diagram is obtained by reversing the movement of the photon (crossed diagram).

(a)   (b)   (c)

Figure 6.19: (a) J. Steinberger, (b) R. Jackiw, (c) W. Bardeen, S. Adler.

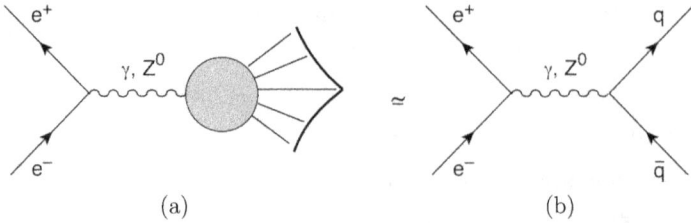

Figure 6.20: (a) Feynman diagram showing the process: $e^+e^- \to$ hadrons with intermediate virtual photon and Z^0 boson; (b) equivalence in the Quark Model.

$$\Gamma(\pi^0 \to \gamma\gamma) = [N_c(Q_d^2 - Q_u^2)]^2 \left(\frac{\alpha^2}{32\pi^3}\right) \frac{m_\pi^3}{f_\pi^2},$$

where, $\alpha = e^2/4\pi$ is the electromagnetic structure constant, m_π is the pion mass and $f_\pi = 130$ MeV its decay constant. Its decay rate $\Gamma(\pi^0 \to \gamma\gamma)$ is predicted from the previous theoretical calculation to be: 7.7 eV, in good agreement with the experimental result of 7.7 (6) eV. This result obtained well before QCD confirms the three colors of quarks. This good agreement with experiment also comes because the $\pi^0 \to \gamma\gamma$ process is not affected by the QCD quantum corrections. This is the so-called *non-renormalization theorem of the anomaly* that was later shown by John Bell (Fig. 1.26) and Roman Jackiw (1939–), Stephen Adler (1939–) and William Bardeen (1941–) (Fig. 6.19).

• $e^+e^- \to$ *hadrons total cross-section at high energy*

According to the Parton Model and the asymptotically free property of QCD, the ratio R of the cross-section for the reaction: $e^+e^- \to$ Hadrons (see Fig. 6.20) over the one of $e^+e^- \to \mu^+\mu^-$ must be equal to $N_c \sum_f Q_f^2$ where the sum runs over quark flavors $u, d, s, \ldots$:

$$R_{e^+e^-} \equiv \frac{\sigma(e^+e^- \to \text{Hadrons})}{\sigma(e^+e^- \to \mu^+\mu^-)} = N_c \sum_f Q_f^2(1 + \alpha_s/\pi + \cdots).$$

We show the measurement of this quantity in Fig. 6.17 as function of the energy. To produce three quarks u, d, s, we shall work in the energy between 1.5 and 3 GeV region where this perturbative QCD result is applicable (validity of the perturbative calculation of QCD above the resonances ρ, ω, ϕ and below the production of resonance J/ψ which is a $\bar{c}c$ bound state of the fourth quark charm c with the anti-charm $\bar{c}$. The dashed (green) line shows the predictions of QCD: $R = 3[(2/3)^2 + (-1/3)^2 + (-1/3)^2] = 2$ which is in very good agreement with experimental data (points with error bars).

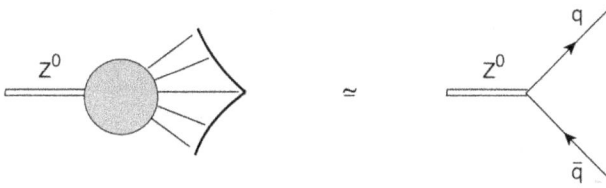

Figure 6.21: Similar to Fig. 6.20 but for $Z^0 \to$ Hadrons.

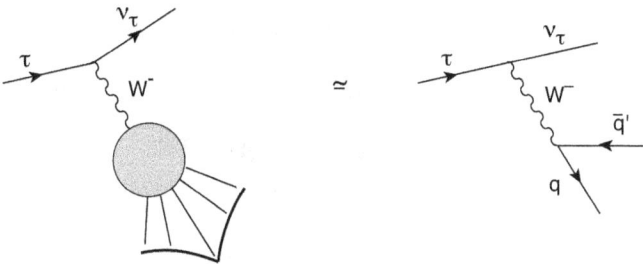

Figure 6.22: Similar to Fig. 6.20 but for $\tau \to \nu_\tau +$ hadrons.

• *The hadronic decays of the $W^\pm$ and Z^0 bosons*

A similar result is obtained in the analysis of the decay rate into hadrons of $W^\pm$ and Z^0 bosons mediators of weak interactions (see Chapter 7). The decay of the Z^0 is represented in Fig. 6.21.

• *Heavy lepton τ hadronic decay*

The heavy lepton τ having a mass 1.8 GeV was discovered by Martin Perl (Fig. 6.8) and its associated neutrino was observed later. The total number of leptons is six as shown in Table 7.1. Under the quark parton model (Fig. 6.22), the branching ratio of the decay rate $\tau \to \nu_\tau +$ hadrons over the leptonic decay one, $\tau \to \nu_\tau + l + \nu_l$, and including QCD corrections is:

$$R_\tau \equiv \frac{\Gamma(\tau \to \nu_\tau + \text{hadrons})}{\Gamma(\tau \to \nu_\tau + l + \nu_l)} = 3(1 + \alpha_s/\pi + \cdots),$$

while experiments give $R_\tau = 3.65(5)$ which is not in agreement with the parton model prediction $R_\tau = 3$. However, this difference of 20% between the theoretical and experimental prediction measures the values of the quantum corrections to the quark parton model which are relatively large at this energy. This deviation has been analyzed by E. Braaten, A. Pich and the author (Fig. 6.8) in 1990 [37] to determine the value of the QCD coupling constant α_s shown in Fig. 6.7 at the value of the τ-mass.

6.6 QCD as a Theory of Hadrons

Understanding of the *properties of hadrons is linked heavily to the problem of confinement of quarks and gluons at low energy*. However, there is still no analytical convincing proposal for the resolution of this problem. All what we know is that the potential of QCD behaves like a $1/r$ Coulomb potential at short distance (high energy) and is linear in r at long distance (low energy). So, I am going to give as follows a (non-exhaustive) list of the approximate analytical approaches and numerical lattice simulation introduced for understanding the hadronic properties.

♣ Why a High Spin Hadron is More Heavy?

This can be understood by analogy with QED. Take the case of a positronium which is an $e^+ e^-$ bound state. This state is called *orthopositronium* (1) when the e^+ and e^- spins are oriented in the same direction (total spin $= 1 = 1/2 + 1/2$), while it is a *parapositronium* (2) if not (total spin $= 0 = 1/2 - 1/2$). The difference between the two states comes from the difference of sign of the magnetic field which involves a small difference in their level of energy (*hyperfine splitting*). In case (1), they repel magnetically, which implies that the magnetic force acts against the electric force. This is not the case of (2) because the two forces are added. Therefore we must provide more energy in (1) as in (2) explaining that the energy level of (1) is higher than that of (2). The same reasoning can be applied in the case of QCD to explain why the Δ^{++} of spin $3/2$ (3 quarks aligned spins) is heavier than the proton of spin $1/2$ (spin of a d quark opposite to the other two quarks u). The fact that the Δ^{++} is heavier than the proton ensures the stability of our matter because otherwise the proton would decay to Δ^{++}. Similarly, one can understand the 600 MeV mass difference between the ρ and π meson. However, if we look at the $\bar{c}c$ meson bound states of the heavy quark c, we see that the difference in mass between the meson of spin 1 (the J/ψ) and spin 0 (the η_c) is only 100 MeV. This may be explained by the fact that each quark has a calculable magnetic moment and the more the quark is heavy, less is the associated magnetic force. So it takes less energy to separate the two hyperfine states of $\bar{c}c$.

◇ Non-Relativistic or Potential Models

We saw in the Quark Model that for explaining the hadronic spectra, one just uses the quark constituent or effective masses: $m_u \simeq m_d \simeq M_P/3 =$

(a)   (b)   (c)   (d)

Figure 6.23: (a) J.M. Richard, (b) M. Neubert (courtesy of Thomas Hartman, Johannes Gutenberg University Mainz), (c) J. Gasser, (d) K. Wilson.

313 MeV and $m_s \simeq M_\Omega/3 = 557$ MeV where, these masses contain mostly the value of the binding energy of these quarks inside the baryon and a tiny contribution of current or perturbative mass that appears in the QCD Lagrangian and in approaches based on current algebras called Chiral Lagrangian approach [17]. Potential models improve the Quark Models by including other corrections (spin-orbit, hyperfine energy interaction,...). The predictions of these models on the spectroscopy of hadrons are very impressive. However, the direct link to QCD of these models for hadrons built with light quarks u, d, s is still mysterious which resulted a neglect of these models to the benefit of other approaches. This is not the case of the heavy quarks where the QCD Coulomb potential $V(r) \simeq \alpha_s(r)/r$ is a good approximation to explain spectra of heavy hadrons as the quark mass where the energy scale is greater than the scale Λ of the QCD coupling [see e.g. the review written by J.M. Richard (1947–, Fig. 6.23) [14]].

♡ Heavy Quark Effective Theory (HQET)

Another alternative is the effective theory called HQET to study the masses and decay of a quark containing a heavy quark and a light quark like the $B(\bar{b}d)$ meson where one uses an expansion in terms of $1/m_b$ which has a meaning because the b quark mass m_b is large relative to the scale Λ of QCD. A comprehensive review on this subject has been written by Matthias Neubert (1962–, Fig. 6.23) [15].

♠ Bag Models

These models have also generated a lot of enthusiasm in the years 1970–1980 but were also gradually abandoned. They consider hadrons as spherical balls (bag) with an energy zero, outdoors (the boundary conditions). The most popular model is the one by the Massachusetts Institute of Technology (MIT) group (R.L. Jaffe, Fig. 6.12c) where hadrons are obtained by exciting different color singlet combinations inside the bag.

♣ Effective Chiral Lagrangian

The approach is based on an effective Lagrangian which reproduces the chiral symmetry of QCD and permits to study QCD at low energy where perturbative calculations in terms of quarks and gluon degrees of freedom are not applicable. One then describes hadrons by the products of the pion fields ϕ and makes the analysis by developing the inverse of the pion coupling constant f_π where the pion field enters via the unitary matrix $U = \exp(i\phi/f_\pi)$. The simplest form of the Lagrangian is $\mathcal{L} = (f_\pi^2/4)\mathrm{Tr}[\partial_\mu U^+ \partial^\mu U]$, to it one can add higher dimension terms. However, *the theory is not renormalizable* (not finite) and faces a growing number of arbitrary low energy constants (counterterms) to be determined by experimental data or from some other approaches (lattice calculations, QCD spectral sum rules) combined with the constraints of chiral symmetry (relation between the left and right component or between the pseudoscalar and scalar components of the field) in order to have a consistent theory. *This approach is a modern version of the algebra of currents* and was developed in 1984 by Juerg Gasser (1944–, Fig. 6.23) and H. Leutwyler (Fig. 6.6) and later on by several authors. A recent review on the subject is given by E. de Rafael (Fig. 4.86) and A. Pich (Fig. 6.8c) in [17].

♦ Lattice Calculations

This approach developed in 1974 by Kenneth Wilson (1936–2013, Fig. 6.23), Nobel Prize of 1982, is to *discretize the action of QCD in the form of lattice points in space-time* for numerical simulations of observables. Here *quark fields are defined at lattice site while gluon fields are defined on the links connecting neighbor sites.* The development of this approach is linked to the progress of the computer because one can work at present with lattices of larger size more close to reality where the action is continuous, but also to the introduction of the quantum corrections due to quarks loops. Several international groups are working on it and get some interesting results. However, the meaning of the errors obtained in some results is often not clear and raises some doubts on their veracity requiring some cross-check of the results by some other groups.

6.7 QCD Spectral Sum Rules (QSSR)

• *Introduction and the Weinberg Sum Rules*

QSSR is based on the duality between observables that can be measured experimentally and are calculable in QCD. This duality arises from the *dispersion relation* that connects the integral of the imaginary part or spectral function $\text{Im}\Pi$ (measurable experimentally) of the observable (two-point correlation function Π being the time ordered product of two hadronic currents) with its real or total part (in principle computable with the QCD at the energy where perturbative calculations can be applied) using the analyticity property of the correlation function. This dispersion relation is written as (see textbooks): $\Pi(q^2) = \int_0^\infty dt/(t-q^2)(1/\pi)\text{Im}\Pi(t)$. Supercon-vergent Weinberg sum rules (see Chapter 5) that have been discussed in the case of the algebra of currents are the first examples of this type of sum rules where, using the asymptotic realization of chiral symmetry at very high energy, he has derived the sum rules: $\int_0^\infty dt \,\text{Im}\Pi_{LR}^{(1+0)}(t) = 0$ and $\int_0^\infty dt\, t\,\text{Im}\Pi_{LR}^{(1,0)}(t) = 0$ where the indices 0 and 1 denote that the hadrons entering into the spectral functions have spin 0 and 1, while the index LR means that the spectral function is the difference between the one of the vector with the one of the axial-vector currents. From these sum rules and assuming a dominance of the lowest mass hadrons in the spectral integral, Weinberg deduced a relation between the mass and coupling of the vector meson ρ with those of its axial partner A_1: $M_{A_1} \simeq \sqrt{2}M_\rho \simeq 1.1$ GeV, which, within the crude approximation used, is in good agreement with the measured mass of 1.26 GeV [5]. Improved versions of these Weinberg sum rules including the effects on non-zero quark mass values and QCD radiative corrections have been studied by E.G. Floratos (Fig. 6.15), E. de Rafael (Fig. 4.8) and the author in [40].

• *The SVZ sum rules*

In 1979, Michael Shifman (1949–), Arkady Vainshtein (1942–), and Valya Zakharov (1940–) (SVZ) [18] (Figs. 6.24 and 6.25) have largely improved the previous dispersion relation by the contribution, on the one hand, of an exponential weight $\exp(-t\tau)$ in the integral which replaces the kernel $1/(t-q^2)$ where τ is a new variable (imaginary time in the language of quantum mechanics) which replaces $1/q^2$. This weight enhances the relative contribution of the low energy region inside the integral and thus optimizes the extraction of the lowest mass resonance parameters. On the other hand, the *SVZ sum rule* improves the perturbative calculation of QCD at relatively low energy of the order of $\tau^{-1/2} \simeq (1-2)$ GeV by the introduction

Figure 6.24: A. Vainshtein, the author, V. Zakharov and M. Shifman at the Zakharovfest, Ringberg castle, Tegernsee, Munich (16–21 May 2005).

of the non-perturbative effects simulated by the average values in the vacuum of the product of quark and/or gluon fields called quark, gluon and quark–gluon mixed condensates. Indeed, it is assumed that the confined part of QCD is approximately parametrized by the sum of the contributions of these higher and higher dimension condensates that appear in the $1/q^{2n}$ Wilson (Fig. 6.24) expansion, where the hadron correlation function can be written in the form:

$$\Pi(q^2) = C_0 + \sum_{n=2,3,\ldots} \frac{C_{2n}\langle 0|O_{2n}|0\rangle}{q^{2n}} \ ,$$

where C_{2n} are Wilson coefficients that can be calculated within perturbation theory and $\langle 0|O_{2n}|0\rangle$ are the non-perturbative condensates. C_0 is the contribution due to the standard perturbation theory. The best-known condensate is the *quark condensate* $\langle 0|\bar{\psi}\psi|0\rangle \equiv \langle \bar{\psi}\psi \rangle$ which breaks chiral symmetry spontaneously as it is related to the pion mass and to its decay constant f_π (which controls its decay) through the *Gell-Mann–Oakes–Renner relation*: $m_\pi^2 f_\pi^2 = -(m_u + m_d)\langle \bar{u}u + \bar{d}d\rangle$. The pion plays here the role of a quasi-Nambu–Goldstone for breaking such a symmetry.

Figure 6.25: M. Shifman, V. Zakharov, H.G. Dosch and the author at QCD 10, Montpellier in 2010 during the celebration of the 30 $\oplus$ 1 years of the SVZ sum rules.

In addition to this quark condensate, SVZ have postulated the existence of the *condensates built from gluon fields*, which, at dimension four, are given by the $\langle \alpha_s G^2 \rangle \equiv \langle 0 | \alpha_s G_a^{\mu\nu} G_{\mu\nu}^a | 0 \rangle$ condensate where $G \equiv G_{\mu\nu} = \partial_\mu A_\nu^a - \partial_\nu A_\mu^a + g f_{abc} A_\mu^b A_\nu^c$ is the strength associated to the eight gluon fields A_μ^a; a, b, c are color and μ, ν are Lorentz space–time indices; f_{abc} is the $SU(3)_c$ gauge group structure constant. Some other more complicated combinations and some condensates built from the gluon and quark fields (mixed condensates) appear when the dimensions of the condensates are larger than four. At dimension five, we have the *mixed condensate* $\langle g \bar{\psi} G \psi \rangle$. At dimension six, we have the *four-quark* $\langle \bar{\psi}\psi\bar{\psi}\psi \rangle$ *and three-gluon* $\langle g G^3 \rangle$ *condensates*. The idea of SVZ is that, using a finite number of these condensates (in practice having dimension smaller or equal to 6–8), one can describe with a good approximation the hadronic correlation function $\Pi(q^2)$ thanks to the convergence of the $1/q^2$-expansion at this value (1–2 GeV) of the energy. Kostya Chetyrkin (1950–) (Fig. 6.26), Valya Zakharov (Fig. 6.25) and the author [41] have added a new term due to a *tachyonic* ($\lambda^2 \le 0$)

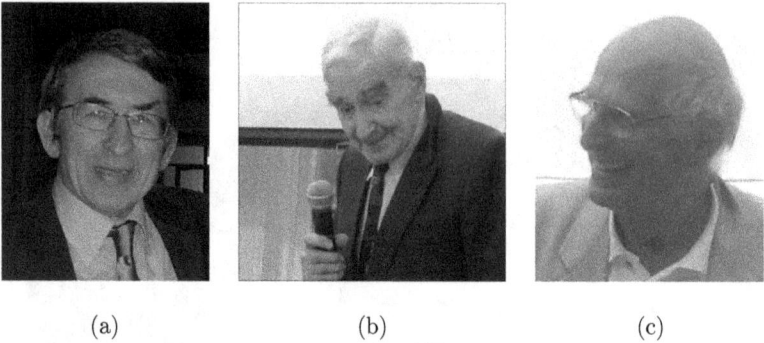

(a)   (b)   (c)

Figure 6.26: (a) K. Chetyrkin, (b) B.L. Ioffe (courtesy of ITEP), (c) H.G. Dosch.

gluon mass squared having a dimension $d = 2$ which is not present in the original Wilson expansion. This is due to the fact that a $d = 2$ condensate cannot be present in the QCD Lagrangian because it violates gauge invariance. Here, *this term is used to parametrize (by duality) the terms of higher order of perturbation theory* for improving the approximation of perturbation series which are of the ultraviolet (UV) high-energy origin. This UV term should not to be confused with an infrared (IR) $d = 2$ term of the low-energy origin appearing in the gluon propagator discussed in the literature.

6.8 QSSR Phenomenology

The QSSR Phenomenology is rich and its different applications for studying the hadron properties are successful. In the early time of the QSSR, the predictions of SVZ on the values of the ρ meson mass and couplings and on its axial-vector partner, the A_1 meson were impressive. In one hand, this analysis has been extended by SVZ to the $J/\psi(\bar{c}c)$ meson family for predicting the charm quark mass and especially the gluon vacuum condensate $\alpha_s\langle G^2\rangle$ with the values 1.26 GeV and 0.04 GeV^4 respectively. On the other hand, QSSR has been used by Boris Ioffe (1925–, Fig. 6.26) and by Hans Guenter Dosch (1936–, Fig. 6.26) *et al.* to give the first predictions of the light baryon spectra [19], which we have improved with H.G. Dosch and M. Jamin (1962–) [42] (for a review, see e.g. [33, 34]).

♣ QCD Fundamental Parameters

Due to the duality between the experimental data and the QCD theory through the improved dispersion relation, *the approach can provide a phenomenological estimate of some QCD fundamental parameters once the properties of the corresponding hadrons are experimentally known and vice-versa.* Among these applications, we have already discussed the determination of the QCD coupling α_s from τ-decay. In the following, we shall mention few other examples.

• *Light Quark masses*

In the early beginning of QCD, the definitions of the quark masses were unclear. This is due to the fact that, unlike the electron, the quark is not freely observed which renders delicate the meaning of its mass. *The mass of the observed electron can be identified with the pole of the propagator* which is commonly named *pole or on-shell mass*. In the quark (constituent) model, the mass of the light quarks u and d, which are almost equal according to the approximate isospin symmetry, are taken to by 1/3 of the proton mass as the proton is composed by 2 up and 1 down quarks. However, these masses are not the ones entering into the QCD Lagrangian, which are the masses generated by the Higgs boson coupling to the quark–antiquark bilinears (see Chapter 7).

— *The masses of the Lagrangian are called current algebra masses* where their ratios are known prior to QCD from the current algebra phenomenology. For instance: $m_u/m_d = 0.50(3)$ and $2m_s/(m_u + m_d) = 24.4(1.5)$. However, their absolute values which are useful for controlling the breaking of chiral and flavor symmetries are unknown. In 1979, E.G. Floratos (Fig. 6.15), E. de Rafael (Fig. 4.8) and the author have clarified the problem by introducing the definition of the *renormalization group invariant* $\hat{m}$ (very similar definition as the QCD scale Λ) and of the *running masses* $\bar{m}(Q)$ function of the energy Q (similar to the running QCD coupling) within the 't Hooft dimensional renormalization $\overline{MS}$ scheme:

$$\bar{m}(Q) = \frac{\hat{m}}{(\log Q/\Lambda)^{2/-\beta_1}},$$

to order α_s, where $-\beta_1 = (33 - 2n/3)/2$ is the lowest order value of the β-function for n number of quark flavors. *The running mass also decreases to some power of* $1/\log$ *with increasing Q but slower than that of α_s.* It enters into the QCD Lagrangian mass term $m\bar{\psi}\psi$ and is related to the so-called current algebra but not to the constituent quark mass.

— In 1981, C. Becchi (1939–), F.J. Yndurain (1940–2008) (Fig. 6.27), E. de Rafael (Fig. 4.8) and the author [43] have been the first to give a lower bound on the sum of the u and d quark masses using the analyticity properties of the two-point correlator associated to the pseudoscalar quark current $(m_u + m_d) : \bar{u}\gamma_5 d$: and by including the α_s corrections to the QCD pseudoscalar correlator (see Appendix E). Series of papers have been written in the literature (as quoted in [44]) for improving this early bound and estimate of the light quark masses. The values of the u, d, s quark masses from [44] are compiled in Table 6.1. They agree with the ones from lattice QCD calculations compiled in [5] where determinations from some other sources can be also found.

— One can notice that the values of these masses are relatively light $(m_u = 2.8$ MeV) compared to the ones from the constituent quark models $(M_u = M_p/3 \simeq 330$ MeV), while it is the first value which is induced by the Higgs mechanism. *This feature indicates that the Higgs mechanism is not the full origin of the proton mass and consequently of the mass of the Universe. Here, the binding energy of quarks inside the proton gives the dominant contribution to the proton mass.*

• *Heavy quark masses*

For the heavy quark sytems, the binding energy is smaller than for the light quarks as they are almost static i.e. non-relativistic. In the case of $J/\psi(3100)$ charmonium state which is a $\bar{c}c$ bound state, the *constituent mass* is $M_c = M_{J/\psi}/2 \approx 1.5$ GeV [14]. In the case of the QCD based approach with a mass of the QCD Lagrangian, one may define here, within perturbation theory (PT), *a pole mass analogue to the electron mass* which is possible as the c-quark mass is much larger than the QCD scale Λ such that perturbative calculations can make sense. Rolf Tarrach (1948–, Fig. 6.27) [20] has provided the expression of the pole mass in perturbation theory. In [45], the author has clarified the connection of the pole

(a)   (b)   (c)

Figure 6.27: (a) C. Becchi, (b) F.J. Yndurain, (c) R. Tarrach.

and running masses in the $\overline{MS}$ scheme. A phenomenological estimate of the pole mass leads to a value approximately similar to the one of the constituent mass [33, 34]. However, this mass definition suffers from large QCD quantum corrections where the PT series converge badly and from some other PT technicalities at larger order (UV renormalons). Therefore, it cannot lead to a precise prediction. On the contrary, the running mass $\bar{m}(Q)$ is well-defined and converges faster. The recent values of the charm and bottom quark masses from [46] are compiled in Table 6.1. Some other determinations using some variants of QSSR and some other methods are quoted by PDG [5].

• *Non-pertubative QCD condensates*

The well-known QCD condensate: $\langle 0|\bar{\psi}\psi|0\rangle \equiv \langle\bar{\psi}\psi\rangle$ can be deduced from the GMOR relation $m_\pi^2 f_\pi^2 = -(m_u + m_d)\langle\bar{u}u + \bar{d}d\rangle$ once the value of the sum of light quark masses is known. Using the result in Table 6.1, its value at the energy of 2 GeV is $\langle\bar{u}u\rangle = -(267\pm7)^3$ MeV^3. The ratio of light quark condensates $\langle\bar{s}s\rangle/\langle\bar{d}d\rangle$ comes from the kaon sum rules [47] and the heavy baryons spectra in a work with R. Albuquerque and M. Nielsen [48]. The value of the gluon condensates $\langle\alpha_s G^2\rangle$ and $\langle gG^3\rangle$ have been re-extracted recently from the $e^+e^- \to J/\psi + \ldots$ charmonium data by the author [46]. The value of the mixed condensate $\langle g\bar{\psi}G\psi\rangle$ has been estimated from the light baryon sytems [19, 42]. The value of the tachyonic (imaginary) gluon mass λ^2 comes from τ-decay and $e^+e^- \to$ hadrons data [49].

◇ Open Charm and Beauty States

• *Pseudoscalar D, D_s, B, B_s Mesons and CP violation*

These states are composed with one light and one heavy quarks or with

Table 6.2 QCD parameters from QSSR.

QCD parameters	Values	Ref.
$\alpha_s(M_\tau)$	0.325(8)	[12, 38]
$\Lambda(n_f = 3)$	(353 ± 15) MeV	[12, 13, 38]
$-\langle\bar{\psi}\psi\rangle(2$ GeV$)$	$(267 \pm 7)^3$ MeV^3	[44]
$\kappa \equiv \langle\bar{s}s\rangle/\langle\bar{u}u\rangle$	(0.74 ± 0.06)	[48]
$-(\alpha_s/\pi)\lambda^2$	$(7 \pm 3) \times 10^{-2}$ GeV^2	[49]
$\langle\alpha_s G^2\rangle$	$(7 \pm 2) \times 10^{-2}$ GeV^4	[46]
$M_0^2: \langle\bar{u}Gu\rangle \equiv M_0^2\langle\bar{u}u\rangle$	(0.8 ± 0.2) GeV^2	[19, 42, 50]
$\langle g^3 G^3\rangle$	(8.2 ± 2.0) GeV$^2 \times \langle\alpha_s G^2\rangle$	[46]
$\rho\alpha_s\langle\bar{\psi}\psi\rangle^2$	$(5.8 \pm 1.8) \times 10^{-4}$ GeV^6	[38, 42, 51]

unequal heavy quark masses. In the pseudoscalar channels ($J^{PC} = 0^{-+}$), we have the $D^+(\bar{d}c)$, $D_s^+(\bar{s}c)$ and $B^+(\bar{b}u)$, $B_s^0(\bar{b}s)$ and the $B_c^+(\bar{b}c)$ mesons where d, s, c, b are the down, strange, charm and bottom quarks. Analogous quark contents occur for the vector $J^{PC} = 1^{--}$ states where the mesons are denoted by D^*, D_s^*, B^*, B_s^* and B_c^*. All of these mesons have been experimentally observed and their masses can be found in the PDG book [5]. Of particular interests are the D, D_s and B, B_s mesons, as the mass difference ΔM between the neutral state B^0 and its anti-particle $\bar{B}^0$ measures *CP*-violation. ΔM is proportional to the quantity $|V_{td}V_{tb}^*|^2\mathcal{M}$ where V_{ij} is the *Cabibbo–Kobayashi–Maskawa (CKM) mixing angle* (see Chapter 7) entering into the vertex of the i–j–W coupling of the W-boson to the quarks i and j. $\mathcal{M} \equiv \langle B^0|\bar{d}(1 - \gamma_5)\gamma_\mu b\bar{b}(1 - \gamma_5)\gamma_\mu d|\bar{B}^0\rangle = \frac{16}{3}f_B^2 M_B^2 B_B$ is the $B^0 - \bar{B}^0$ matrix element where B_B is the so-called bag parameter which measures the validity of the vacuum saturation estimate (it is expected to be around one for the heavy quark open states). f_B *is the decay constant analogue to* f_π *of the pion* and defined as: $\langle 0|(m_d + m_b)\bar{d}\gamma_5 b|B\rangle = f_B M_B^2$. It controls the leptonic decay: $B \to \tau\nu_\tau$ which is proportional to the square of the τ-lepton mass. Thus, for determining the CKM mixing angles, one needs to evaluate the decay constant f_B and the bag parameter B_B. Using QSSR, we have estimated the combination $f_B\sqrt{B_B}$ and compared the results with the estimated value of f_B. We (with K. Hagiwara, D. Nomura, A. Pivovarov) found that the deviation from the vacuum saturation estimate due to the non-factorizable contributions to the matrix elements is only of about 10% [52]. The estimate of f_B using QSSR have been earlier improved in [53] where it has been found that $f_D \approx f_B$ which is unexpected in the very heavy quark mass limit where a $1/\sqrt{M_b}$ behavior is predicted. This result indicated large quark mass corrections and the charm quark mass is not sufficiently heavy for applying this heavy quark prediction. Later on, this result has been confirmed by lattice calculations and experimental results. Recent estimates of the decay constants have been done and reviewed in [54]. The resulting value of the decay constants from the analysis are summarized in Table 6.3 and can be compared with some other results compiled in [5].

• *Vectors $D_{(s)}^*$, $B_{(s)}^*$ and pseudoscalar B_c mesons*

An updated estimate of the corresponding decay constants has been done in [52]. The results are summarized in Table 6.3.

• *Scalar D_0^*, D_{s0}^* and axial-vector D_{s1}^* mesons*

The masses and decay constants of the scalar and axial-vector open states have been estimated in [56]. The results are compiled in Tables 6.3 and 6.4 where one can notice a good agreement with the data when available.

Table 6.3 Decay constants of the pseudoscalar and vector open charm and beauty systems in units of MeV from QCD spectral sum rules. f_P are normalized as $f_\pi = 130.4$ MeV and $f_K = 156.1(9)$ MeV.

Charm	Bottom	Ref.
Pseudoscalar		
f_D	f_B	
$204(6) \equiv 1.56(5)f_\pi$	$206(7) \equiv 1.58(5)f_\pi$	[54]
$\leq 218.4(1.4) \equiv 1.68(1)f_\pi$	$\leq 235.3(3.8) \equiv 1.80(3)f_\pi$	[54]
$207(9)$	–	Data [5]
f_{D_s}	f_{B_s}	
$246(6) \equiv 1.59(5)f_K$	$234(5) \equiv 1.51(4)f_\pi$	[54]
$\leq 253.7(1.5) \equiv 1.61(1)f_K$	$\leq 251.3(5.5) \equiv 1.61(4)f_K$	[54]
f_{B_c}		
$436(40) \equiv 3.34(31)f_\pi$		[55]
Vector		
f_{D^*}	f_{B^*}	
$250(11) \equiv 1.92(8)f_\pi$	$209(8) \equiv 1.60(6)f_\pi$	[52]
$\leq 266(8) \equiv 2.04(6)f_\pi$	$\leq 295(15) \equiv 2.26(12)f_\pi$	[55]
$f_{D_s^*}$	$f_{B_s^*}$	
$270(19) \equiv 1.73(12)f_K$	$225(10) \equiv 1.44(6)f_K$	[55]
$\leq 287(18) \equiv 1.84(12)f_\pi$	$\leq 317(17) \equiv 2.03(11)f_\pi$	[55]
Scalar		
$f_{D_{s0}^*}$ $\quad$ $f_{D_{s0}^*}/f_{D_0^*}$		[56]
$217(25)$ $\quad$ $0.93(2)$		

Table 6.4 Masses of the scalar and axial-vector open charm and beauty systems in units of MeV from QCD spectral sum rules. f_P are normalized as $f_\pi = 130.4$ MeV.

Charm	Bottom	Ref.
$M_{D_{s0}^*}$		
$2297(113)$		[56]
2317		
$M_{D_{s1}^*}$		
$2440(113)$		Data [5]
2460		
$M_{D_{s0}^*} - M_{D_0^*}$	$M_{B_0^*} - M_B$	
25	$422(196)$	[56]

♡ Exotic Hadrons of QCD

We have seen that the mesons and baryons are respectively bound states of quark–antiquark and of three quarks and are uncolored (white) or in

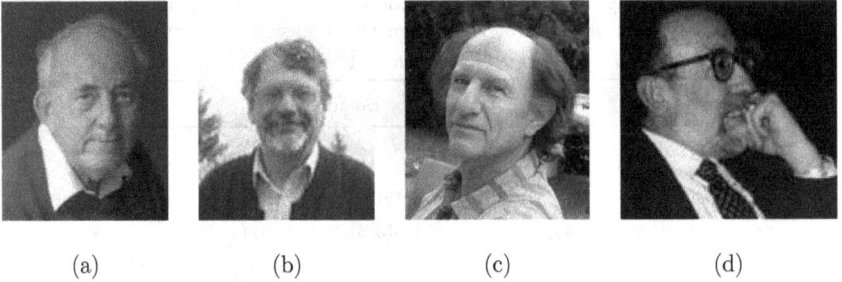

Figure 6.28: (a) P. Minkowski, (b) W. Ochs, (c) G. Mennessier, (d) P. Pascual.

color blind states. As long as this non-colorful (color blindness) property of hadrons is respected, nothing prevents to have more exotic hadronic states beyond the usual quark model of mesons and baryons:

• *Glueballs (ball of glue) or gluonia*

These bound states of two or three gluons were proposed for a long time in 1975 by H. Fritzsch and Peter Minkowski (1941–, Fig. 6.28) [21]. Using QCD spectral sum rules approach combined with some low-energy theorems [57], the author, G. Veneziano and Alberto Bramon in 1987–1989 studied the gluonium content and the quark–gluon mixing of the light spin zero scalar mesons and concluded that *the light $\sigma/f_0(600)$ meson is an "hemaphrodite meson" which contains a lot of gluonium but likes to decay into pair of pions* which are quark–anti-quark mesons (so-called *OZI rule violation* of non-perturbative origin), while the $f_0(1.6)$ *meson* found by the GAMS group at CERN is *a pure gluonium which decays predominantly into pairs of $U(1)$ singlet pseudoscalar particles $(\eta\eta, \eta\eta', \eta'\eta')$* which contain gluon in their wave functions as predicted by perturbative QCD expected to be valid at this higher energy. A parallel study has been done by P. Minkowski and Wolfang Ochs (1943–, Fig. 6.28). In 2010, analysis of the experimental results of π–π and photon–photon scatterings data done by R. Kaminski, G. Mennessier (Fig. 6.28), W. Ochs and X.G. Wang [58] supported the previous conclusion where an *analytic K-matrix model* introduced by Gerard Mennessier [22] has been used. In this model, a summation of pion loops entering into the scalar propagator and contributions of different vector and tensor meson poles in the $\pi\pi \to \pi\pi$, $\gamma\gamma$ amplitudes have been considered. This gluonium interpretation of the nature of the light scalar mesons $\sigma/f_0(600)$ and $f_0(980)$ is an alternative to the four-quark explanation proposed by R.L. Jaffe (Fig. 6.12) and many others on the still obscure nature of these scalar mesons.

• *Hybrid quark–gluon mesons*
They are bound states of quarks and gluons and are described by the (axial)
vector mixed quark–gluon operators $\bar{\psi}\gamma_\mu(\gamma_5)\lambda_a G_a^{\mu\nu}\psi$. They are exotic in the
sense that a vector particle can have a positive charge conjugate C quantum
number, while the standard vector particle like the photon has a negative
C. Among different papers in the literature reviewed in the books [33, 34],
the correct QCD expression using the QSSR predictions of the light hybrid
meson masses, couplings and decay widths (1/2 of the particle lifetime)
have been done by the author, J.I. Latorre and Pedro Pascual (1934–2006,
Fig. 6.28) in [59]. The mass prediction for the vector particle $J^{PC} = 1^{-+}$ (J:
spin, P: parity and C: charge conjugate) has been improved recently in [60]
and has been confronted with the experimental candidate $\pi_1(1600)$. For the
hybrid mesons containing heavy quarks and gluon fields, the early QSSR
calculations have been made in [23] which have been improved recently in
[24]. At present, there are no serious hybrid meson candidates.
• *Meson–antimeson molecules and tetraquarks*
*Meson–antimeson molecules are bound states which are assumed to bind
via weak Van der Vaals like-forces. Tetraquarks are bound states of four-
quarks or of a diquark (Qq)–antidiquark $(\overline{Qq})$.* Several mesons that are not
predicted by the standard model of quarks for charmonium states were
recently discovered in e^+e^- experiments by BELLE (KEK, Japan), BABAR
(SLAC, USA) and BES (Beijing, China) and in $\bar{p}p$ experiment by Tevatron
(Chicago) from the analysis of their decays into $J/\psi\pi^-$ and $J/\psi\pi^+\pi^-$.
These exotic experimental candidates were called X, Y, Z and are *expected
to be molecules or/and four-quark states*. Different QCD-based approaches
have predicted the masses and widths of such exotic states (see, e.g., reviews
by F. Navarra (1959–), M. Nielsen (1954–, Fig. 6.29) and collaborators [25]).
To the lowest order of QCD PT series including large numbers of QCD condensates, masses, couplings and widths of these exotic mesons have been
estimated within QSSR. For instance, the $X(3872)$ looks to be a good
candidate for a four-quark state [62]. However, a confirmation of these LO
results, where the definition of the quark mass used is ambiguous, requires
the inclusion of quantum (radiative) corrections. More recently, QSSR estimates including these perturbative QCD radiative corrections have been
investigated [61] which then improve existing lowest order (LO) results in
the literature. These improved results show that LO results using as input
the running perturbative heavy quark masses are less affected by radiative corrections. Therefore, one may state that the observed $X(3872)$ can
mostly be a four-quark state, while the observed $Z_c(3900)$ and $Z_b(10610)$

Figure 6.29: The author, M. Nielsen, F. Navarra.

almost coincide with the masses of a pure $\bar{D}D^*$ and $\bar{B}B^*$ spin one axial-vector meson–antimeson molecule states. However, the $Y_c(4260)$, $Y_c(4360)$ and $Y_c(4660)$ experimental candidates are too light compared with the mass prediction around 5 GeV for a pure $\bar{D}^* D_0^*$ spin one vector meson–antimeson molecule, therefore suggesting strong mixings of the molecule/four-quark states with some other vector states of the J/ψ family.

• Meson–baryon molecules and pentaquarks

Another possible exotic state is composed of 5 quarks assembled either as meson–baryon molecule or as 5-quark (pentaquark) state (see Fig. 6.30). Recently in July 2015, the LHCb group at CERN has discovered two such candidates: $P_c(9380)$ and $P_c(4450)$ through the decay $\Lambda_b \to J/\psi p K$ in the $J/\psi p$ invariant mass spectrum. This new discovery goes in line with the recent ones of the meson–antimeson molecules and/or four-quark states, which signal new physics beyond the standard quark model of hadrons.

6.9 Hot Plasma Soup of Quark–Gluon

If we refer to the Big Bang (Fig. 2.4), it is expected that immediately after the explosion, we have a very hot soup of quarks and gluons. Several

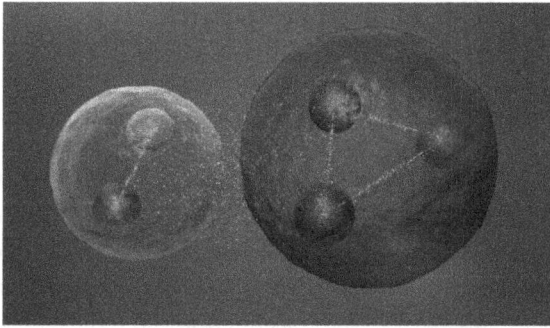

(a)

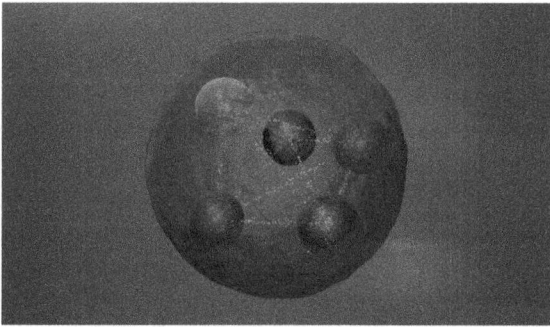

(b)

Figure 6.30: (a) Meson–baryon molecule, (b) Pentaquark.

theoretical approaches based on QCD have been (and continue to be) devoted to understand this phenomenon which is a new area of QCD because the medium is at the thermodynamic phase (finite temperature, pressure,...). Experiments to test this phase transition was conducted at CERN SPS and Brookhaven (New York) where collisions of heavy ions (instead of the proton) have been used to get more important energies near the early Universe. At the LHC (Large Hadron Collider) experiment that we shall discuss later, the ALICE group is dedicated to these experiments at energies not yet reached by previous experiments. This field of research is very active in QCD and is expected to develop within the next few years.

6.10 AdS/QCD Correspondence

We pursue this chapter of QCD on the method of *AdS/QCD correspondence or Holographic QCD* which has been introduced for string theory (see Chapter 9) and which is *the correspondence between quantum field theories such as $N = 4$ super Yang–Mills theory and the 5-dimensional anti-de Sitter space formulated in terms of string or M-theory.* Its application to the quark–gluon plasma [26] where the prediction that *the ratio of shear viscosity η to the volume density of entropy S has universal value:* $\eta/S = \hbar/4\pi k_B \simeq 6.1 \times 10^{-13} k_B S$, where $\hbar$ is the reduced Planck constant and k_B the Boltzmann constant, has been confirmed in 2008 by the RHIC experiment at Brookhaven. Some other applications of this approach for studying hadron spectra and wave functions have been done in the literature such as the *light-front holography* approach by Stanley Brodsky (1940–) and Guy F. de Teramond (Fig. 6.31) and collaborators [27]. The approach has been used to understand some other non-perturbative aspects of QCD and to check the SVZ expansion where some crude estimate of the QCD condensate values have been done by different groups in the literature (see e.g. [28]). In particular, using some model, *Andreev and Zakaharov found naturally the presence of the two-dimension term due to the tachyonic gluon mass in the OPE* [29] supporting the previous result of [41], while a systematic study of different QCD two-point correlators in various holographic models by Frederic Jugeau, Hery Ratsimbarison and the author [63] confirmed this result.

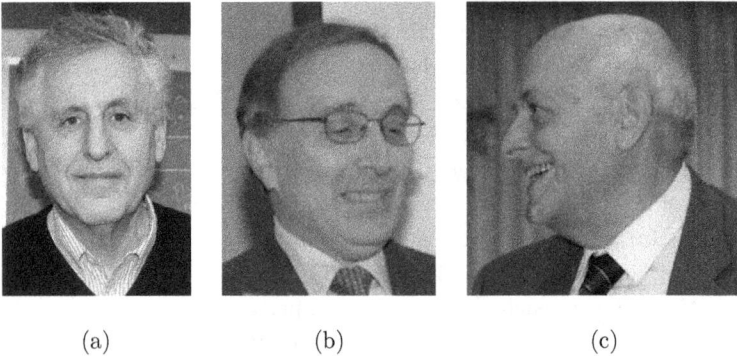

(a) (b) (c)

Figure 6.31: (a) S.J. Brodsky, (b) G.F. de Teramond, (c) A. Di Giacomo.

6.11 QCD Confinement

Some activities are investigated by different groups for understanding the confinement of quarks and gluons and the complicated structure of the QCD vacuum. In particular, Adriano Di Giacomo (1936–, Fig. 6.31) *et al.* and V.I. Zakharov *et al.* (Fig. 6.25) use 't Hooft's proposal where a monopole could condense in the vacuum and becomes dual to a superconductor which confines chromoelectric charges which are the quarks. This happens because, it is the only way to have an extra symmetry in QCD which comes from the degrees of freedom on the boundary of the physical space (dual variables). However, QCD is far from superconductivity as the latter does not have confinement and is an Abelian (described by a $U(1)$ group like QED). Another 't Hooft's proposal at the QCD02 Montpellier conference is that confinement can be treated perturbatively by a suitable choice of the Lagrangian which reproduces this effect (QCD Coulomb potential $\oplus$ linear term) at the low-energy infrared (IR) regime, though this Lagrangian cannot be used to describe the ultraviolet (UV) region of the theory because it is not renormalizable. However, we do not yet have any convincing analytic explanation of confinement.

6.12 QCD is in a Good Shape

Despite our lack of formulating the true analytic form of confinement, we have seen in the previous discussions that QCD is in a good shape. Indeed, *QCD gives a precise description of the high-energy phenomena thanks to asymptotic freedom, while different approximate QCD-based models fit with a good accuracy the non-perturbative features of hadron physics.* These facts indicate that we are on the right track. The QCD theory of strong interactions in terms of quarks and gluons and based on an unbroken gauge theory like QED is the modern version of nuclear physics.

Chapter 7

Electroweak Standard Model

7.1 Weak Interactions in the 1930–1960s

We have shown in previous chapters how we went from the concept of atoms (which has taken 25 centuries to be understood scientifically) to the notions of quarks and leptons that we understood in few decades. This rapid speed of our knowledge is due to complementary and alternative results between theory and experiment that quickly tests the validity of the theoretical predictions. Essential progress comes from quantum mechanics and relativity which gave birth to the quantum theory of relativistic fields and which, in turn, gave birth to gauge theory. We took as an example of these gauge theories the case of *QED and QCD describing with great success the electromagnetic and strong interactions that are described by gauge groups which have, respectively, the Abelian $U(1)$ rotation symmetry and non-abelian $SU(3)_c$ special unitary symmetry. These symmetries are unbroken because the gauge fields (photon for QED and gluons for QCD) are massless.* However, one can ask the question if the force responsible for the β decay of nucleus (the weak interaction) can also be described in the same way as QED and QCD.

♣ First Experimental Facts

• *The electron neutrino (ν_e)*
We saw previously that the experience of Chadwick in 1914 on the β decay radioactivity of a nucleus transmuting to another nucleus with electron emission has highlighted the existence of a new particle associated with the electron (the neutrino). Indeed, the momentum of the observed electron did not reproduce the difference between the one of initial and final nucleus (principle of conservation of energy) such that the missing energy would be taken by a (invisible) neutral particle of very small mass. Incorrectly called neutron by Pauli, Fermi has renamed (with the permission of Pauli) this new particle neutrino (small neutron in Italian) after the discovery of the true neutron.

• *The muon (μ) and muon neutrino (ν_μ)*
We have also seen that the search of the π meson (pion) by Anderson in 1937 (the particle predicted by Yukawa as responsible for the strong interaction of the nucleus), led to the discovery of the muon (μ) and the associated antiparticle (the muon antineutrino: $\bar{\nu}_\mu$), resulting from the decay of the pion by the reaction: $\pi \to \mu\bar{\nu}_\mu$. The muon will then give rise to the electron and its associated neutrino plus the antineutrino of the electron by the reaction $\mu \to \nu_\mu + e^-\bar{\nu}_e$. Note that because of the Pauli exclusion principle, one cannot have in the same state the μ and ν_μ. This selection rule corresponds to the *lepton number conservation*.

◇ Intensity and Range of the Weak Interaction

In contrast to strong and electromagnetic interactions, it was first thought that *the weak interaction is universal* because it seems to act with the same intensity between leptons (weak lepton process), between leptons and hadrons (weak semi-leptonic process) and hadrons (weak hadronic process). In addition to the difference in intensity between the three forces, there is also the difference in range: *the strong interaction with the Yukawa theory has a short range of the order of 1 Fermi (the proton size) because it is mediated by the pion, while the electromagnetic interaction is infinite in range due to the zero mass of the photon.* For the weak interaction, the constant which controls the intensity of the reaction $n \to p + e + \bar{\nu}_e$ would be $(G_F M_P^2)/4\pi = 10^{-6}$ where $G_F = 10^{-5}/M_p^2$ in GeV^{-2} unit with $M_p = 0.938$ GeV being the mass of the proton. G_F is called Fermi constant which is very small compared with the electromagnetic fine structure constant: $\alpha \equiv e^2/4\pi = 1/137$.

Figure 7.1: E. Fermi.

♡ Fermi Effective Theory and Universality of Weak Interactions

These types of reactions involving the production of neutrinos are new because they are not the strong interactions that happen inside the nucleus. These are not also electromagnetic interactions because neutrinos are neutral and therefore cannot interact with the photon. To explain this phenomenon, Fermi (Fig. 7.1) introduced in 1934 an *effective theory of local current–current (CC) four-fermion interaction* due to the contact term (from zero range). This interaction is controlled by the Fermi constant G_F. This value was obtained by measuring the lifetime of the muon via the reaction: $\mu \to \nu_\mu + e^- \bar{\nu}_e$. In the case of the neutron β-decay: $n \to p + e^- \bar{\nu}_e$, this interaction is described by the four-fermion Hamiltonian of the form: $\mathcal{H} = G_F \sum_{i=1}^{5} [\bar{\psi}_p \Gamma_i \psi_n] [\bar{\psi}_e \Gamma_i \psi_\nu]$ where Γ_i are combinations of Dirac matrices (scalar S, pseudoscalar P, vector V, axial-vector A and tensor T) that characterize each type of interactions. ψ_f is the spinor (fermion field) which describes each particle. Experimental data $\pi \to \mu \bar{\nu}_\mu$ and $\mu \to \nu_\mu + e^- \bar{\nu}_e$ do not favor the existence of derivatives in the $\mathcal{H}$ expression and of interaction with the charge retention: $(\bar{\psi}_p \Gamma_i \psi_e) \bar{\psi}_n \Gamma_i \psi_\nu$. While the reaction $n \to p + e + \bar{\nu}_e$ favors the combinations S, T or V, A and that of Gamow–Teller $He^6 \to Li^6 + e + \bar{\nu}_e$ favors T instead of A, one may conclude that the combination S and T is responsible of the β-decay. By his side, the decay $\pi \to \mu \bar{\nu}_\mu$ indicates an interaction A or P because the pion is a pseudoscalar particle, while the precise analysis of the electron spectrum in the reaction: $\mu \to \nu_\mu + e^- \bar{\nu}_e$ clearly favors the V and A interactions. These facts show that the situation is not yet clear. However, in 1947, B. Pontecorvo

(a) (b) (c) (d)

Figure 7.2: (a) B. Pontecorvo, (b) T.D. Lee, (c) C.C. Wu, (d) R.L. Garwin.

(1913–1993) (Fig. 7.2) has suggested that the weak interaction between pairs (n, p), (ν_e, e) and (ν_μ, μ) is universal.

♠ Parity (P) Violation and Chiral Invariance

In 1956, T.D. Lee (1926–) and C.N. Yang (Fig. 6.2) (Nobel Prize of 1957) have suggested that the violation of parity P or mirror symmetry (change of a vector in space by its opposite), which explains the contradictory decays of K (kaon) meson into 2 and 3 pions observed in 1949, may also apply to leptonic decay $\mu \to \nu_\mu + e^- \bar{\nu}_e$. Several experiments have been undertaken to test this hypothesis in the various weak processes with production of a neutrino by measuring the asymmetry and polarization of the produced electron or muon. In 1957, C.S. Wu (1912–1997), R.L. Garwin (1928–) (Fig. 7.2), L. Lederman (Fig. 6.16) and M. Weinrich discovered a parity violation in the β decay of Cobalt-60. These results have led to *describe the neutrino by a two-component field called Weyl spinors* (Fig. 1.29) but not by four components as in Dirac's theory (Dirac spinors). Thus, the neutrino is expressed by its two chiral projections $\psi_{R,L} = (1 \pm \gamma_5)\psi/2$ respectively right (R) and left (L). γ_5 is the 4×4 Dirac matrix which labels the parity violation. However, if the neutrino is massless, it can have only a single component while the electron which is massive can have two components (Dirac spinor). Further experimental searches were made for knowing what is the true chirality of neutrinos. The decisive experience for the choice of the chirality of the neutrino and the different types of coupling in the earlier reactions is the relatively small value 10^{-4} of the branching ratio $\pi \to e\bar{\nu}_e$ over $\pi \to \mu\bar{\nu}_\mu$ which favors the V–A coupling and the left chirality

(a)   (b)   (c)   (d)

Figure 7.3: (a) R. Marshak (courtesy of AIP Emilio Segrè Visual Archives), (b) E. Sudarshan, (c) J. Sakurai, (d) N. Cabibbo.

of neutrinos. This result also showed that *all weak processes are universally described by this V–A type of coupling* and the Hamiltonian can be written as: $\mathcal{H} = G_F/\sqrt{2}[\bar{\psi}_1\gamma_\mu(1-\gamma_5)\psi_2][\bar{\psi}_3\gamma_\mu(1-\gamma_5)\psi_4)]$ for the process: $1+3 \rightarrow 2+4$ where $\bar{\psi}_1\gamma_\mu\psi_2$ is the vector current (V) and $\bar{\psi}_1\gamma_\mu\gamma_5\psi_2$ the axial-vector current (A). This universality has been named chiral invariance in 1957 by R.E. Marshak (1916–) and E.C.G. Sudarshan (1931–) (Fig. 7.3), taken over later by Feynman, Gell-Mann and J.J. Sakurai (1933–1982, Fig. 7.3).

♣ Cabibbo Theory and Angle

The currents entering the earlier-mentioned Hamiltonian can be divided into subgroups: the lepton current L_μ and the hadronic current H_μ. This hadronic current is itself divided into part H_μ^0 which does not change the hypercharge ($\Delta Y = 0$) and H_μ^1 which changes the hypercharge ($\Delta Y = 1$). Despite the success of the theory to explain existing data, a new experimental fact shows that all semi-leptonic process $\Delta Y = 1$ are systematically smaller by a factor of 20 over similar processes with $\Delta Y = 0$. This experimental result cannot be explained by quantum corrections because the ratio is too small. Nicolas Cabibbo (1935–2010) (Fig. 7.3) proposed in 1963 that the hadronic current H_μ^1 must be reduced over H_μ^0. The total hadronic current then takes the form: $H_\mu = H_\mu^0 \cos\theta_c + H_\mu^1 \sin\theta_c$ where: $\theta_c = 0.23$ radian is the Cabibbo angle in order to explain the results of previous experiments. With this proposal, one can explain all the leptonic and semi-leptonic processes with $\Delta Y = 0$ and 1 like the transition of the hyperon $\Sigma^+ \rightarrow \Lambda + e^+ + \nu_e$.

◊ P and CP Symmetries Violation

• *P in the decay of $K^+ \rightarrow 2\pi$ and 3π*

We mentioned that the parity violation explains the coexistence of the decays of K^+ to $\pi^+ + \pi^0$ and to $\pi^+ + 2\pi^0, \pi^+ + \pi^- + \pi^+$ which are purely hadronic processes with $\Delta Y = 1$. However, it is impossible to measure directly the violation of parity from these decays of K because it did not have a spin and it also decays into 2 or 3 particles that have no spin. Indeed, to do so, then one should consider a pseudoscalar quantity which one could not form with the momenta of the particles involved in these reactions. This measure is possible in the reaction $\Lambda^0 \rightarrow p + \pi^-$ where particles have spins because one can form a pseudoscalar measurable quantity which is the *asymmetry*.

• *CP in the decay of $K^0 \rightarrow 2\pi$ and 3π*

Instead of K^0 and its antiparticle $\bar{K}^0$, one can construct the states: $|K_S\rangle = N[p|K^0\rangle - q|\bar{K}^0\rangle]$(S as short lived) and $|K_L\rangle = N[p|K^0\rangle + q|\bar{K}^0\rangle]$ (L as long lived) which are their combinations where $N = 1/\sqrt{p^2 + q^2}$ is a normalisation factor. $p = q = 1$ if CP symmetry is conserved. In this particular case, K_S and K_L are the eigenstates of CP with proper values +1 and −1. It can be shown[1] that boson–anti bosons states $(\pi^+\pi^-)$ and $(\pi^0\pi^0)$ have a CP= +1. Then the transitions $K_L \rightarrow \pi^+\pi^-$ and $K_L \rightarrow \pi^0\pi^0$ must be strictly forbidden. K_S and K_L have very different lifetimes. The observation of $K_L \rightarrow \pi^+\pi^-$ and $K_L \rightarrow \pi^0\pi^0$ with a rate of order 10^{-6} times lower than the analogous decays of K_S, but which is not zero, indicates *a violation of CP in the weak interaction*. With similar reasoning, the *processus $K_S \rightarrow 3\pi^0$* must be strictly prohibited if CP is preserved because CP of $K_S = + 1$ whilst that of $3\pi^0$ is −1.

• *CP in the $K^0 - \bar{K}^0$ mass difference*

As K_S and K_L are different in the presence of the weak interaction, they must have a mass difference. The transition of K^0 to $\bar{K}^0$ corresponds to a change of 2 units of strangeness ($\Delta S = 2$). The experimental measurement of this mass difference is $\Delta m_K = (3.484 \pm 0.006) \times 10^{-12}$ MeV. Modern approaches based on the methods of effective lagrangians, QCD spectral

---

[1] For a boson–antiboson system, the associated field (wave function) must be symmetric in space, spin and charge conjugation (C) which is defined as: $C = (-1)^l(-1)^s = +1$ where l and s are the orbital angular momentum and spin. Indeed, for $(\pi^+\pi^-)$, $C = (-1)^l$, and as s $= 0$, CP $= (-1)^l(-1)^l$, which is always positive. For $(\pi^0\pi^0)$, C should be positive implying CP= $(-1)^l$. However, as the spin of K^0 is zero, CP should be also positive.

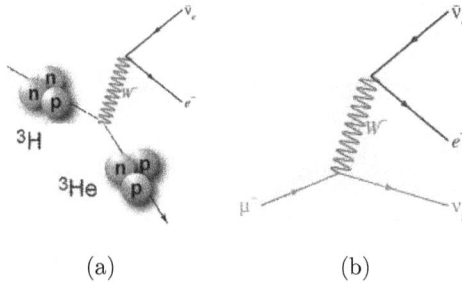

(a)   (b)

Figure 7.4: IVB Model: (a) β radioactivity of tritium to helium-3: $n \rightarrow p + e^- + \bar{\nu}_e$; (b) muon decay: $\mu \rightarrow \nu_\mu + e^- + \bar{\nu}_e$.

rules and lattice calculations have been used to explain this experimental result (see e.g. Chapter 15 of the book in [34]).

♡ Hypothesis of the Intermediate Vector Boson (IVB)

Before 1960, the *phenomenological model of Fermi* described successfully several experimental results at low energy. However, it is *not renormalizable* because it is an effective theory with a constant that has the dimension of the inverse square of energy, while a renormalizable field theory must have a coupling constant that has no dimension. Thus, in the theory of Fermi, we cannot improve predictions including the quantum corrections because they will bring infinite quantities.

To try to get closer to the theory of fields, and by analogy with the Yukawa theory where the strong interaction is mediated by the pion, *Yukawa postulated that there is an intermediate vector boson (IVB) mediator of the weak interaction* which connects the two Fermi interaction currents CC. Although quickly abandoned, this idea has been pursued by Schwinger. Thus, the CC interaction is no longer local (point-like) but is a function of the square of the energy entering the propagator of the IVB: $G_F/2$ is replaced by $g^2/8(M_W^2 - q^2)$ and at $q^2 = 0$, the two theories are identical. To explain these decays (change of the charge, nature of the coupling), the IVB must be charged and it must have a structure to allow a V–A coupling. It must be also very massive for not affecting successful predictions of the Fermi model and for example prohibit the reaction $K \rightarrow W^+\gamma$ which is not observed. That has served (at that time) to give a 2 GeV lower limit to the mass of the W-boson. However, the experience of neutrino–hadron scattering gave a stronger lower limit of the order of 4 GeV. In this approach, the interaction Lagrangian is written: $\mathcal{L}_{\text{int}} = g J_\mu W^\mu$ where W^μ is the field of W and J_μ is the weak current consisting of a lepton part L_μ and a hadronic part $H_\mu = H_\mu^0 \cos\theta_c + H_\mu^1 \sin\theta_c$ where θ_c is the Cabibbo angle, while H_μ^0 and H_μ^1 correspond to the hypercharge transition $\Delta Y = 0$ and $\Delta Y = 1$.

♠ Weak Isospin $SU(2)$ Symmetry

The success of QED to describe the electromagnetic interaction by the $U(1)$ Abelian gauge theory stimulated to seek a similar theory for the weak interaction. Recall that the weak interaction has a universal coupling governed by the Fermi constant G_F as is QED by the electric charge. We also saw that this coupling may be connected to the square of the inverse of the mass of the W boson, which would mean that the strength of the weak interaction would become more smaller when the mass of W increases. We have also seen that *there is always a well defined pair of particles* $(n, p), (e, \nu_e), \ldots$ *that couple to* W and that the particle transforms into its partner by earning one unit of charge after an emission of a virtual W: $n \rightarrow p + W^-$ and $e^- \rightarrow \nu_e + W^-$. One can call these pairs: *doublets*. That leads to think, by analogy with the symmetry of isospin of hadrons of the strong interaction, that one can possibly describe the weak interaction by the simplest gauge group $SU(2)$ where the fermions are in doublets. Considering the left chirality of neutrinos, assuming it is only the left component that is useful, we will call this group $SU(2)_L$. Like the $SU(3)_c$ group of QCD which has $8 = 3\times3-1$ bosons of gauge, *the $SU(2)_L$ must contain* $2\times2-1$ *gauge bosons, namely the* W^+, W^- *and a new neutral boson* Z^0, the effect of which has not yet been seen. However, *these gauge bosons have to be massive to explain the experimental results* while local invariance of gauge theory is only satisfied if the bosons are massless like the photon and 8 gluons. Therefore, we are still in a deadlock.

7.2 Higgs Mechanism and Higgs Boson

♣ Spontaneous Symmetry Breaking

In 1964, F. Englert (1932–) and R. Brout (1928–2011), Peter W. Higgs (1929–), Gerald S. Guralnik (1936–2014), C.R. Hagen (1937–), and Tom W.B. Kibble (1932–) (Fig. 7.5) showed that *one can spontaneously break the local symmetry of a gauge theory by introducing spin zero scalar fields that*

(a)

(b)

Figure 7.5: (a) Kibble, Guralnik, Hagen, Englert and Brout (Sakurai Prize of 2010); (b) Englert and Higgs (Nobel Prize of 2013) (courtesy of Maximilien Brice, CERN).

are neither matters nor gauge fields. From now, *we shall call this the Higgs mechanism* to simplify the notation. One can illustrate this mechanism by considering the rotation group $U(1)$ of QED and by introducing a neutral complex scalar field $\phi = (1/\sqrt{2})(\phi_1 + i\phi_2)$ which has two degrees of freedom: its real and its imaginary component.

◇ The Higgs Potential

To break the $U(1)$ symmetry, one considers the simplest *scalar potential* of the form $V(\phi) = \mu^2 \phi^+ \phi + \lambda(\phi^+ \phi)^2$ where λ is positive because we want the lowest (vacuum) energy state. One takes a negative μ^2 such that

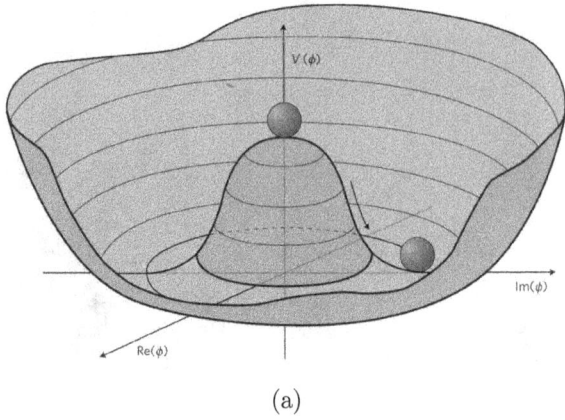

(a)

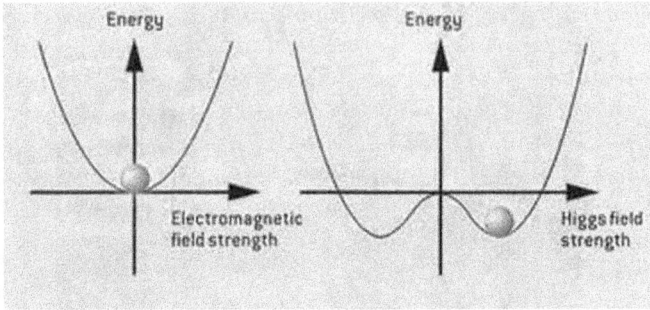

(b)

Figure 7.6: (a) Higgs potential in the form of a mexican hat. The stable vacuum is minimum (position of the ball after his fall) (courtesy of CMS); (b) comparison of the electromagnetic potential and that of Higgs.

we can break the symmetry where the vacuum is no more invariant by rotation. The potential will then take the form of a Mexican hat with two minima (Fig. 7.6) instead of a single minimum at the center which is the average value in a vacuum of the field by the rotation. After the breaking of symmetry, the average value in the vacuum of the scalar field: $\langle 0|\phi_1|0 \rangle = \nu_\phi/\sqrt{2} \neq 0$ (where $\nu_\phi = \sqrt{-\mu^2/\lambda}$) *corresponds to the stable minimum of the potential*, which is not the case of the unstable minimum (false vacuum) at the center of the figure because if you put a ball there, it descends to reach the minimum. The movement of vibration (radial oscillation) of the ball in the valley of the minimum is that of a physical particle which is the Higgs boson carrying a weak charge (sum of the weak isospin and hypercharge quantum numbers). However, this choice of the minimum is not unique because we can move along the channel without spending any energy (*vacuum degeneracy*). *This little shift (oscillation) along the channel corresponds to a small change in energy due to massless particles of spin zero known as Nambu–Goldstone bosons.*

♡ Vector and Higgs Boson Masses

These spin zero bosons will then be absorbed by spin 1 gauge bosons and appear as their longitudinal components. *After the breaking of the symmetry, the spin 1 (vector) bosons acquire their masses $M_W^{\pm} = g(\nu_\phi/2)$ where g is the $SU(2)_L$ charge or coupling.*

By summarizing, we start, before symmetry breaking, with a massless W field with two degrees of freedom and the scalar field, which has two degrees of freedom (its real and imaginary components). What makes a total of four degrees of freedom. After symmetry breaking, the spin 1 vector boson has acquired a mass and has now three degrees of freedom where it has stolen one from a scalar field that disappears in the operation. It remains therefore one free degree of freedom left corresponding to a physical scalar field that we will call the *Higgs field H*. By rewriting the scalar field as: $\phi = \nu_\phi + H/\sqrt{2}$, we also see that the terms of $V(\phi)$ will induce the mass squared term: $(1/2)(-2\mu^2)H^2$ corresponding to the mass of the Higgs boson: $M_H = \sqrt{-\mu^2}$ where μ^2 is negative. However, $\mu^2 \equiv -\lambda\nu_\phi^2$ is an arbitrary number because of λ, such that the Higgs boson mass is *a priori* undeterminate. The only constraint which one can have comes from *unitarity* where *the Higgs width should be smaller than its mass*. In this way, a precise calculation provides the upper bound: $M_H \leq 1.2$ TeV which is somewhat a loose bound.

♠ Fermion Masses

A similar mechanism can also generate masses to fermions. This can be achieved by adding a new interaction term $g_f\bar{\psi}_L\phi\psi_R$ in the Lagrangian which respects gauge invariance. g_f is the coupling (so-called *Yukawa coupling*) of the scalar field to fermions; ψ_L and ψ_R correspond to right and left chirality of the fermion fields with opposite chirality. It results in the Yukawa coupling behaving like the fermion mass m_f which indicates that heavy fermions (τ lepton or b and t quarks) couple strongly to the Higgs field. This property will be used when detecting experimentally the Higgs boson which we shall see in Chapter 8.

♣ Higgs Mechanism and Higgs Boson Imaging

The Higgs mechanism has been much publicized after the Higgs boson discovery at LHC which we shall see in Chapter 8.

One can imagine that the vacuum is the filled range of the Higgs fields that are grains of sand. A person who does not interact with (does not walk on) the beach will cross it faster (like the massless photon). Now, suppose that you walk on the beach which is the analogue of the Higgs field: if you have a racket, you seem lighter because you interact less with the grains of sand, while if you put regular shoes, you seem heavier because you press and you interact more with the grains of sand. Thus, more the particle interacts with the Higgs field, more heavier it is. A Higgs boson is the analog of a grain of sand in this beach and it also has a mass. It is this particle that we are extracting out of this field (ejecting a grain of sand from the beach) and what we will seek in the experiments.

Another popular image is the one by Gerard Miller, where he compares the Higgs field with different persons chatting and sitting regularly in the meeting room. When a famous person enters the room, everybody wants to approach him. This famous person has then to increase her resistance when moving. It looks like she acquires a mass just like a particle which acquires a mass when moving through the Higgs field.

Now, if the rumor crosses the room, it creates the same kind of clustering but only among the persons themselves which transmits the information. These clusters are analogue to the Higgs bosons.

7.3 Electroweak Standard Model

♣ Status of the Progress Done Until 1967

Development and success of QED is based on the $U(1)$ group which is a renormalizable gauge theory (because you can add finite quantum corrections) obtained by the perturbation theory stimulated to seek theories in this direction in order to describe other weak and strong interactions.

On the other side, knowledge of subnuclear particle quarks which brings us a bit more to the heart of the matter, the success of the quark model by Gell-Mann to describe the observed hadrons (mesons and baryons), Feynman parton model and the discovery of asymptotic freedom have been essential in the development of the theory of the strong interaction which led later to QCD.

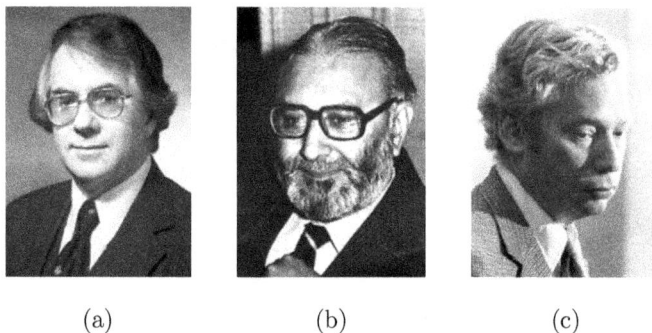

(a) (b) (c)

Figure 7.7: (a) S.L. Glashow, (b) A. Salam, (c) S. Weinberg.

Phenomenological success of the Fermi model (highlights of V–A coupling, of the left chirality of neutrinos and of the assumption of the BVI), the discovery that the particles are in a doublet of the weak isospin group $SU(2)_L$ (the L index takes into account the fact that neutrinos are left-handed) as well as the proposal of Higgs mechanism (which generates the mass of the W boson in accordance with gauge invariance) were big steps in the development of the theory of weak interactions.

◇ The Aspects of the Standard Model

Based on empirical and theoretical facts, S.L. Glashow (1932–), A. Salam (1926–1996) and S. Weinberg (Fig. 7.7) Nobel Prize, 1979, proposed *a model that puts on the same level (unified) electromagnetic and weak interactions. This standard electroweak model called for simplicity the Standard Model, is the product of the isospin weak group $SU(2)_L$ with the $U(1)$ describing QED, while the Higgs mechanism is supposed to give mass to the gauge bosons and fermions.* The model is developed at the scale of the infinitely small components which are quarks and leptons. Thus, *the model postulated that the matter in our Universe came from assembly of quarks and leptons that can be divided into three pairs (doublets) of leptons and three pairs of quarks* (see Fig. 5.6). Recall that the quarks and leptons that we already discussed previously are infinitely small particles called fermions which each have a spin 1/2 (one of the internal intrinsic quantum numbers that characterize them) and obey Fermi–Dirac statistics. Each couple differs by their masses and their intrinsic quantum numbers (Fig. 5.6). Each particle has its own image called the antiparticle. *However, there are more*

Table 7.1 Types of leptons and their intrinsic quantum numbers. L_l is the lepton number. Q is the charge in e unit. Leptons are fermions of spin 1/2 like quarks but does not participate to strong interactions.

Lepton	L_e	L_μ	L_τ	Q	Mass [MeV]
e	+1	0	0	−1	0.54858(0)
ν_e	−1	0	0	0	≤2 eV
μ	0	+1	0	−1	105.6584(0)
ν_μ	0	−1	0	0	≤2 eV
τ	0	0	+1	−1	1776.82(16)
ν_τ	0	0	−1	0	≤1 keV

particles than antiparticles (CP violation) in nature which ensures matter stability in addition to proton stability.

♡ Leptons

We summarize the properties of leptons observed in Table 7.1. *The best-known family member of leptons is the electron* which has a negative charge −1 in unit of 1.6×10^{-19} coulombs, that we have extensively discussed previously. *His wife the neutrino of the electron (ν_e) has a zero charge* and is very light with a mass inferior to 2 eV. Thus it is invisible because it does not interact with matter (zero charge) although this cosmic ray bombards us incessantly everywhere we are. *The 1st brother of the electron is called the muon (μ)* which is 200 times more heavier than the electron. He is married to the *neutrino of the muon (ν_μ)* which is neutral and having a mass less than 2 eV. *The 2nd brother of the electron is called the tau (τ)* which is 18 times as heavy as the muon and his wife the *tau neutrino (ν_τ)* has a mass which does not exceed 1 keV. All these particles have been discovered experimentally. Measurement of the width of the Z^0 boson at LEP and SLC allowed to limit the number of neutrinos (without or with a very small mass) to $N = 2.92$ (5) from PDG average [5] (Fig. 7.8), while the mass of any new charged heavy lepton must be greater than 102.6 GeV, a result obtained from its decay to $W + \nu$.

♠ Quarks

Now let us look more closely at the nucleus around which the electrons revolve to form atoms (Fig. 5.6). By breaking the core, we will discover the nucleons (protons and neutrons). At this time, we are performing nuclear

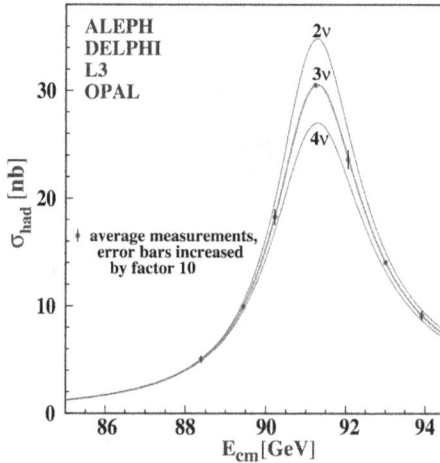

Figure 7.8: Measure of the number of neutrinos from Z^0-decay at LEP.

physics. *By breaking the proton, we discover that it is composed of three elementary particles called quarks by Gell-Mann.* In the Standard Model (SM), the family of quarks resembles that of leptons (see Fig. 5.6). The lightest is called u (*up*) of (current perturbative) mass equal to 2.8 MeV but, unlike the electron, it *has a fractional charge 2/3 and three colors* referred to as blue, green and red. His wife the d *(down)* quark has a charge $-1/3$ and a mass of 5 MeV. The first brother (2nd family) of the u is called c (*charm*), which has a mass of 1.3 GeV and is married to the s (*strange*) quark of a mass 96 MeV. They differ from the 1st couple (1st family) by their masses and internal quantum numbers but have similar charges. The second brother of the u (third and last family?) is called t (*top*) who is married to the b (*bottom*) quark and respectively have masses of 170 and 4.2 GeV. These different types of quarks are also called *flavors*. Unlike leptons, *quarks masses are not measurable directly* because the quarks do not exist in the free state as they are glued together to form matter (properties of confinement of quarks). Thus, we need theoretical approaches to determine their masses. These different theoretical determinations are explained for example in the author's books [30–32]. However, as mentioned in Chapter 6, the case of the heavy t quark is exceptional because it cannot hadronize as it is not sensitive to the confinement. That is why we can directly measure its mass by the experience. The properties of quarks have been summarized in Table 6.1.

♣ Higgs Mechanism and Electroweak Mixing Angle θ_W

The structure of the group $SU(2)_L \otimes U(1)$ shows us that there are $2 \times 2 - 1 = 3$ gauge bosons of the group $SU(2)_L$ and 1 gauge boson of the group $U(1)$ called hypercharge group. Using the Higgs mechanism, one gives masses to three bosons, W^+, W^- and W^0 of $SU(2)_L$ but one must maintain the mass of the boson B associated with group $U(1)$ equal to zero to preserve the $U(1)$ symmetry of QED. After symmetry breaking, two neutral fields W^0 and B will mix to form the physical Z^0 heavy neutral massive boson and the massless photon observed in experiments. The mixing angle θ_W is characteristic of the Standard Model and is measured by different experiments. It connects the weak charge to the electric charge by the relationship: $g = e/\sin\theta_W$ and may be expressed in terms of the masses of the boson Z^0 and $W^\pm$: $\sin^2\theta_W = 1 - M_W^2/M_Z^2$ (see more in Part VIII). Its experimental value is: $\sin\theta_W = 0.213\,16(12)$ [5].

♦ Neutral Current

One of the first consequences of the Standard Model is the existence of the *neutral current mediated by the new massive Z^0 boson* which was not predicted by the model of Fermi. The search for the existence of these neutral currents was very difficult as one cannot use all experiments involving charged leptons like for example $Z^0 \rightarrow \mu^+\mu^-$ because these processes are dominated by the exchange of a photon $\gamma \rightarrow \mu^+\mu^-$ which is also neutral. The experiment of the SPS at CERN with the Gargamelle bubble chamber to look for neutral current began under the leadership of André Lagarrigue (1924–1975, Fig. 7.9) and Paul Musset (1933–1985). It consists of bombarding the proton with beams of neutrinos issued from the primary beams

| | | | | |
| (a) | (b) | (c) | (d) | (e) |

Figure 7.9: (a) A. Lagarrigue, (b) J. Illiopulos, (c) L. Maiani, (d) C. Bouchiat, (e) P. Meyer.

of protons and filtering the neutrinos emitted by several meters of earth and metal arranged in front of the target. In the reaction with Z^0 exchange, the incident neutrino will produce a new neutrino that leaves no trace in the bubble chamber, while in a reaction with charged current with a charged $W^\pm$ boson exchange, there is a production of a charged lepton member of the same weak isospin doublet that will leave traces in the bubble chamber. In 1973, the Gargamelle group published the discovery of neutral currents challenged by Americans but reconfirmed immediately by CERN.

♡ GIM Mechanism and the Charm Quark

In 1970, Glashow (Fig. 7.7), Illiopoulos (1940–) and Maiani (1941–) (GIM) (Fig. 7.9) propose to explain the difference in mass of 3.48×10^{-6} eV between the K_L–$\bar{K}_S$ transition ($\Delta S = 2$: S is the strangeness) and the branching ratio: $\Gamma(K_L \rightarrow \mu^+\mu^-)/\Gamma(K_L \rightarrow \text{all}) = 6.87 \times 10^{-9}$ by the need to have a fourth quark, *the charm c of charge* $+2/3$. The charm would be the partner of the strange quark (Fig. 5.6) in the context of the Standard Model and restores the symmetry between the two families (u, d) and (c, s). This proposal generalizes the hadronic current proposed by Cabibbo which has explained the semi-leptonic transitions $\Delta Y = 1$, where Y is the hypercharge. Thus, in terms of quarks, the hadron current becomes: $J_H^\mu = \bar{u}\gamma^\mu(1 + \gamma_5)d_c - \bar{c}\gamma^\mu(1 - \gamma_5)s_c$ where d_c and s_c are the orthogonal components after rotation of d and s: $d_c = \cos\theta_c d + \sin\theta_c s$ and $s_c = -\sin\theta_c d + cos\theta_c s$ where θ_c is the Cabibbo angle. Note that only the 1st term of J_μ^H has been used by Cabibbo for explaining the β decay. In the scheme of GIM, $K_L \rightarrow \mu^+\mu^-$ reaction proceeds by the Feynman box diagram of Fig. 7.10a with virtual boson exchange $W^\pm$ and a light u quark while Fig. 7.10b shows the additional effect of the c quark that compensates the dominant log-terms coming from the contribution of Fig. 7.10a.

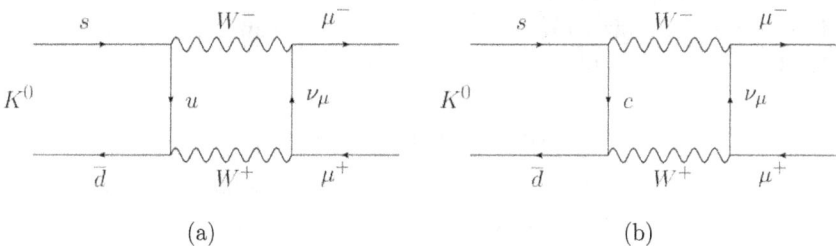

(a)　　　　　　　　　　　　(b)

Figure 7.10: Diagrams showing the $K_L \rightarrow \mu^+\mu^-$ process in the Standard Model: (a) exchange of a u quark; (b) exchange of a c quark.

Thus, the sum of the two diagrams gives a branching ratio of the order of $g^4(m_c^2 - m_c^2)/M_W^2 \times 10^{-8}$ in agreement with experiment for m_c of the order of 1.3 GeV. Similar diagrams explain the mass difference $K_L - \bar{K}_S$ obtained by changing leptons by quarks in the diagrams.

♠ Renormalizability of Standard Model

• $U(1)$ *Axial anomaly*
The presence of the c quark also restores the symmetry between quarks and leptons (two families of quarks and two families of leptons: Fig. 5.6). This allows to have a theory without anomaly. The anomaly comes from the contribution of the triangle diagram of the type in Fig. 6.18 (called *triangle anomaly*) which is due to the divergence (derivative) of the axial current. It does not depend on the mass of the fermion circulating on the triangle. However, the presence of this term is not desirable in theory because it breaks the gauge invariance. Claude Bouchiat (1932–), John Illiopoulos and Philippe Meyer (1925–2007) (Fig. 7.9) showed in 1972 that this contribution disappears into the Standard Model if one has the same number of families of leptons and quarks with three colors which circulate along the triangle loop.

• *Renormalizability*
In 1973, *G. 't Hooft and M. Veltman demonstrated the renormalizability of the Standard Model* by using the method of dimensional regularization, which was also, independently, introduced in 1972 by Bollini and Giambiagi (see [32]). Here, the spacetime has an arbitrary dimension n during the computations of the Feynman diagrams and one takes $n = 4 - \epsilon$ once calculations are completed and then lets $\epsilon \to 0$. In this approach, infinities are becoming poles (singularities) in power of $1/\epsilon$ and non-local terms of type $(1/\epsilon) \log Q^2$ which may be present during the calculations would automatically disappear (locality). These poles in $1/\epsilon$ will be absorbed by those from the renormalization constant of the bare quantities and give a finite final result which is physically observable.

♣ Discovery of Charmonium

The important role of the charm to validate the Standard Model has motivated his experimental research. We have seen that the explanation for the K_L decay predicted a charm quark mass of the order of 1.3 GeV. If the theory is correct, one must therefore find a bound state of charm–anticharm ($\bar{c}c$) in the region of (3–4) GeV. The accelerator e^+e^- at SLAC, Stanford that comes to a total energy of 9 GeV energy of 4.5 GeV per beam therefore must produce it. In 1974, SLAC (B. Richter group) and BNL, Brookhaven (S. Ting Group) (Fig. 6.16), where the second experiment was conducted with a 30 GeV proton synchrotron, announced the discovery of a particle of mass of 3.1 GeV decaying into $\mu^+\mu^-$ which has an anomalously narrow width. They called it respectively ψ and J. Theorists have speculated that it is a $\bar{c}c$ bound state (charmonium) that they checked the width by perturbative QCD calculations. A new dynamic has emerged because *the mass of the charm is relatively large compared to other light quarks (u, d, s) and to the scale $\Lambda \simeq (200-300)$ MeV of QCD.* Then, one can (*a priori*) safely use the QCD perturbation theory for predicting its mass and decay widths. Thus, by analogy with QED in the calculation of the decay of positronium into 3 photons, one can calculate the decay of J/ψ into light hadrons which is done by an annihilation process $\bar{c}c \to 3$ gluons $\to$ hadrons built from the light quarks. By a simple counting of coupling, this approach predicts the hadronic width of the J/ψ of the order of α_s^3 where $\alpha_s = g^2/4\pi = 0.2$ is the QCD coupling constant evaluated at the mass of J/ψ. The discovery of the meson D (a $\bar{c}d$ bound state) having the mass of 1864 MeV heavier than $M_\psi/2$ supports this result because the J/ψ is not heavy enough to decay to a pair D^+D^-, so all its hadronic width comes from light hadrons. Since the J/ψ decaying into light hadrons is suppressed as α_s^3, one then expects that the J/ψ is narrow as indicated by experiments. The discovery of the J/ψ also opened a new path that is the *physics of heavy quarks*. As the charm is relatively heavy, it may be considered as a non-relativistic particle where approaches using the Schrödinger equation with a potential inspired by QCD models can be applied. As we have mentioned in Chapter 6 on QCD, the predictions of these models of potential on the spectra of hadrons are impressive although a closer link with the perturbative aspects of QCD remains unclear.

◇ Third family of Leptons and Quarks

• *Heavy lepton τ*

However, the adventure for the search for new leptons and new quarks continues. The same year, 1974, as mentioned in Chapter 6, Martin Perl discovered a new charged heavy lepton that he baptized *tau* (τ: third) with mass 1784 MeV in the same experiment of SLAC where the J/ψ were discovered. Its identification has led to confusion because it was thought that it was the $D(\bar{c}d)$ meson, one was actively looking for. A more detailed analysis is based on the universality of the weak coupling of τ on the pair (e, ν_e) and (μ, ν_μ) in the purely leptonic process: $e^+ e^- \rightarrow \tau^+ \tau^-$, where each τ will decay by the reaction: $\tau^+ \rightarrow \nu_\tau + \mu^+ \nu_\mu$ and $\tau^- \rightarrow \nu_\tau + e^- \bar{\nu}_e$. Experimentally, one will detect the pair (e, μ) called $(e\mu)$ *events* which allowed to differentiate the τ from the D because the D prefers to decay to $\mu^+ \mu^-$ but not in $e^+ e^-$ as the branching ratio is relatively suppressed [of the order of $(m_e / m_\mu)^2$] because of chirality. *The discovery of the τ and later its associated neutrino ν_τ completes to three the number of the families of leptons.*

• *Heavy quarks b and t*

However, if the Standard Model is correct, one must find the third family of quarks such that the theory does not contain an axial anomaly. We have already anticipated the discovery of quark b having charge $-1/3$ and of quark t of charge $+2/3$ in Chapter 6 and we have summarized their properties in Table. 6.1. The b quark was discovered in 1977 by L. Lederman of the Fermilab group in the pp experiment at 400 GeV, while the t quark was discovered in 1995 again at Fermilab by the D$\oslash$ and CDF groups in the $\bar{p}p$ experiment of 1 TeV energy. *The discovery of the pair (t, b) thus completes the three families of quarks of the Standard Model.*

♡ Three Families and only Three!

The discovery of the boson Z^0 (see following discussion) and the precise measurement of its width in the LEP experiments allowed to determine the number of possible lepton families if the Standard Model is correct. Assuming that all the neutrinos are massless or have negligible masses, Fig. 7.8 results show that the number of families is $N = 2.92(5)$. This indicates that *if the Standard Model is correct, one has already found all elementary particles constituents of matter including six quarks and six leptons.*

♠ Naturalness of the CP Violation for Three Families

In 1972, M. Kobayashi (1944–) and T. Maskawa (1940–) (Nobel Prize of 2008, Fig. 7.11) showed that *for three families of quarks or leptons, CP violation is natural.* This result is based on the mathematical properties of the 3×3 unitary matrix that mixes the three families and which is a generalization of the Cabibbo rotation matrix and GIM mechanism for two families. In the case of two families, one can always redefine the fields and absorb their phase by a redefinition of the 2×2 unitary matrix that

(a) (b)

Figure 7.11: (a) M. Kobayashi, (b) S. Maskawa.

remains real. This matrix has a single mixing angle (of rotation) which is the Cabibbo angle θ_c. This is not the case of a 3×3 unitary matrix which must have three mixing angles and a phase.[2] It is this phase $e^{-i\delta}$ which will naturally cause CP violation. These mixing angles and phase, measured with precision (see PDG[5]), meet the constraints required by the Standard Model. In addition to the already known result of decays of the K, CP violation has been seen in e^+e^- accelerators (B-factory) by BaBar (SLAC, Stanford) and Belle (KEK, Japan) in the decays of D and B in 2001 and confirmed in 2012 by CERN LHCb.

♣ Discovery and Properties of the $W^\pm$ and Z^0 Bosons

• $W^\pm$ and Z^0 in the $S\bar{p}pS$

The Standard Model predicts the masses of the $W^\pm$ and Z^0 bosons respectively at 80 and 92 GeV [5]. They were seen for the first time in 1983 by UA1 and UA2 (Underground Area 1 and 2) experimental groups SPS (Super Proton Synchrotron) at CERN transformed into a $\bar{p}p(S\bar{p}pS)$ collider with 450 GeV beam energy. This transformation of the accelerator's proton from the SPS on a circumference of 7 km, in hadron $\bar{p}p$ is technically difficult because one must make antiprotons which will be gathered in an intermediate storage ring and will be rendered more uniform by cooling in order to inject them in a coherent way in the collider. This technological

---

[2]In general $n \times n$ unitary matrix of a $SU(n)$ group must have $n(n-1)/2$ angles and $(n-1)(n-2)/2$ phases.

Particles and the Universe

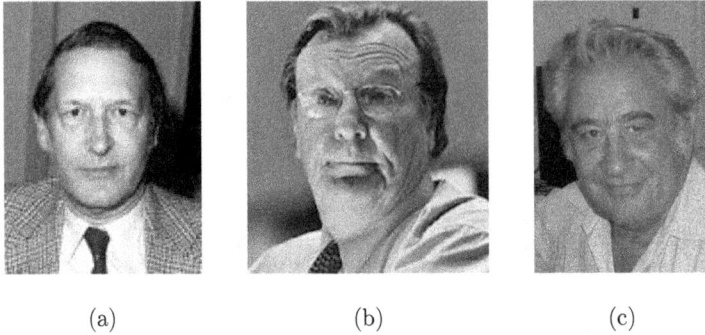

Figure 7.12: (a) S. Van Der Meer, (b) C. Rubbia, (c) P. Darriulat.

challenge, which was a first for CERN, has been accomplished by Simon Van Der Meer (1925–2011, Fig. 7.12). The discovery of the $W^{\pm}$ bosons and Z^0 at the mass predicted by the Standard Model earned S. Van Der Meer and Carlo Rubbia (1934–) — initiator of the project and leader of the UA1 experiment — the Nobel Prize in 1984; UA2 was led by Pierre Darriulat (Fig. 7.12).

• Z^0 *boson at LEP1*

However, as early as 1981, before the discovery of the $W^{\pm}$ and Z^0 bosons, one has already planned to construct a new e^+e^- collider LEP (Large Electron Positron: see Chapter 8) by using the SPS infrastructure to improve the detection of the $W^{\pm}$ and Z^0 bosons. Their discovery has accelerated the implementation of the project to refine the analysis of their properties. LEP's phase 1 of 107 GeV energy just above the mass of the Z^0 boson was built in a tunnel of 27 km in circumference and started in 1989 where millions of Z^0 have been produced by the mechanism of Fig. 6.20. Four groups ALEPH, DELPHI, OPAL, L3, each affected in a specific detector participated in this experiment. It helped measure with high accuracy the mass of the Z^0, $M_Z = 91.188(2)$ GeV, and its total width $\Gamma(Z \to \text{all}) = 2.495(2)$ GeV. In particular, the measurement of its decay width into invisible particles: $\Gamma(Z \to \bar{\nu}\nu) = 499.0(1.5)$ MeV allowed to count the *number of light neutrino*, that can be produced by decay of the Z^0, to be $N = 2.92(5)$ (Fig. 7.8) confirming the three families of the Standard Model. The democracy of the couplings of the Z^0 to the pairs $\bar{q}q$ and l^+l^- was also verified as well as a precise measurement of the QCD coupling $\alpha_s(M_Z)$. Some other properties of the Standard Model have been also measured (see PDG [5]).

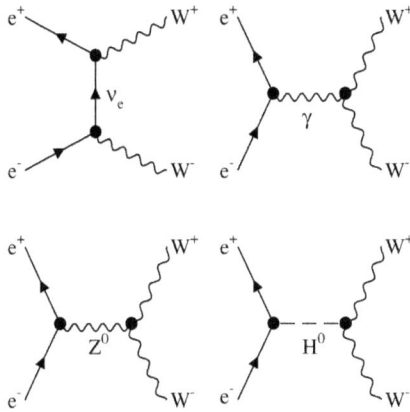

Figure 7.13: Production of a pair of W at LEP2.

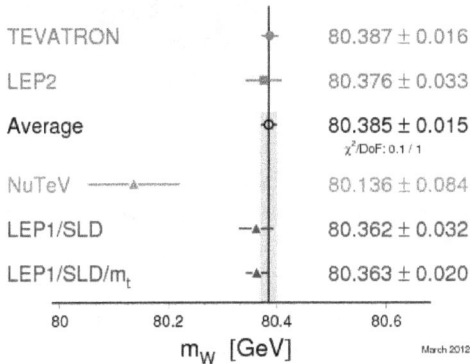

Figure 7.14: Values of the mass of the W in GeV by different experiments.

• $W^{\pm}$ *bosons at LEP2 in 1996*

Encouraged by the success on the Z^0, one is passed to the phase 2 of LEP to search the W in the reaction $e^+e^- \rightarrow W^+W^-$ through the reactions in Fig. 7.13 where H^0 is the Higgs boson. From 1996 to 2000, the total energy of the accelerator was increased gradually from 161 to 209 GeV well beyond the planned performance. Each experimental group was able to collect 11 000 pairs of W^+W^- thanks to the excellent luminosity of the machine. As the total energy is very close to the threshold of the W^+W^- pair production, it is expected that the total cross section is very sensitive to the mass of the $W^{\pm}$ leading to a precise measurement of the value of this mass. Figure 7.14 shows the results obtained by different experiences: the CERN LEP and SLD of SLAC, Stanford e^+e^- colliders, Tevatron which is a $\bar{p}p$ collider of 1 TeV energy at Fermilab and NuTEV which is a neutrino–proton collider at Fermilab. The average of these results lead to $M_W = 80.385(15)$ GeV and its corresponding total width is $\Gamma(W \to$ all) $= 2.085(42)$ GeV [5].

7.4 Uncomplete Success of the Standard Model

Along the lines of this chapter, we have seen that the *Standard Model of weak interactions passes all experimental precision tests, while the particles which it has predicted have all been seen (six quarks, six leptons and the* $W^\pm$, Z^0 *bosons*). We shall see in Chapter 8 that *the missing Higgs boson giving masses to these particles has been recently discovered*. This success is completed by the electromagnetic part of the SM where *QED describes with unprecedented accuracy the anomalous magnetic moment* $a_l \equiv (g_l - 2)/2$ *of the electron and muon as well as the value of the fine structure constant* $\alpha = e^2/4\pi$ as we have discussed in Chapter 4.

We have also seen in Chapter 6 that *QCD describes accurately the hadron phenomena* despite our present inability to understand analytically the difficult confinement problem.

Though on the right track, we still remain with many unanswered fundamental problems which we shall list in Chapter 9 and to which some tentative answers will be given.

Part III
Higgs Boson Discovery and Beyond

Chapter 8

Higgs Boson Discovery

8.1 The Hunt for the Higgs Boson

♣ Constraints on the Higgs Boson mass from e^+e^-

As we saw in Chapter 7, the mass of the Higgs boson is not fixed theoretically in the Standard Model except the loose upper bound of about 1.2 TeV. One then expects that experiments can constrain and fix it. The Higgs boson can be produced in e^+e^- from the two reactions illustrated in Fig. 8.1. The first corresponds to the decay of a real Z^0 produced at LEP1 and would allow to detect a Higgs mass less than that of the Z^0. The second reaction is the production of a virtual Z^0 at LEP2 and will produce a real Z^0 and a Higgs H^0. On Fig. 8.2, we show the variation in the variable $\Delta\chi^2$ (*which quantifies the difference between the theoretical prediction and experimental measurement*) as a function of the mass of the Higgs boson assuming that the Standard Model is the true electroweak theory. The curve results from the combination of precise measurements in the e^+e^- collisions at LEP1, LEP2 of CERN and SLD at SLAC, Stanford with those coming from the $\bar{p}p$ collider Tevatron at Fermilab. The optimal outcome should be at the minimum of the curve. Tolerant values of $\Delta\chi^2$ below 1 (68% confidence level or $\pm 1\sigma$), one would have a mass of the Higgs in the area of 94^{+29}_{-24} GeV while that, for a confidence level of 95% ($\pm 2\sigma$), which corresponds to $\Delta\chi^2 < 2.7$, precision measurements provide a mass of the Higgs $M_H < 152$ GeV fixed by the blue band of the curve. By adding the

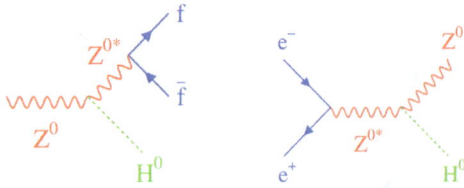

Figure 8.1: Production of the H Higgs in e^+e^- via the ZZH coupling.

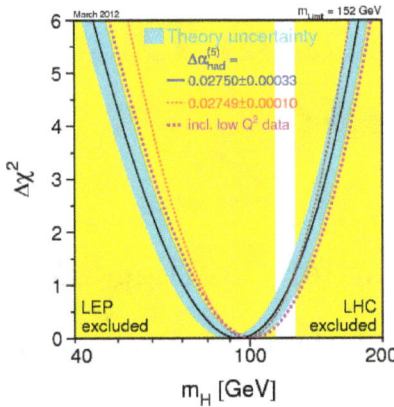

Figure 8.2: Constraint on the mass of the Higgs.

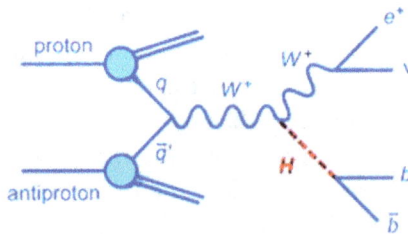

Figure 8.3: Production mechanism of the Higgs boson at the Tevatron.

direct measurement of LEP2 which gives a mass $M_H > 114$ GeV (yellow region), this result affects the upper bound which becomes $M_H < 171$ GeV.

Direct search on the production of the Higgs was also made at the Fermilab Tevatron by the D⌀ and CDF groups at the $p\bar{p}$ beam energy of 1 TeV which circulates in 6.86 km ring. *The dominant production is done via Fig. 8.3 by the production of a virtual W via a u quark and an antiquark $\bar{d}$*

issued respectively from proton and antiproton where the Higgs will decay into a pair of $\bar{b}b$. The experimental result in July 2011 excludes the area between 156 and 177 GeV (95% confidence level).

◇ Constraints on the mass of the Higgs Boson at LHC

Unlike the Tevatron (Fig. 8.3) case, the direct production of the Higgs at the LHC is dominated (88%: see Fig. 8.4) at an energy of 7 TeV of the beam of the LHC (Large Hadron Collider: see Chapter 9), by the two gluons fusion (Fig. 8.5) coming from the proton structure functions. However, *massless gluons cannot couple directly to the Higgs boson. This coupling must be done through a loop of quarks that is dominated by the heavier top quark* (the coupling of the Higgs to a pair of quarks is linearly proportional to the mass of the quark). *Similarly, the coupling of the two massless photons to the Higgs can be done only through a loop.* This loop contains a t quark or

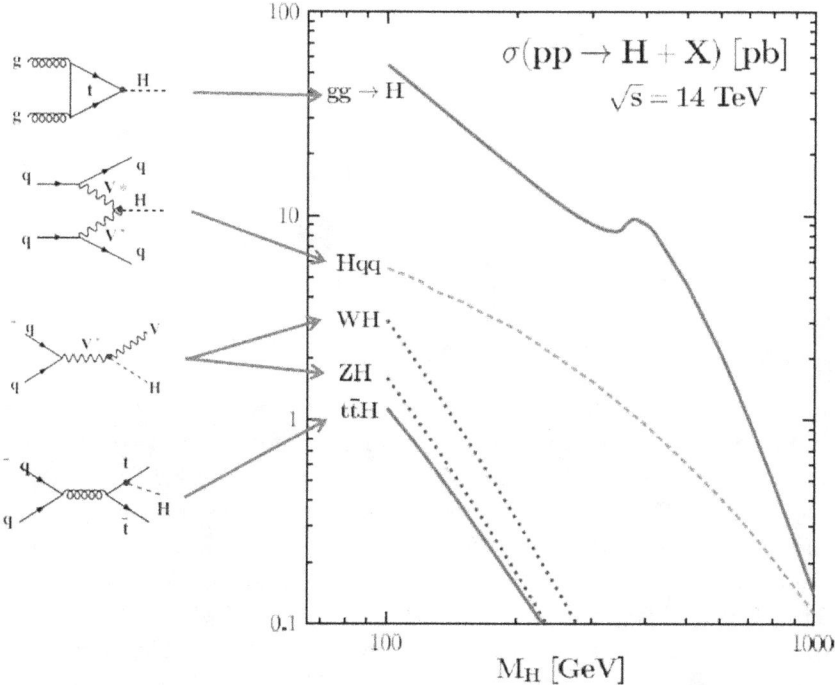

Figure 8.4: Different cross-sections for the Higgs production at the LHC.

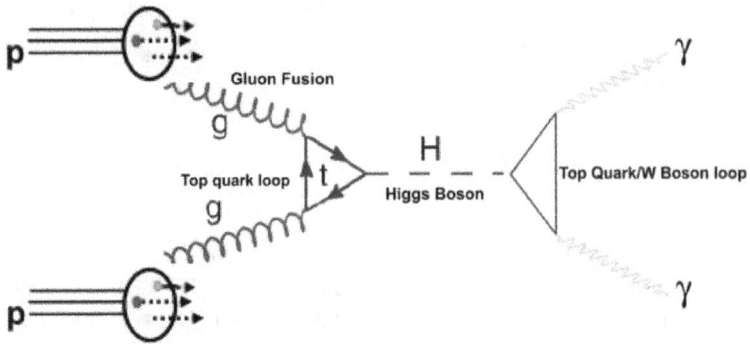

Figure 8.5: Mechanism of production and best signal of the Higgs at the LHC.

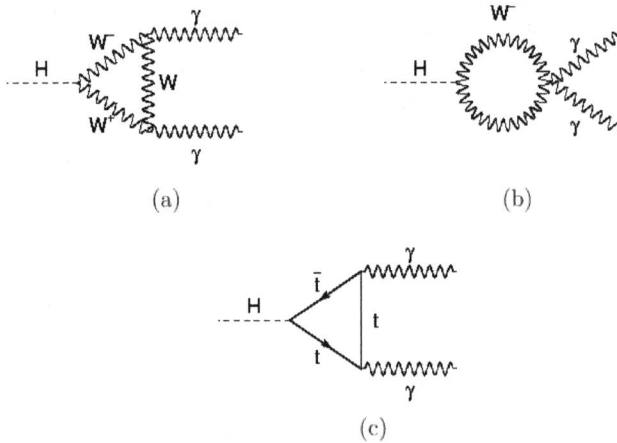

(a) (b)

(c)

Figure 8.6: Higgs mechanism decay to $\gamma\gamma$.

a boson W (Fig. 8.6). Cross-sections of different productions are compared in Fig. 8.4, while the branching ratios for the different decays of the Higgs in the context of the Standard Model are given in Fig. 8.7. In December 2011, the CMS of the CERN LHC group excludes the mass of the Higgs of the Standard Model above 127 GeV with a confidence level of 95% (2σ) while the ATLAS excludes at the same confidence level a mass lower than 115.5 GeV and larger than 131 GeV. However, a small window between 237 and 251 GeV is not excluded by ATLAS, while the excluded higher value of the mass is 468 GeV for ATLAS and 600 GeV for CMS. The lower bound of 115.5 GeV is consistent with that obtained by LEP2. The CMS and ATLAS

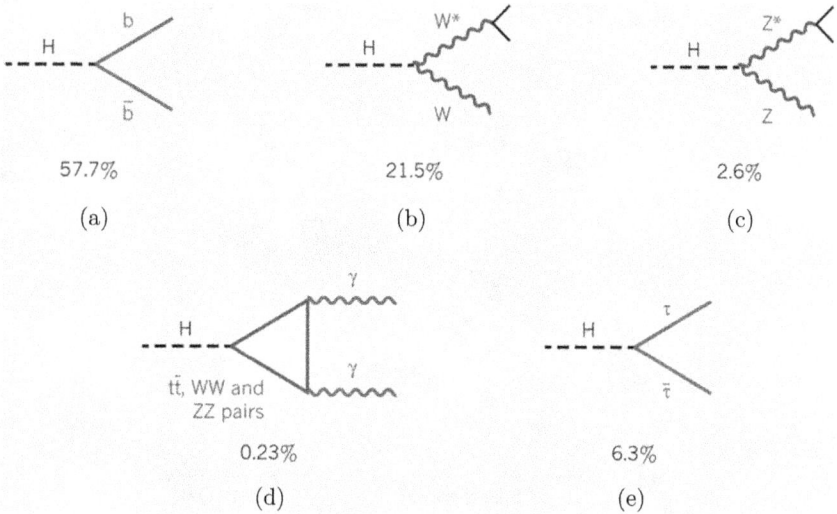

Figure 8.7: Higgs mechanism decay and branching ratio for a mass $M_H = 125$ GeV.

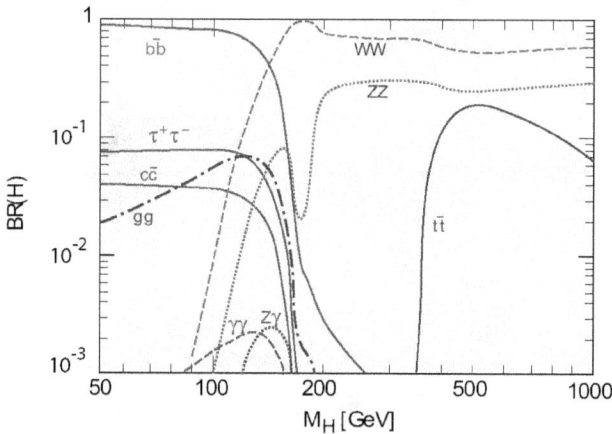

Figure 8.8: Branching ratio of different Standard Model Higgs decays as function of its mass.

results for the experimental data collected in 2011 are presented in Fig. 8.9. Does the appearance of a bump in each experiment indicate the presence of the Higgs or is it just a statistical fluctuation because the confidence level of 2σ is still low to validate a discovery?

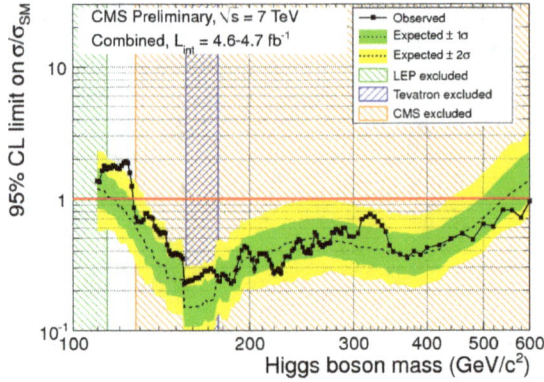

(a)

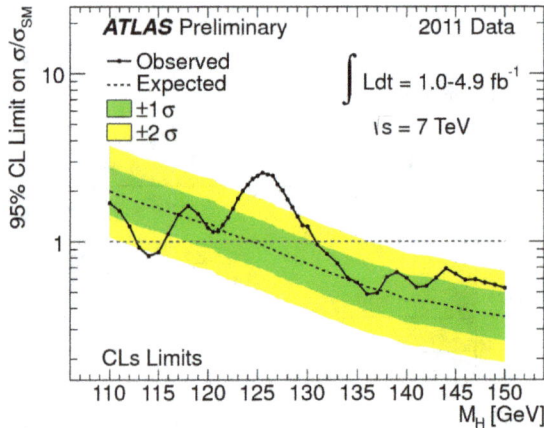

(b)

Figure 8.9: Ratio between the total cross-section observed by CMS (a) and ATLAS (b) (black dots at 95% confidence level) in 2011 and that predicted by the Standard Model (green band to $\pm 1\sigma$ and yellow $\pm 1\sigma$). CMS also shows the areas excluded by his experience.

The previous constraints have shown that: *if the Standard Model is the correct theory of the unified electroweak interaction, the Higgs boson must be found in the region between 115.5 and 127 GeV (Fig. 8.2 white band).* In fact, an indication of the signal has appeared in this window, as well as at LEP2 around 115 GeV. However, LEP2 experiment has reached its maximum energy of 209 GeV, which is no longer sufficient to continue and improve this observation. There was no question to increase the energy of

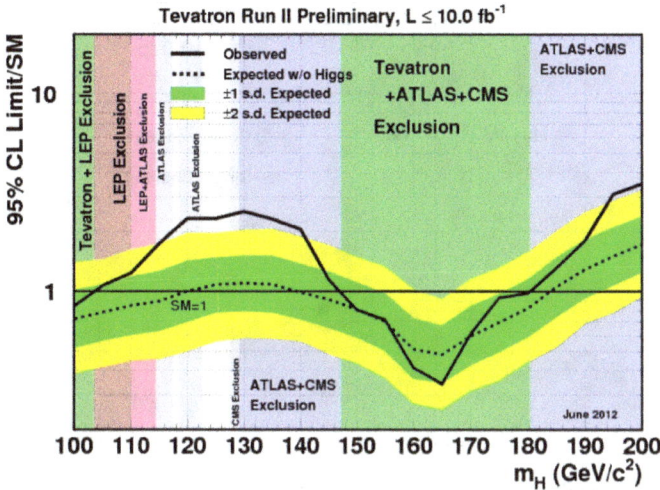

Figure 8.10: Constraint on the mass of the Higgs from Tevatron data. Those from LEP and LHC are also shown.

LEP2 because the project to build the LHC (see Chapter 9) has already been accepted. LEP2 was closed (bitterly) in 2000 such that one can start the LHC facilities whose experiment was expected to start in 2005.

♡ Higgs Boson at the Tevatron

Despite the stopping of the experiment of the Tevatron in September 2011 for budgetary reasons, Tevatron has continued to pursue the Higgs boson search by improving the analysis of already accumulated data. *The dominant mechanism of production comes from Fig. 8.3 and the decay channels analyzed are $H \to \bar{b}b, W^+W^-$ and $\gamma\gamma$.* In June 2012, the two groups D⊘ and CDF jointly published the results shown on Fig. 8.10 (black line) where the mass of the Higgs would be around 120 GeV with 2.9σ if one combines results from the $H \to \bar{b}b$ decay. The constraints coming from various experiments are also shown. Figure 8.11 shows the determination of the H couplings with particles emitted for $M_H = 125$ GeV.

8.2 The Higgs Big Show at CERN

The existence of this small vacant window between 115 and 127 GeV, the appearance of signal seen at 115 GeV LEP2 and the false alarm at 144 GeV

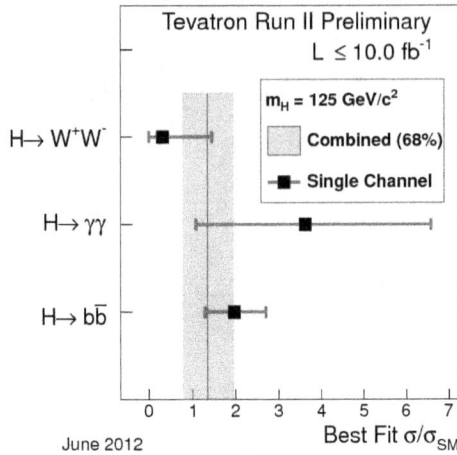

Figure 8.11: Coupling of the Higgs obtained at the Tevatron. The band is the prediction of the Standard Model.

given by ATLAS and CMS (statistical fluctuation) in 2011, as well as the signal on Fig. 8.9 encouraged the continuation of the hunt for the Higgs boson.

♣ Rise in Power of LHC

In spring 2012, thanks to their technical expertise, the LHC engineers have decided to move the beam from 7 to 8 TeV to increase the likelihood of production. Added to this is the increase in luminosity.

◇ Gold Reactions

The LHC will focus on gold reactions (less bacground noise) of the type in Fig. 8.7c and d to search for the Higgs boson.

• *The* $H \to \gamma\gamma$ *channel*

Although we expect a very small branching ratio (0.2% in the Standard Model: Figs. 8.6 and 8.7) because the cross-section behaves as α^2, this channel is very clean because there is no contamination due to hadronic decays. Also the measurement of 2 photons processes is well under control. In fact, *the mechanism of decay of the Higgs to 2-photons proceeding by diagrams of Fig. 8.6 shows that the decay and production (Fig. 8.5) occur at the level of the quantum corrections (quark and W bosons loops). This reaction is also useful because it tests at the same time the field theory aspect of the non-abelian group* $SU(2)_L$ *of the Standard Model where 3 and*

4 bosons self-couplings are present in the Lagrangian (Fig. 8.6).

• *The $H \to l^+l^- + l^+l^-$ channel*

This channel comes from the decay of Higgs to pairs of $Z^0(Z^*Z)$ where one Z^0 is virtual (Z^*) and an another real (Z). These bosons will in turn decay each to l^+l^- pair. To identify this reaction, one rebuilds the invariant mass of a pair l^+l^- coming from the real Z which should reproduce the mass of the Z. *This reaction is also clean because it is not contaminated by the hadronic backgrounds.*

♡ Preparation of the Announcement of the Results

In early July 2012, there was great excitement at CERN as ATLAS and CMS announced their results that had, from each side, remained secret. Rolf Heuer, Director-General of CERN, has summoned Fabiola Gianotti (spokesperson of ATLAS) and Joseph Incandela (CMS spokesperson) (Fig. 8.12). They have planned to present on July 4 the results at CERN and disseminate them by videoconference to the large International Conference of Physics of High Energies (ICHEP) from Melbourne which was at the same time as our biannual International Conference on Quantum Chromodynamics (QCD 12) of Montpellier (2–7 July 2012) (see Fig. 8.13) where speakers of ATLAS and CMS were already invited to present their findings on July 3 in our conference. However, despite our insistence to their spokesmen, these speakers had only the right to present the old data by 2011 if their presentations were made before the official announcement in Melbourne on July 4. We then changed our program to include the live videoconference from CERN on July 4 and then postpone to July 7 the presentations of speakers of ATLAS and CMS on data freshly gathered in 2012. During the two sessions, the room was packed and journalists were moved. Participants' impatience to know these important news and their satisfaction at the end of presentations were largely visible.

♠ Discovery of the Higgs Boson in the $H \to \gamma\gamma$ Channel

In Fig. 8.14, we show the events seen in the CMS detector and simulated in ATLAS. In Fig. 8.15, we show the number of events of the two photon spectra based on the mass of the Higgs measured by CMS and ATLAS with the combined data of 2011–2012. The mass spectrum is based on the decay into two photons of hadrons and it decreases according to the QCD

Figure 8.12: F. Gianotti (ATLAS), R. Heuer (DG, CERN), J. Incandela (CMS) during the presentation of the discovery of the Higgs at CERN.

CERN Courier   November 2012

Faces & Places

CONFERENCE

QCD 12 and Higgs-like happiness

Amid the beauty of Montpellier in France, the 16th International Conference in Quantum ChromoDynamics, QCD 12, took place on 2–6 July. It brought together theorists and experimentalists from laboratories around the world, all working actively within QCD. Mid-way through the conference, a live webcast from CERN thrilled participants with the news of the discovery of a Higgs-like boson.

Participants pose in the Montpellier sunshine. (Image credit: Alize Photo.)

Figure 8.13: QCD 12 International Conference Participants (2–7 July 2012) of Montpellier (from the November 2012 CERN Courier article).

predictions. The appearance of the little bump indicates the presence of a resonance at 125 GeV. Now, there arises the question of the statistical significance of this bump. The two plots in Fig. 8.15 show that for CMS the bump is 4.1σ above the background noise while for CMS it is 4.5σ, which is not far from the statistical threshold of 5σ to validate a discovery. In any case, the sum of two results is significantly above the threshold for the discovery.

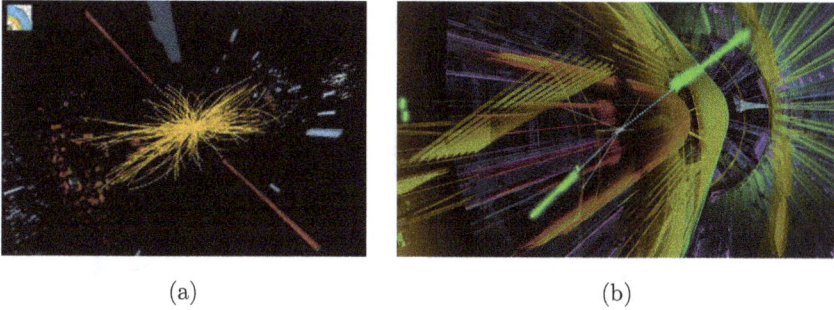

(a) (b)

Figure 8.14: Higgs decay to $\gamma\gamma$ (red jet): (a) seen by CMS (b) simulation by ATLAS (green jet).

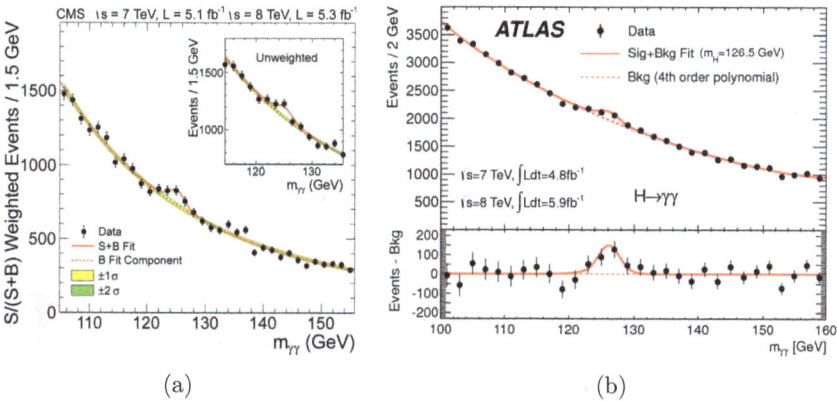

(a) (b)

Figure 8.15: Number of events of the spectrum in $\gamma\gamma$ by interval of 1.5 GeV for CMS (a) and 2 GeV for ATLAS (b) based on the mass of the Higgs boson obtained by combining 2011–2012 data. The sub-curve of ATLAS shows what remains once as background noise has been removed rendering more visible the bump at 125 GeV. Bottom of the CMS figure shows values of the corresponding standard deviations.

♣ Discovery of the Higgs Boson in $H \rightarrow l^+l^- + l^+l^-$

The two groups then present the results of the channel $H \rightarrow l^+l^- + l^+l^-$ via the decay of the Z^*Z pairs. As we mentioned before, it is a perfect channel to eliminate the hadronic background noise of QCD but the price to pay is the lost in the numbers of observable events. It is also cheaper than via the pair W^+W^- because the $W^\pm$ will decay each to $l^\pm + \nu_l$ and identifying of ν_l is more difficult than that of the charged lepton $l^\pm$. We show in Fig. 8.16 a visualization of these events in the detector. We show

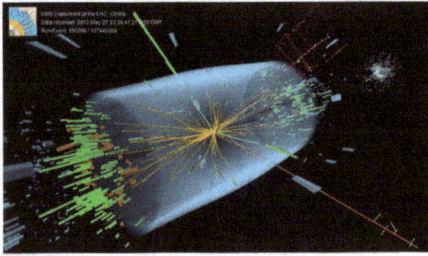
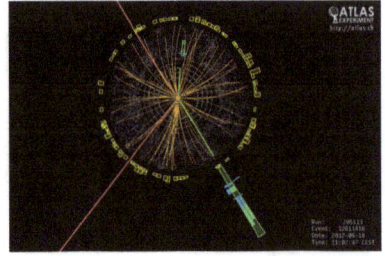

(a)   (b)

Figure 8.16: Decay of the Higgs to 2 electrons (green) in CMS (a) and to 2 muons (red) in ATLAS (b) via W^+W^- pair.

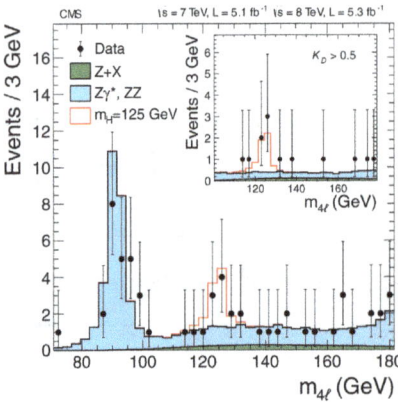
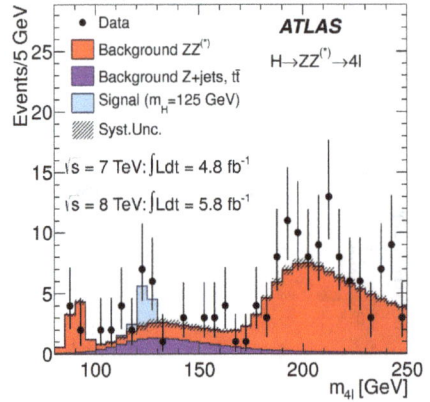

(a)   (b)

Figure 8.17: Decay into 4 leptons of the Higgs in CMS (a) and ATLAS (b) as a function of the invariant mass $2(l^+l^-)$. The CMS sidebar shows what remains after subtraction of the backgrounds.

in Fig. 8.17 the observed spectra on the basis of the invariant mass l^+l^-. The corresponding standard deviations to the 2011–2012 data are given in Fig. 8.18. The excess observed at 125 GeV is the mass of the Higgs boson. The standard deviations are respectively 3.1σ and 3.6σ (99.7% probability that is the correct event) for CMS and ATLAS. These values are further enhanced in the last update made by the two groups (6.5σ for ATLAS). These results confirm independently the discovery made in the $\gamma\gamma$ channel. *Thus the sum of each experience gives 5σ results (99.9999% that it is correct) and 6σ for combined CMS and ATLAS data.* The latest combined

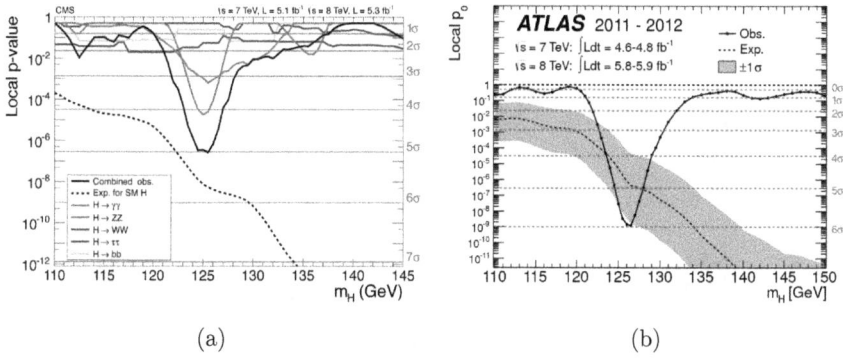

Figure 8.18: Deviation of the observation of the Higgs boson for the 2011–2012 LHC data which reaches more than 5σ i.e. more than 99.9999% confidence level that observed events are not statistical fluctuations: (a) CMS: measurement of five-channel decays; (b) ATLAS: measurement of decay to $\gamma\gamma + ZZ + WW$.

Higgs boson mass results from ATLAS is [31]: $M_H = (125.36 \pm 0.41)$ GeV, while the CMS combined result is [32]: $M_H = (125.03 \pm 0.30)$. Considering that the two results are independent, *one can take the weighted average of these two results, leading to the Higgs boson mass*:

$$M_H = (125.15 \pm 0.24) \text{ GeV.}$$

◇ Consolidation of the Discovery by Other Measurements

Some other decay channels were also measured, which allowed to increase the values of standard deviation but also measure the couplings of the Higgs to other particles. These results are summarized in Fig. 8.18 where one sees a good agreement with the theoretical predictions of the Standard Model (vertical dotted line).

♡ Is it the Standard Model Higgs?

It is, now, too early to know the exact nature of this particle.

● *Spin*
Observations from the decays of the Higgs in a pair of photons (probability of 0.2%) and ZZ^* (2% probability) which give the final state of four leptons indicate that *this particle is not a fermion of spin 1/2 because the photon and Z have spin 1* (C.N. Yang theorem). Other tests have confirmed that *it is a scalar boson of zero spin but not a spin 2 particle.*

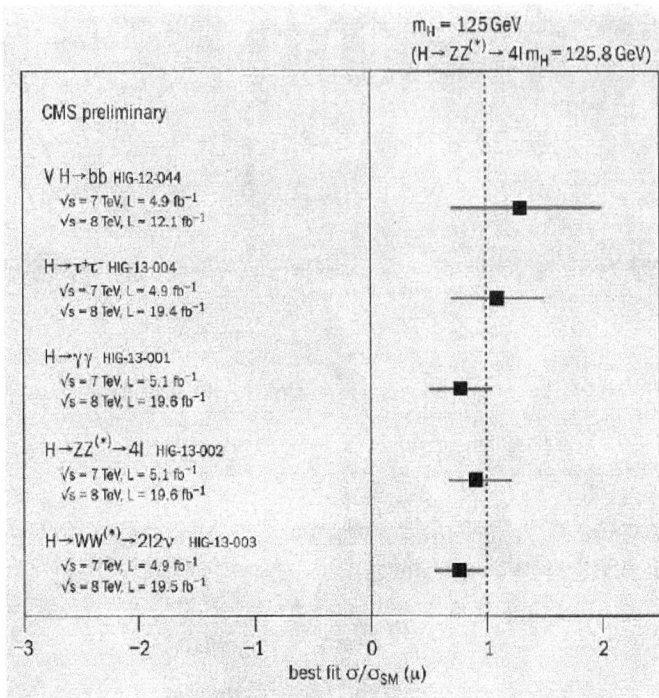

Figure 8.19: Couplings of the Higgs to the quarks, leptons and bosons gauge obtained by combining measures of CMS and ATLAS. The vertical dotted line is the prediction of the Standard Model. The full vertical line is what would have been obtained if there is no Higgs.

- *Couplings*

Measurements of its couplings to the quarks, leptons and gauge bosons were just made and showed perfect agreement (within the limits of accuracy of experience) with the predictions of the Standard Model (Fig. 8.19).

- *Quantum corrections*

Observations of the Higgs production through two gluons and its decay into two photons also showed a very good agreement with the prediction of the Standard Model. As gluons and photons have no mass, the Higgs cannot couple directly with them. The coupling can only be done via quantum corrections which are dominated by the heavier top quark loop (Fig. 8.5). The two-photon coupling can only proceed through a loop of W bosons (Fig. 8.6) which is another type of quantum correction specific of the Standard Model.

♠ Consequences of the Higgs Boson Discovery

The Higgs discovery is a milestone in our understanding of the origin of the Universe. However, as can be seen in Table 6.1, if one adds the masses (perturbative masses) of the quarks *uud* induced by the Higgs boson, that make up the proton having an experimental mass of 938.27 MeV, *one finds that the Higgs boson gives only a minuscule portion of 10.6 MeV of the proton mass, which indicates that the Higgs mechanism is not the origin of the whole proton mass.* In fact, it is the binding energy between these quarks which is most responsible for the mass of the proton. In general, the same conlcusion applies to the masses of hadrons composed of light quarks *u, d* and *s*. On the other hand, the masses of heavy quarks, gauge bosons and the Higgs itself seem to be all issued mostly from the Higgs mechanism. *Thus, identifying the Higgs Boson particle with God seems exaggerated!*

Chapter 9

The Era After Higgs

9.1 Theoretical Status After Higgs

♣ Standard Model

The discovery of the boson having a mass of 125 GeV which is very similar to the Higgs of the Standard Model (SM), the missing link of the chain, is an important step for our understanding of the origin of the Universe. Until now, the SM overcame all experimental and precision tests that he has had to face, while all the particles he predicted were observed.

◇ QED

Quantum electrodynamics (QED) as a part of the electroweak unified SM is tested with an impressive precision of 1 part per billion from the calculations and measurements of the fine structure constant α and of the electron and muon anomalous magnetic moments $a_l = (g_l - 2)/2$ (Chapter 4). However, at this high-precision, the confrontation of the measured and the SM prediction of the muon anomalous magnetic moment reveals a discrepancy of about 2.4–3.6σ (Fig. 4.10) which may indicate either there is a new phenomena beyond the SM or the present theoretical estimate of the hadronic contribution from e^+e^- and τ decay data needs to be improved. Understanding this discrepancy is the subject of the present active researches.

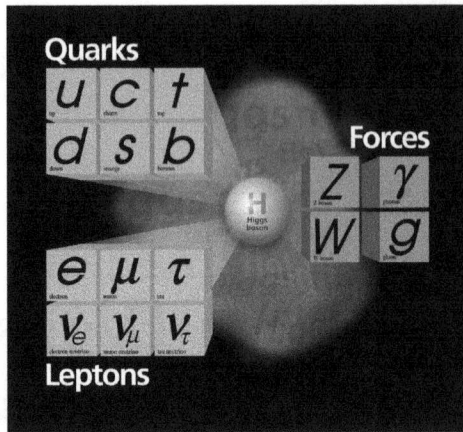

Figure 9.1: Observed particles of the SM including QCD classified as fermions (quarks and leptons) and bosons (gauge and Higgs bosons).

♡ QCD

Not less, Quantum Chromodynamics (QCD) describing the strong interaction and which is an inherent part of the SM is known with great precision: evaluation to order α_s^5 of the decay of the heavy lepton τ and of the e^+e^- cross-section by Chetyrkin *et al.* (Fig. 6.26), calculation of jets of hadrons to order α_s^2, analysis of the hadronic properties by non-perturbative approaches (lattice simulations, spectral sum rules, ...), precise determinations of α_s, quark masses, hadron couplings, ... from the various processes which are the basic parameters of QCD and of the SM.

♠ Summary

The present status of the SM including QCD is best summarized by the artistic figure in Fig. 9.1 representing the discovered SM particles.

9.2 Some Open Theoretical Problems

However, despite the success of the SM and QCD, many fundamental questions remain unanswered (see also the review of John Ellis in [30]):

♣ Are There Other Higgs Bosons?

Models with a second doublet of scalar bosons exist in the literature (two-higgs doublet model). These models predict that a second Higgs boson should be found in the region between 200–600 GeV accessible by the LHC.

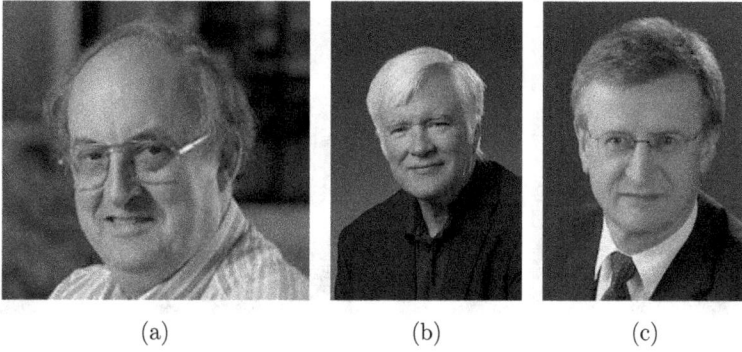

Figure 9.2: (a) W. Bardeen, (b) C.T. Hill, (c) M. Lindner.

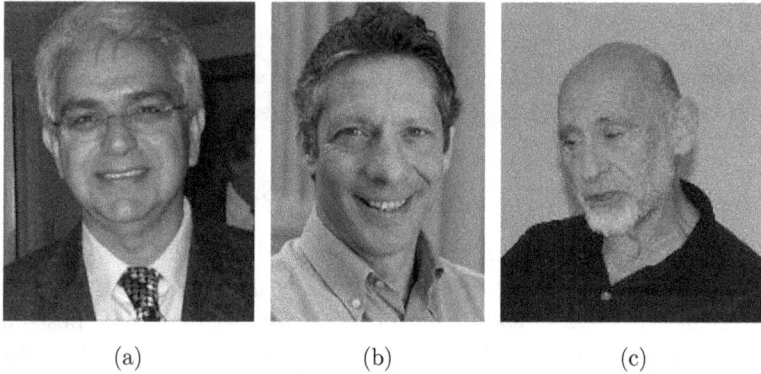

Figure 9.3: (a) S. Dimopoulos, (b) E. Fahri, (c) L. Susskind (courtesy of Jonathan Maltz, wikipedia.org).

◇ Are There Some Other Alternatives to Generate Masses?

Although the Higgs boson is discovered, the question on the true mechanism of this electroweak symmetry breaking is still unresolved. Alternatives to the Higgs mechanism as the *dynamical breaking by non-perturbative top quark condensate* [W. Bardeen (1941–), C.T. Hill (1951–) and M. Lindner (1957–) (Fig. 9.2)] or *technicolor models* [Savas Dimopoulos (1952–), Edward Fahri and Leonard Susskind (1940–) (Fig. 9.3)], ... are available on the market. For technicolor, the Higgs boson that was discovered at the LHC would be the lightest part of a rich spectrum of technicolor particles which may be found at energy reached by LHC.

Figure 9.4: (a) J. Pati, (b) H. Fritzsch, (c) H. Harari.

♡ The Problem of Mass Hierarchy

We do not yet understand why the electron is very light and why the top quark is heavy. The *SM also predicts that neutrinos are massless.* However, *experiments show that neutrinos are massive* because they oscillate as they mix together.

♠ Are There More Fundamental Particles?

At this stage, the possibility that quarks and leptons are bound states of more elementary particles called *preons* is not yet necessary although several models have already been proposed in the past. The best-known are the models of J. Pati (1937–), Abdus Salam, H. Fritzsch-Mandelbaum and Haim Harari (1940–) (Fig. 9.4). *These models have been motivated for trying to understand the above mass hierarchy and its origin.*

♣ SM Answers Only Partly the Baryon Asymmetry

Charge–parity (*CP*)*-violation appears naturally in the SM through the phase of the Cabibbo–Kobayashi–Maskawa* (*CKM*) *mixing matrix.* However, *the size of the corresponding CP-violation is to weak for explaining the matter–antimatter asymmetry* (large dominance of matter) leading to the stability of the Universe.

◇ SM Does Not Answer Some Cosmological Problems

As we shall see in Part V, *cosmology requires the existence of dark matter and dark energy which constitutes 96% of the Universe* (Fig. 13.4).

However, the SM does not answer to this question. *The problems of baryon asymmetry and of the cosmological inflation also remain unanswered by the SM.*

9.3 Grand Unified Theory (GUT)

If our present approach for understanding the Universe is correct, *these three forces (electroweak and strong) can be unified within a Grand Unified Theory (GUT) at the unification scale of about* 10^{16} *GeV* (Fig. 9.7). *GUT is mainly based on gauge theories where particles and gauge bosons are described as point-like objects* (local wave functions) and where the gauge couplings are dimensionless which guarantee the renormalizability (subtraction of infinities appearing in the S-matrix) of such (relatively) low-energy gauge theories.

♣ How to Unify the Three Microscopic Forces?

One has succeeded in unifying the electromagnetic and weak forces with the SM. There arises also the question of the unification of this electroweak force with the strong force. In response, Howard Georgi (1943–, Fig. 9.5) and S.L. Glashow (Fig. 9.5) proposed the $SU(5)$ group in 1974, later named GUT by Andrzej J. Buras (1946–), John Ellis, Mary K. Gaillard and Dimitri Nanopoulos (1948–) (Figs. 6.15 and 9.5), which includes the product of the electroweak group $SU(2)_L \otimes U(1)$ with the $SU(3)_c$ group of QCD. However, the model predicts an unstable proton which is not favored by the experience. Other unification models built on larger groups: $SO(10), SU(5) \otimes U(1),\ldots$ have been proposed in the literature.

(a)   (b)   (c)

Figure 9.5: (a) H. Georgi, (b) A.J. Buras, (c) D. Nanopoulos.

(a)　　　　　　　　　　　　　(b)

Figure 9.6: (a) J. Wess (courtesy of Renate Schmid, MFO), (b) B. Zumino.

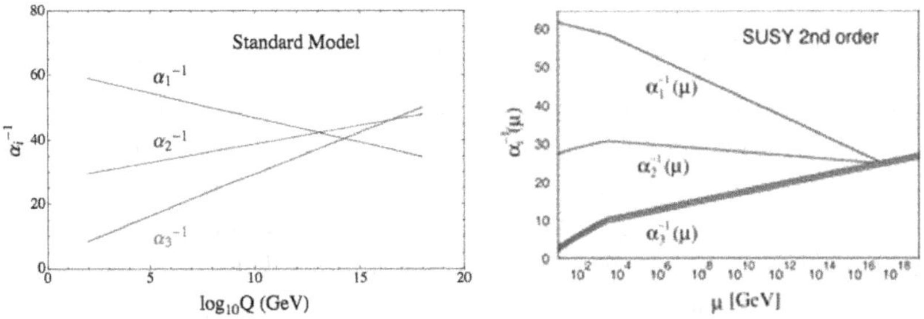

(a)　　　　　　　　　　　　　(b)

Figure 9.7: (a) Evolution of three microscopic forces as function of energy in the SM (b) their unification by SUSY.

By studying the energy dependence of the three couplings related to the strong, electromagnetic and weak forces, it is expected that the three forces meet approximately around 10^{15} GeV (see Fig. 9.7a).

◇ Why Supersymmetry?

Supersymmetry (SUSY) was introduced by Julius Wess (1934–2007) and Bruno Zumino (1923–2014) (Fig. 9.6) in 1974. *It encompasses intrinsic and spacetime (spin, isospin, color) symmetries and puts on the same footing the fermions and bosons.* Thus, each fermion of spin 1/2 has a scalar spin zero partner (called sfermion), while a vector boson of spin 1 has a fermionic partner of spin 1/2 (called ...ino) (see Fig. 9.8). *These SUSY particles should have higher mass values as they have not yet been seen at low energy*

SUPERSYMMETRY

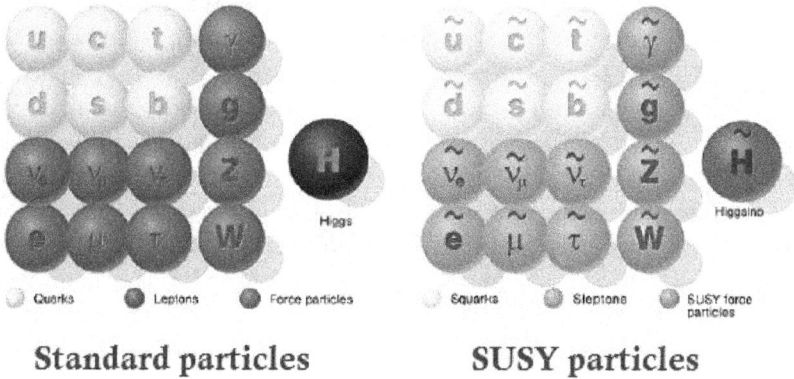

Standard particles   **SUSY particles**

Figure 9.8: SM particles and their SUSY partners.

experiments, implying that supersymmetry should be broken. However, there are not yet any convincing methods (perturbatively or non-perturbatively) to break SUSY. Therefore, there are no accurate predictions of the masses of these new particles which have not yet been seen. It is expected that LHC should either see some of them, or imposes lower bounds on their masses. The minimum version of SUSY called MSSM (minimal supersymmetric SM) is based on the group $SU(5)$ by Georgi–Glashow. It allows to solve theoretical problems appearing in the $SU(5)$ non-SUSY model which are:

• *Unification of the three microscopic couplings*
The strict meets of three couplings at 10^{16} *GeV* (Fig. 9.7b), which is not the case of non-supersymmetric $SU(5)$ model (Fig. 9.7a). This fact suggests that the more we increase the energy, more nature becomes simpler and beautiful: is that Heaven?

• *Hierarchy problem*
The problem of hierarchy (naturalness) by the cancellation of the infinite terms between the fermion and the sfermion loops in the renormalization of the mass of the Higgs boson.

• *R parity*
The achievement of R parity which is equal to 1 for SUSY particles and -1 for particles of the SM and which *prohibits couplings that violate baryon and lepton numbers which then ensures the stability of the proton and then of the Universe.*

• *Neutralinos and dark matter*

Four neutralinos denoted by $\tilde{\nu}$ of spin $1/2$ come from the mixing of the SUSY partners of the Z^0(Zino), of the photon (photino) and of the Higgs (Higgsino). Their masses are expected to be in the range of the weak scale from 100 GeV to 1 TeV depending on their nature (more Zino,...) and they couple to other particles with strengths characteristic of the weak interaction scale. These neutralinos interact weakly with matter and are called *Weak Interacting Massive Particle (WIMP)* which make them good candidates for explaining the origin of dark matter. *R parity prevents the decay of the lightest neutralino* called lightest supersymmetric particle (LSP) *which is stable* and then can explain *cold dark matter*. If the SUSY breaking occurs at the weak scale of about 100 GeV, one also expects that *gravitino is a* candidate for dark matter. *It can be a very light particle having a mass of the order of eV.*

• *MSSM faces the Higgs boson discovery*

Unfortunately, *the discovery of the Higgs boson around the mass of the SM prediction causes some difficulties on the natural and minimal SUSY extension of the SM called MSSM* as the model favors a lower value of the Higgs mass around 115 GeV instead of the observed value of 125 GeV. Therefore, some peoples start to be less enthusiastic on the naturalness of the SUSY approach!

9.4 Theory of Everything and Gravitation

Our dream is to find a consistent theory, *the Theory of Everything (TOE)*, which unifies the macroscopic gravitation with the three microscopic electroweak and strong forces (see Fig. 9.9) at the *Planck scale.*

♣ Graviton–Graviton Scattering

Like the electroweak and strong forces which are mediated by spin 1 vector gauge bosons (photon, $W^\pm$ and Z bosons and gluons), *gravitation is supposed to be mediated by a spin 2 particle called graviton.* However, if we try to evaluate the graviton–graviton scattering process, one finds that its amplitude behaves like $G_N q^2$ where the Newton constant G_N is the inverse of the square of the Planck length of about 10^{19} GeV such that for values of the energy q^2 much bigger that G_N, the amplitude increases and violates unitarity. Similar problems have been encountered in the early days of weak interactions where Fermi has parametrized the four-fermion interaction amplitude by the Fermi coupling $G_F \simeq 10^{-5}/M_P^2$. At low-energy

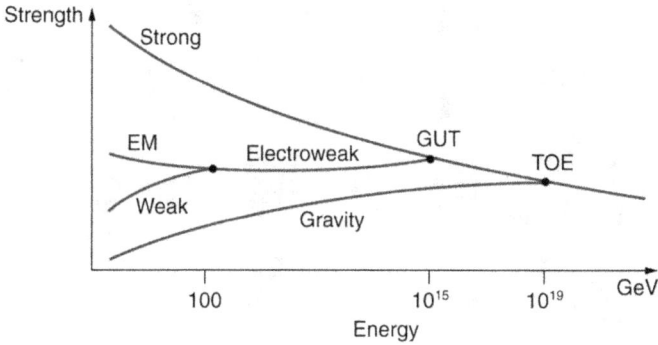

Figure 9.9: Different steps for the unification of three forces in a Grand Unified Theory (GUT) and with a gravitational macroscopic force in the Theory of Everything (TOE).

(a)   (b)   (c)

Figure 9.10: (a) S. Fubini, (b) G. Veneziano, (c) C. Lovelace.

much lower than $M_W = 80$ GeV, Fermi theory describes quite well the weak interactions processes but fail at larger energies. Within the advent of gauge theories, the Fermi coupling has been replaced by the W-propagator $1/(q^2 - M_W^2)$ times a dimensionless weak gauge coupling, which has cured this default.

◇ String Theories

In the absence of a gauge theoretical description of gravity, *string theories (models) propose a solution to the previous problem by treating particles*

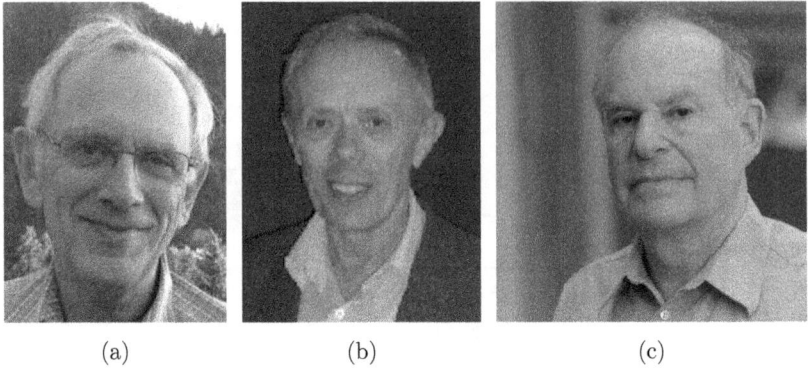

(a)   (b)   (c)

Figure 9.11: (a) P. Ramond, (b) A. Neveu, (c) J.H. Schwarz.

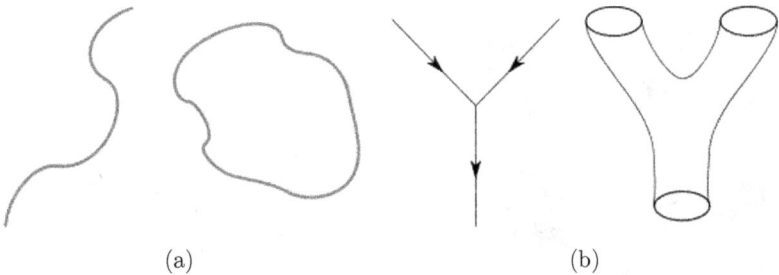

(a)   (b)

Figure 9.12: (a) open and closed strings; (b) vertex interaction in gauge and string theories.

as non-local but not point-like objects. These particles appear as quantum states of a surface (1-dimensional object) called string. There can be open and closed strings (Fig. 9.12a). A point-like three particles vertex of gauge theories is replaced by a non-local one in string theories as shown in Figs. 9.12a and b. *Its generalization to higher dimensions is called brane*: a point is a brane with zero dimension; a string is a brane with one dimension; a membrane is a 2-dimensional brane,...

The ideas of strings are issued from the Veneziano (Fig. 5.3) *amplitude used for hadron states like* $\pi\pi$ *scattering* (see Chapter 5) in 1968, where the amplitude can be expressed in terms of the Euler Gamma and Beta functions after the summation of resonance poles and then does not violate unitarity, while Veneziano and Sergio Fubini (1938–2005) (Fig. 9.10) have introduced an operator formalism for computing scattering amplitudes.

(a)   (b)   (c)   (d)

Figure 9.13: (a) W. de Sitter, (b) E. Witten, (c) J. Maldacena, (d) I. Antoniadis.

However, the drawbacks of this approach for hadron physics is the existence of a massless spin 2 particle which has never been observed, while *the self-consistency of the approach needs the presence of tachyons (particle with an imaginary mass)*. In this period, a lot of activities have been performed for eliminating such a wrong state leading to the introduction of *a bosonic string with a critical dimension $d = 26$ after a loop calculation of the amplitude by Claud Lovelace* (1934–2012) in 1971. Just after, *Pierre Ramond (1942–), André Neveu (1946–) and John Henry Schwarz (1941–)* (Fig. 9.11) *have introduced fermions for eliminating these unwanted tachyons with a string theory having a critical dimension $d = 10$ which was the beginning of superstrings. In 1995, Edward Witten* (1951–, Fields Medal in 1990) (Fig. 9.13) *showed that the five existing different versions of superstring theories in $d = 10$ dimension space–time* known as type I, type IIA, type IIB and two heterotic $[E_8 \otimes E_8$ and $SO(32)]$ strings *are particular limits of a unique so-called M-theory which can be mapped to one another by dualities. All of them possess SUSY (equal number of bosons and fermions) which makes them free of tachyons.* Type I possesses open strings (segments with end points) and closed strings (closed loops), while the other types contain only closed strings.

♡ AdS/CFT Correspondence or Holographic Duality

Introduced by Juan Maldacena (1968–, Fig. 9.13) in 1997, the *anti-de Sitter/conformal field theory (AdS/CFT)* (also called gauge/gravity duality) correspondence *implies that the M-theory is equivalent to a quantum field theory.* For instance, *there is an equivalence between an $N = 4$ super Yang–Mills theory and the 5-dimensional anti-de Sitter* [Willem de Sitter

(a)   (b)

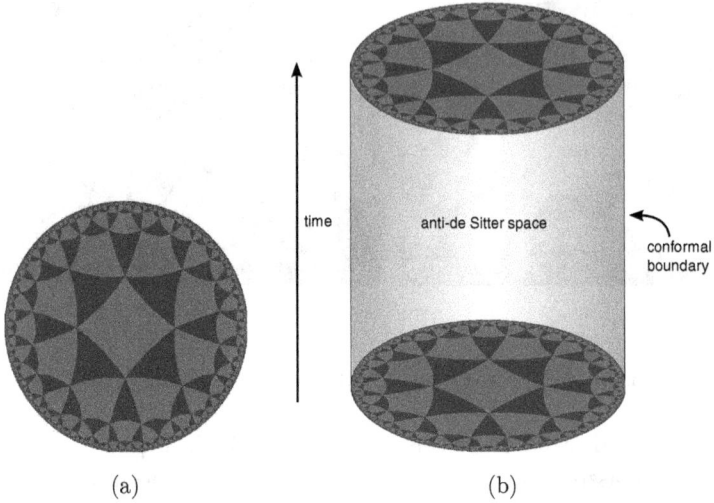

Figure 9.14: (a) Tessellation of the hyperbolic plane by triangles and squares; (b) Three-dimensional AdS space is like a stack of hyperbolic disks, each one representing the state of the Universe at a given time. The resulting space-time looks like a solid cylinder (courtesy of Alex Dunkel, wikipedia.org).

(1872–1934, Fig. 9.13)] space formulated in terms of string or M-theory,[1] where *the AdS space is a mathematical model of space–time in which the metric (distance between two points) is related to hyperbolic space which differs from the one in ordinary Euclidean geometry.* The hyperbolic plane can be illustrated in Fig. 9.14a by the tessellation of a disk by triangles and squares. One can define the distance between points of this disk such that the triangles and squares have the same size and the circular outer boundary is infinitely far from any point inside the disk. Now, one can imagine a stack of hyperbolic disks which represent each the state of the Universe at a given time. The resulting geometric object is a 3-dimensional AdS space. It looks like a solid cylinder in which any cross-section is a copy of the hyperbolic disk.[2] The cylinder which is the boundary plays an important role as it looks like the Minkowski space, the space–time of quantum field theory. Then, one claims that this quantum field theory on the boundary is equivalent to the gravitational theory on the bulk AdS

---

[1]This result confirmed G. 't Hooft's observation in 1974 that in the large color number limit, QCD calculations in gauge theory ressembles to the ones in string theories.

[2]This construction describes an hypothetical Universe with only two space and one time dimensions, but it can be generalized to any number of dimensions.

space in the sense that there is a "dictionary" for translating calculations in one theory into calculations to the other. *The conformal field theory is like a hologram which captures information about the higher-dimensional quantum gravity theory.* In this sense, the AdS/CFT correspondence is also called *holographic duality.*

♠ M-Theory for Gravitation

As a massless spin 2 particle appears naturally in superstring theories, these theories are good candidates for unifying the gravitational (macroscopic) force with the three microscopic (electroweak and strong) forces, which are expected to be realized around the Planck energy of 10^{19} GeV (see Fig. 9.7c). Although a complete formulation of M-theory is not known,[3] the theory should describe 2- and 5-dimensional objects called branes and should be approximated by $d = 11$ *dimensional supergravity* at low energies. *Modern attempts to formulate M-theory are typically based on matrix theory or the AdS/CFT correspondence discussed previously. This holographic duality renders the M-theory as an example of quantum theory of gravity.* It also provides a deeper non-perturbative picture of string theory.

♣ Compactification of Extra-Dimensions

According to the early work in 1919–1920 of Theodor F. E. Kaluza (1885–1954) and Oskar Klein (1894–1977) (see Fig. 9.15) (Kaluza–Klein), *electromagnetism can be derived from gravity in a unified theory if there are four space dimensions instead of three, and the fourth is curled into a tiny circle.* This approach is generalized to string theories, where the extra dimensions are assumed to close up on themselves to form circles. In the limit where these circles become very small, one obtains a space–time having lower number of dimensions. Intuitively, at large distances, a 2-dimensional surface with one circular dimension appears as a 1-dimensional object (Fig. 9.16).

◇ Phenomenological Tests of M-Theory

• *Particle physics accelerators*
A direct test of string theories with particle physics accelerators of the type of the LHC is out of reach due to the huge energy needed which is of the

---

[3]According to Witten, the name M has no precise meaning due to the absence of the real understanding of the structure of M-theory. It can be matrix, magic, mystery, . . .

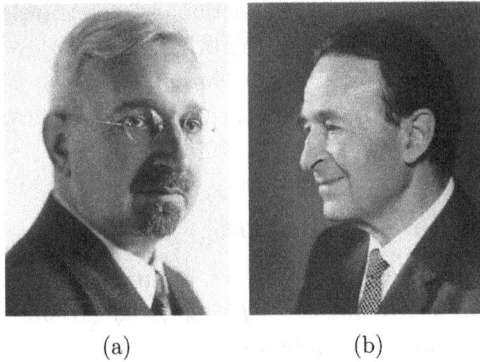

(a)   (b)

Figure 9.15: (a) T. Kaluza, (b) O. Klein

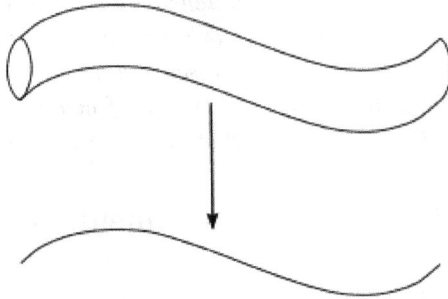

Figure 9.16: Compactification of 1- to 2-dimensional surface.

order of the Planck mass of 10^{19} GeV. The hope is that some low-energy consequences of these models will be observable. Instead, a low scale version of this approach has been investigated by Ignatios Antoniadis (Fig. 9.13) and collaborators where the string scale is lowered to the TeV scale and then the theory can manifest at lower energies. Thus, it can be tested at LHC through the measurement of dijets and/or γ+jets which can be affected by corrections due to stringy resonances poles. A resonance mass less than 4.5 TeV is excluded by present ATLAS and CMS data.

• *Gravity test experiments*

The model can also be tested in some experiments probing the structure of gravity in the sub-millimeter range, which could reveal deviations from Newtonian gravitation at distances smaller than a millimeter.

• *Big Bang*

However, one of the main characteristic prediction of string theory is that it does not require the initial time $t = 0$ of the Big Bang where the Universe is reduced to a point of density and infinite temperature (initial singularity). *For string theory this time $t = 0$ is the phase transition between contraction and expansion of the Universe that was already there before the time $t = 0$ of the Big Bang.* In these theories, the Universe is not necessarily point-like, while its temperature and its density do not exceed a maximum value that is not infinite. These predictions may be eventually tested by future experiments in astrophysics.

• *Black hole entropy*

Using Einstein's relativity and Hawking's radiation, there were hints in the past that black holes have thermodynamic properties that need to be understood microscopically. *A microscopic origin for black hole thermodynamics is finally achieved in string theory.* String theory sheds light on the subject of black hole quantum mechanics.

• *QCD*

Some applications of the M-theory within the AdS/CFT correspondence in QCD have been discussed in Chapter 6.

9.5 Concluding Remarks

There are still many phenomena to understand before arriving at the time $t = 0$ of the origin of the Universe.

♣ Big Bang

It seems unclear whether it is through the mechanism of the Big Bang with an initial singularity that the Universe was created because string theories predict that this time $t = 0$, also called *wall of Planck*, is none other than the Universe going through phase transition in contraction, that has been already there, to an expanding Universe that formed our current Universe.

♦ Supersymmetry

It is not also clear if SUSY describes our Universe as, until now, there is no manifestation of this theory at present energies.

♡ Superstrings/M-theory

The same question is addressed to superstrings or M-theory, though it is a promising approach for unifying gravity with the three other microscopic forces, for explaining the small value of the cosmological constant which may control the future of our Universe, for explaining dark matter and dark energy, and for providing a mechanism for a cosmic inflation.

♠ Observed Higgs Mass

However, the Higgs mass, discovered at 125 GeV by the LHC group, is too heavy for the minimal (natural) version of SUSY which expects a mass around 115 GeV, while the observed mass is also too light for superstring-based theories and/or for a multiverse (our Universe is part of it), which expect a mass around 140 GeV.

At present, we are waiting for the near future data from LHC which will certainly shed light on the future direction of the researches in particle physics .

Part IV
Experiments to Go Back in Time

Chapter 10

High-Energy Colliders

Before LHC

Experiment in high energy physics is to explore the infinitely small constituents of matter and test the related theories by the particle accelerators or by using cosmic rays.

Thus, opening a door without knowing what is behind! $\cdots$

(John Ellis, CERN)

♣ TV as a Linear Electron Accelerator

Our old-fashioned TV is a particle accelerator composed of an electron gun and magnet which deflects the electrons of their path (Fig. 10.1). These deviated electrons will form the image that we see on the screen.

◇ The LEP e^+e^- Accelerator at CERN, Geneva

The principle of accelerators is the same as the one of the TV except that the energy of the electron emitted (85 GeV) is five million times more powerful in the case of the experiment at LEP (Large Electron Positron e^+e^-) which has operated at CERN from 1989 to 2000. Instead of having a linear accelerator such as Stanford (SLC: SLAC Linear Collider), *at LEP, one accelerates the electron in a circular ring 27 km in circumference at 100 m below the ground* in order to preserve the environment (Fig. 10.2). The

Particles and the Universe

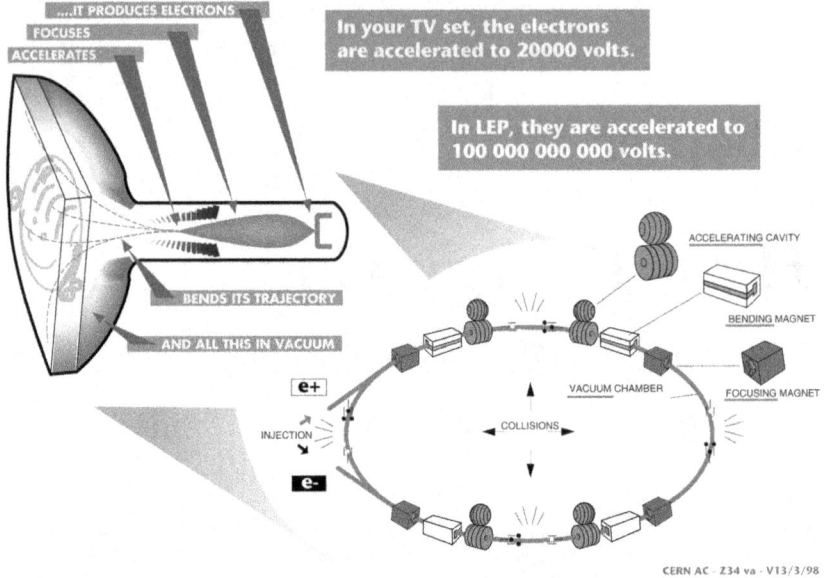

Figure 10.1: Schematic view of a TV and of the LEP accelerator.

positron (antielectron) circulates in the opposite direction and at the same speed. They will collide inside the four collision boxes shown in Fig. 10.1 called detectors which replaces the television screen. *The most used particle physics detectors are wire detectors invented by G. Charpak* (Nobel Prize of 1992, Fig. 6.16) and which were later applied in the seaports (le Havre, Toamasina, ...) to control the goods inside large containers. Each white box represents an international experimental group each with about 100 physicists. In the case of LEP, these groups are ALEPH, DELPHI, L3 and OPAL. The results of the e^+e^- collisions are shown schematically on Fig. 10.3a and traces of these particles in the detector are shown in Fig. 10.3b.

♡ Particles Discovered in e^+e^- Accelerators

• *Meson spectra*
In Fig. 6.17, we show the spectrum of mesons (bound states of quarks and antiquarks) and the Z^0 boson mediator of weak interactions, discovered in different e^+e^- accelerators via the reaction: $e^+e^- \rightarrow \gamma, Z \rightarrow$ hadrons. The discovery of *the rho (ρ) and omega (ω) mesons that are bound states*

Figure 10.2: The site of CERN in Geneva showing the circular tunnel of LEP on horseback between the France and the Switzerland. The first plan is Geneva airport. The Jura is in the background.

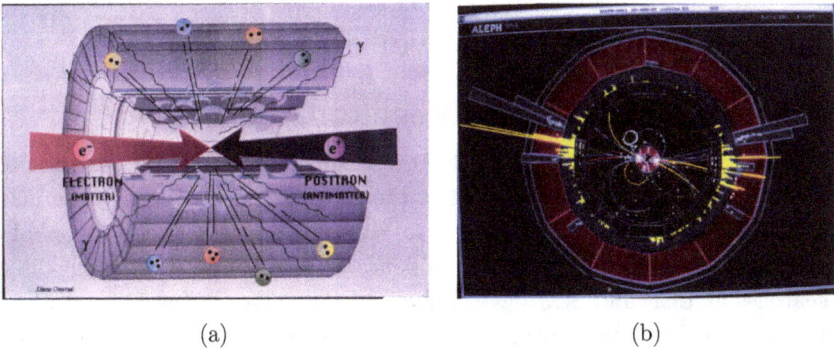

(a)   (b)

Figure 10.3: (a) Schematic view of an e^+e^- collision; (b) tracks in the detector.

of $\bar{u}u - \bar{d}d$ and $\bar{u}u + \bar{d}d$ is an indirect observation of the quarks u and d. The phi (ϕ) meson is an $s\bar{s}$ bound state. The J/ψ meson is a $\bar{c}c$ bound state. Its discovery (B. Richter and S. Ting, Nobel Prize of 1976, Fig. 6.16) was an important step for the Standard Model because it completes the second couple of the family of quarks to have similarity with the two pairs of already discovered leptons. The presence of the fourth quark c is also necessary to explain some rare decays of K meson and the mass difference between K^0 and its antiparticle $\bar{K}^0$ [mechanism of Glashow–Iliopoulos–Maiani (GIM) in 1969 (Fig. 7.9)]. The Υ discovered, by L. Lederman of the E288 collaboration at Fermilab in 1977, is a $\bar{b}b$ *bound state.*

• *The heavy lepton* τ

It was discovered in 1974 by M. Perl (Nobel Prize of 1995, Fig. 6.8) in e^+e^- accelerator via the reaction $e^+e^- \to \tau^+\tau^- \to e^{\pm}\mu^{\mp}$ + neutrinos though it is not shown in Fig. 6.17.

• *The* t *quark*

It is not represented on Fig. 6.17 but was discovered at the Tevatron (Fermilab, Chicago) in 1994 in a proton–antiproton collision where each particle has an energy of 1 TeV. *It differs from other quarks because of its high mass* (173.2 ± 0.9) *GeV. It has a very short lifetime* $(10^{-25}\ s)$ *which prevents it from hadronizing to form a* $\bar{t}t$ bound state.

• *Particles of the Standard Model*

All components of the Standard Model have been discovered since the electron in 1897 by J. Thomson (Fig. 1.15) to the tau neutrino in 2000 at Fermilab, Chicago and SNO (Canada) in 2001. *The only lacking particle before 2012 was the Higgs boson* that is supposed to give masses to these particles.

10.1   The LHC: A Jewel of Technology

♣ Search for the Higgs Boson or God Particle!

The Higgs boson searches until 2000 were unsuccessful. However, the predictions of the Standard Model after the accurate observations of the mass of the $W^{\pm}$ and the t quark helped to give an upper value of 153 GeV for the mass of the Higgs with a confidence level of 95% (2σ), while direct searches at LEP and SLC have eliminated the values below 114 GeV. So if the Standard Model is correct, we should find the Higgs boson at the LHC (Large Hadron Collider) in this window of 114–153 GeV.

◇ The LHC of CERN, Geneva

• *From LEP to LHC*

The LHC uses the LEP tunnel but the electron–positron beams have been replaced by the ones of protons circulating in the opposite direction. However, the proton has an energy 70 times higher (7 TeV) to that of the

Figure 10.4: Interior view of the tunnel of CERN.

LEP and can melt 500 kg of copper in its passage or the equivalent of 80 kg of TNT.

• *The LHC tunnel*

A view of the interior of the tunnel is given on Fig. 10.4 where one can note the tons of concrete surrounding the tunnel and the curvature of the ring despite the *9 km in diameter or 27 km in circumference*. Bending protons of 7 TeV energy is a big technological challenge since one should have a 8 Tesla magnetic field that is possible only with magnets without iron and formed coils of metal in a superconducting state (in blue on Fig. 10.5). To do this, one bathes the entire magnet in *superfluid helium* (in grey pale around the blue) *at a temperature of* $-271°$ *Celsius* or $1.9°$ near absolute zero kelvin (see Fig. 10.5). *This LHC facility is the biggest fridge of the Universe that is colder than the sidereal spaces.*

• *Cost of LHC*

The cost of the installation of the LHC is *2.6 billion euros and 1.2 billion for the salary* of the personnel, technicians and physicists. 20% of the cost of installation was supported by non-member states of CERN and the remaining 80% was collected for several years on the annual budget allocated by the 20 member countries.

• *Experimental groups*

The LHC experience involves *five international groups with thousands of members* in several countries of the world and gathered around four types of detectors:

— ATLAS (A Toroidal LHC ApparatuS) and CMS (Compact Muon Solenoid) were dedicated to *search directly and independently the Higgs and new particles beyond the Standard Model.*

— LHCb aim to analyze accurately the *B meson decays and measure the violation of the CP symmetry.*

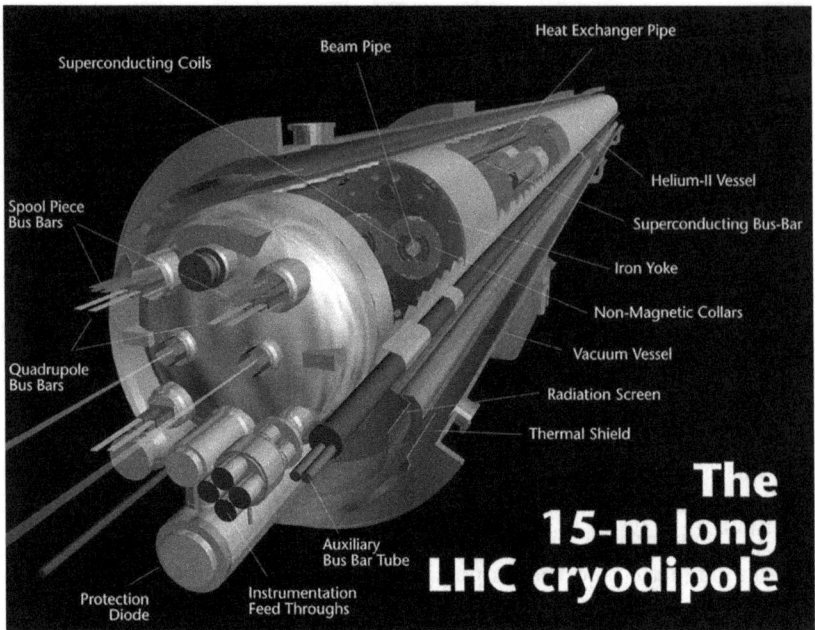

Figure 10.5: Cup of one of the 1232 magnets dedicated to the curvature of the beams of protons in the LHC. There are altogether 9600 magnets to set the path that each measure 15 m with a diameter of 1 m. The magnetic field created is opposed in each of the two small channels in yellow where circulating protons in the opposite direction.

— ALICE (A Large Ion Collider Experiment) mission is to analyze, in heavy ion collisions, *hot soup (plasma) of quarks and gluons* (deconfinement) at high energy and high-temperature densities that had to happen just after the explosion of the Big Bang before the formation of hadrons.
— TOTEM (TOTal cross-section, Elastic scattering and diffraction dissociation Measurement) detector is a small scale LHC experiment dedicated to the *precise measurement of the proton–proton interaction cross-section*, as well as to the study of the proton structure at the LHC energy.

• *Schematic view of LHC*

Figure 10.6 shows the locations of these detectors. You can see especially the mechanism of operation of the LHC. We start with a bottle of hydrogen from which is extracted the proton by removing the electron. The proton began to accelerate linearly (LINAC) and is injected in turn into the injector

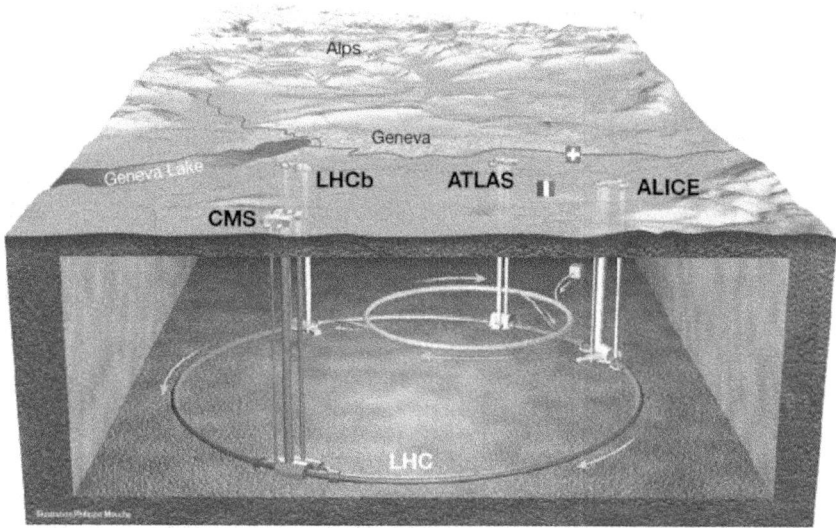

Figure 10.6: Schematic view of the LHC experiment and location of different detectors.

synchrotron of the *Proton Synchrotron (PS) (booster) of 630 m in circumference* with an *energy of 26 GeV* and *the Super Proton Synchrotron (SPS) of 7 km in circumference* with an *energy of 450 GeV* before arriving at the LHC. Once *in the LHC, proton circulates for 20 min to reach the maximum speed close to the speed of light.*

♡ Why a Circular *pp* Accelerator Instead of e^+e^-?

One should notice that in LEP one needs a circular ring of 9 km in diameter for an e^+e^- accelerator each with beam energy of 100 GeV. The reason is that the lightness of the electron/positron implies a strong synchrotron radiation (electromagnetic waves by radial acceleration) of the order of 10^{13} times that of proton causing a loss of energy proportional to $(E/m)^4$ which discriminates against the electron relative to the proton which is 2000 times as heavy. However, *as the proton is a composite state of three quarks, only a fraction (proton structure functions) of 15% (on the order of 1 TeV) of its energy of 7 TeV is useful for the LHC experiment* but it is already very powerful compared to the LEP of 2×100 GeV energy. An e^+e^- collider is useful for experiments of precision because it is very clean as the projectiles are elementary particles. In the case of the proton, the reactions are very complicated, because the simplest physics lies at the level of quarks that make up the proton, but *thanks to its higher energy, a proton machine is more conductive to new discoveries.*

♠ The LHC Detectors

As mentioned previously, wire sensors invented by G. Charpak are used to analyze the *particles produced in 600 million proton shocks per second* at the LHC which correspond to about one hundred thousand CD. Their mission is to identify the nature of these particles and accurately measure their path and their energy. For example, we show the interior of the ATLAS detector (Fig. 10.7) and its schematic (Fig. 10.8).

- *ATLAS*

It is a gigantic underground cathedral 46 m long and 25 m in diameter with the weight of 7000 tons being comparable to that of the Eiffel Tower. It is below the Meyrin village in Switzerland. It involves about 3000 personals (physicists, engineers, technicians, students and support staff) from *38 countries and 174 institutes.* Its large size is related to the choice of having a second toroidal magnetic system. It is the hub of several sub-detectors. The central part of the detector (*tracker*) is used to reconstruct the trajectories of charged particles due to a magnetic field induced by a superconducting coil. Above the tracker, the electrons and photons produce a shower in the *electromagnetic calorimeter.* However, they can be distinguished because, due to its electrical charge, the electron leaves traces in the tracker. Above the electromagnetic calorimeter, there is the *hadronic calorimeter.* To separate the electrons and photons with hadrons, we must choose materials of absorbers. Thus, 18 cm of lead stops a photon energy of 1 TeV while it takes 2 m of iron to stop a hadron of the same energy. These materials bathe in liquid argon at −183°C which allows great stability of the equipment in time. Only neutrinos and muons arrive to escape from these two calorimeters. As neutrinos do not interact with matter, they will not leave a trace in the detectors. Thus, only muons can emerge from calorimeters. As

Figure 10.7: Interior view of the ATLAS detector.

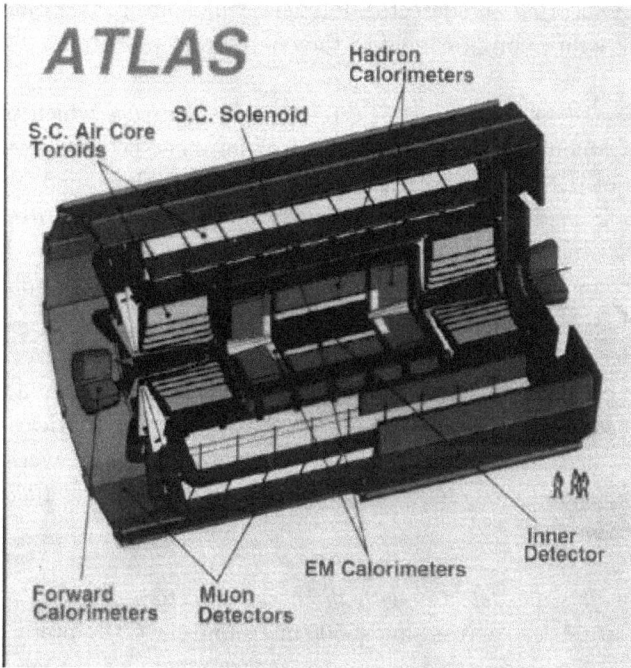

Figure 10.8: Schematic of the ATLAS detector.

they are charged, they can be detected by an another tracker. The particle's flux being low, because one has only muons, one can use simple machines and the least expensive detector wire type to identify them.

• *CMS*

It is a more compact detector 21 m long, 15 m in diameter, but it is heavier (12 000 tons). It is near Cessy in France and involves about *4300 personals* (the biggest size LHC group) *from 42 countries and 182 institutes.* Its superconducting solenoid magnet produces a magnetic field of 3.8 Teslas two times more intense than that of the LHC. *Here, we want to measure the decay into two photons of the Higgs boson, while at ATLAS we want to identify the four muons from the decays of the two intermediate Z^0 bosons produced by the Higgs.* Thus, the electromagnetic calorimeter technologies are very different in both experiments. While ATLAS uses (absorbers) plates and sheets of lead containing 175 000 sectors of detection, all bathed in liquid argon at −183°C, CMS uses 75 000 bars trimmed in crystals more transparent than glass lead tungstate but denser than iron, which will serve

as both an absorber and detector. Particles will emit light scintillation. This
amount of light is proportional to the energy of the photon.

- *LHCb*

It is dedicated to search for CP violation in B decays which produce an
enormous amount of pairs of $\bar{b}b$. This experiment needs a very different
geometry of the apparatus and a lesser luminosity. It is made up of a for-
ward spectrometer and planar detectors. *It is 21 m long, 10 m high and
13 m wide and 5600 tons weight* and sits 100 m near below the Ferney-
Voltaire commune in France. It involves *700 scientists from 66 institutes.*

- *ALICE*

*It is designed to study collisions of heavy ions at high-energy and high-
temperature densities (quark–gluon plasma phase of QCD)* and is still different because of the enormous multiplicity of produced particles. *It is 26 m
long, 16 m in diameter and 10 000 tons weight.* It sits in a cavern below the
Saint-Genis-Pouilly commune in France. It involves about *1000 personals
from 30 countries.*

- *TOTEM*

It is a small scale detector with three sub-detectors of about 3 m. These
detectors are spread across almost 500 m around the CMS interaction point.
TOTEM is dedicated to the *precise measurement of the proton–proton
interaction elastic and diffractive cross-section, as well as to the study of
the proton structure and ISR physics at the LHC energy.* It involves about
100 personals (the smallest size LHC group) *from 16 institutes.* TOTEM
and CMS will coordinate the use of their detectors to perform combined
measurements which will lead to very accurate measurements.

♠ Computer High-Technology at the LHC

Among the particles produced during 600 million shocks per second, most
come from hadronic processes of low momenta and have nothing to do with
the searched new physics. For example the probability of production of a
Higgs boson is 10^{-10} times the total number of produced particles. Thus, a
very severe selection criterion is applied. *From the 600 million events, one
keeps only 200 interactions per second, which corresponds approximately to
the archiving of a flow of 300 megabytes per second or 25 million gigabytes
annually.* CERN has an archiving capacity of 100 million megabytes which
is equivalent to 100 pentaoctets.

Analysis of the data, such as the reconstruction of an event (trajectory, energy deposit, ...), simulation of the studied process and background

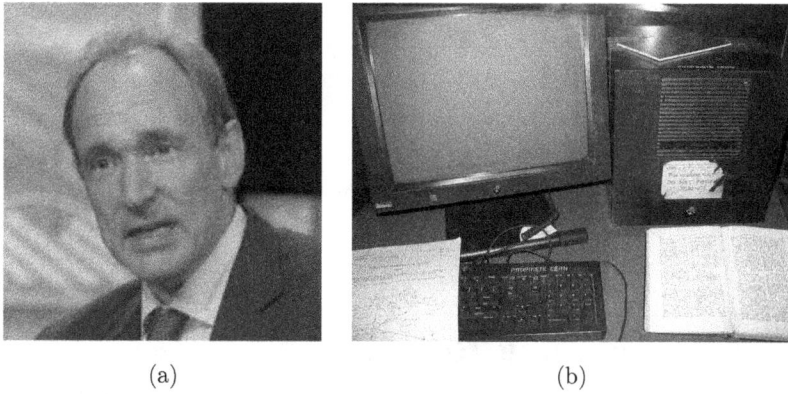

(a)   (b)

Figure 10.9: (a) Tim Berners-Lee; (b) First web server NexT computer used by Tim Berners-Lee at CERN.

(a)   (b)

Figure 10.10: (a) Robert Cailliau and his first Macintosh computer where he has developed the www page at CERN; (b) His original www logo design.

noise, requires large means of calculations. This problem was anticipated for a decade by the technique of *grid calculations inherited from web technology invented at CERN by Tim Berners-Lee in 1990* (Fig. 10.9) and which we know as the origin of the public and commercial www pages on the internet. With Robert Cailliau, they have co-authored a proposal for funding the project. Robert Cailliau (Fig. 10.10) developed later the web browser for the Mac OS operating system. *The LHC computing grid is composed of CERN, 11 national computing centers, 150 centers in laboratories and thousands of users with their own computers.* Thus connected in series as in an electrical circuit, one could have 200 000 processors in 34 countries. The local user can then run computer tasks in a very powerful virtual machine without knowing in what processor these tasks will be executed on the planet. This is the so called *Cloud* in the new language from Microsoft and computer scientists.

10.2   Concluding Remarks

We have seen from the example of LEP and LHC that fundamental researches lead (indirectly) to some progresses in high technology. Most of these innovative materials are already available as prototypes which we may use in the near or far away future for our everyday life. We shall come back to these discussions in Part VI.

Chapter 11

Fishing the Invisible Neutrinos

11.1 From Reactors to Cosmic Rays

Historically, *the neutrino has been introduced by Pauli in 1930 for explaining the missing energy in the β-decay radioactivity* $n \rightarrow p + e^- \bar{\nu}_e$ where the momenta of the initial neutron is larger than the one of the observed states proton and electron. We have also seen in Chapter 7 that *the neutrino* ν_e *only participates in weak interactions and does not interact with matter because it is neutral. In fact, the solar and cosmic neutrinos are incessantly bombarding us without any effects* rendering their detection difficult.

However, *the nature of the neutrinos remains mysterious.* We have seen in Chapter 7 that *there are three different types of neutrinos but there is only an upper limit on their mass* (Table 7.1). From the Standard Model, they are massless and their number is $N_\nu = 3$. However, the observation of the deficit of solar electron neutrino flux issued from solar nucleosynthesis when they arrive to the Earth because they are not be able to transmute an atom of ^{37}Cl into ^{37}Ar indicates that the ν_e transforms into ν_μ at a large distance. That is due to the *neutrino oscillation* which is predicted to be periodic (sinusoïdal) and which *can only happen if, at least, one of these neutrinos is massive.* Massive neutrinos should also affect the expansion of the Universe after the Big Bang as one may expect that there could be a critical density where the gravity can stop this expansion. This critical density corresponds to about 3 protons per m^3 which is about 300 millions of neutrinos. Therefore, *a neutrino mass larger by about* 10^{-8} *times the proton mass is enough* for making the density larger than the critical density and then *stopping the infinite expansion of the Universe.* Thus, measurements of the neutrino mass and oscillation are necessary steps for our understanding of the origin of the Universe.

Another still unanswered question is the nature of neutrino and anti-neutrino. In Dirac's theory like the Standard Model, they are different while for Majorana, a neutrino is the same as an antineutrino like in supersymmetric models.

We shall discuss below some experiments dedicated to study the neutrino properties.

11.2 Reactor and Accelerator Experiments

♣ Electron Neutrino (ν_e)

Nuclear reactors are potential sources of neutrinos (10^{13} neutrinos per square centimeters per second). Thus, Clyde Cowan (1919–1974) and Frederick Reines (1918–1998) in Fig. 11.2, Nobel Prize of 1995 shared with Martin Perl (Fig. 6.8) initiated the experiment at the Hanford nuclear reactor in 1953 (Fig. 11.1) but the experiment had large background due to cosmic rays. They moved to the new Savannah River nuclear reactor in 1955 and put the detectors 11 m from the reactor and 12 m underground. *Neutrinos are detected as initiators of the inverse β-decay reaction:* $\bar{\nu}_e + p \rightarrow n + e^+$ where one measures the produced positron e^+ which annihilates with the e^- to produce 2 photons of 0.5 MeV energy in opposite directions and also the neutron. e^+ and n are slowed down by the water, while the neutron is captured by the cadmium some microseconds after the positron capture. *This experiment enabled them to confirm the existence of the anti-electron neutrino ($\bar{\nu}_e$) in 1956.*

◇ Muon Neutrino (ν_μ)

For detecting the muon neutrino in 1962, L. Lederman (Fig. 6.16), M. Schwartz (1932–2006, Fig. 11.2) and J. Steinberger (Fig. 6.19) (Nobel Prize of 1988) have bombarded a matter target by a proton beam. Thus, one creates the pion (π), after the interaction of the proton on the nucleons target, which decays later into muon and electrons (note that the production

Figure 11.1: F. Reines and C. Cowan at the control center of the Hanford experiment for a first detection of the muon neutrino in 1953.

| (a) | (b) | (c) |

Figure 11.2: (a) C. Cowan, (b) F. Reines, (c) M. Schwartz.

of electron is relatively suppressed by a factor $(m_e/m_\mu)^2$, i.e. of the order of 10^{-5}). These charged muons will be stopped by an absorber. The analysis of the emitted neutrino indicates that the pions decay subsequently only to a muon but not to an electron which indicates that *the produced neutrino is not the same as the one in β decay. This new neutrino is called muon neutrino.*

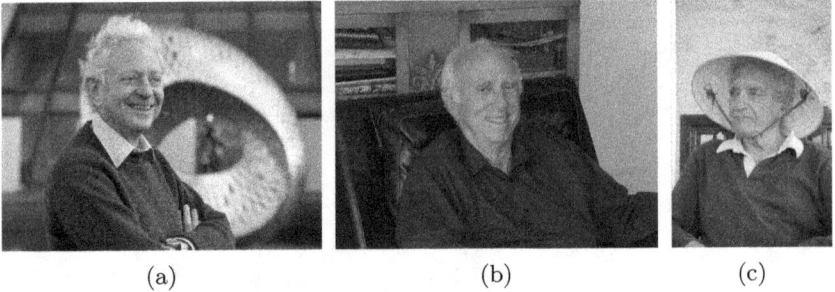

(a) (b) (c)

Figure 11.3: (a) L. Lederman, (b) M. Perl, (c) J. Steinberger.

♡ Tau Neutrino (ν_τ)

Discovering the third charged lepton τ in the e^+e^- collision at the SLAC, Stanford in 1975, *from the decay process*: $e^+e^- \to \tau^+\tau^- \to e^\pm\mu^\mp+$ *missing energy*, M. Perl (Nobel Prize of 1995, Fig. 11.3) *highlighted the existence of the neutrino associated to the* τ by an analysis of the energy conservation of the τ and the produced particles. A direct signature of ν_τ has been done recently at Fermilab (2000) where a neutrino interacting with the nucleus produces the charged heavy lepton τ confirming that this neutrino is the (ν_τ) associated to the τ lepton.

♠ Neutral Current and the Z^0 Decay

The discoveries of the neutral current by Gargamelle at CERN in 1973 (one of the essential steps to validate the Standard Model) and the heavy Z^0 boson by the UA1, 2 groups at CERN in 1983, decaying into a neutrino and antineutrino $Z \to \nu\bar\nu$, and which has been used by LEP to limit the number of massless or light neutrinos to $N_\nu = 3$, have highlighted the importance of neutrinos in our understanding of particle physics.

♣ Neutrino Masses

The principle for measuring the neutrino masses is based on the conservation of the energy which has also been used for its discovery. As they are produced associated to the charged lepton, it is enough to study the mass spectrum of the charged lepton. At the end of the spectrum, where the charged lepton has maximal energy and then the neutrino is at rest, one can deduce an upper bound on the neutrino mass by comparing the data with the theoretical prediction for a massless neutrino. A similar method has been used for giving an upper bound on the mass of the tau neutrino by comparing the spectrum of hadron produced in the hadronic decays of the charged heavy lepton τ.

◇ Neutrino Oscillations

Reactor and accelerator experiments can also be used to study neutrino oscillations.

• *Reactor experiments*

Similar to the original experiments of Cowan and Reines, one can use the electron antineutrino produced by the reactors which is about 10^{21} per second and with an energy of a few tens MeV. However, one should find the optimal position of the detector because if it is placed too near or too far from the neutrinos source, one cannot see the phenomena. *The best result was obtained from the Nuclear Power Station Chooz (Ardennes, France)* (Fig. 11.4) where the detector was placed 100 m underground and 1 km from the source. After optimizing the experimental parameters (intensity of the flux as function of the produced electricity, age of the combustible, backgrounds,...), one has concluded after 2 years of data collection that the experiment is not adequate for detecting the neutrino oscillation. However,

Figure 11.4: Chooz Nuclear Power Station at les Ardennes, France (copyright EDF/Didier Marc).

CERN to Gran Sasso Neutrino Beam

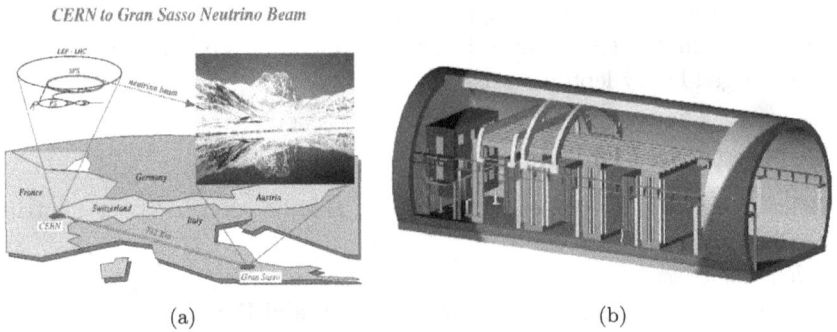

| (a) | (b) |

Figure 11.5: (a) Sketch of the Gran Sasso experiment; (b) Interior of the Opera detector.

it has provided a lower bound on the probability of oscillation which was necessary for setups of future experiments.

• *Accelerator experiments*

These experiments are technically similar to the one used by L. Lederman, M. Schwartz and J. Steinberger for discovering the muon neutrino (see section on muon and tau neutrino) where one uses an accelerated proton beam for producing the muon neutrino except that one placed the detector far from the source for enabling the neutrino to oscillate. The detector is at 250 km for the K2K experiment in Japan, 730 km for MINOS in the US and for OPERA in Europe. *For K2K and MINOS, one is analyzing the disappearance of the muon neutrino after oscillations. For OPERA, one studies the production of tau neutrino after oscillations.* The signature is the observation of a charged τ from a neutrino-induced reaction in the detector. For OPERA, the muon neutrino flux comes from proton produced by the Super Proton Synchrotron (SPS) at CERN, Geneva and directed toward the OPERA detector located at 732 km in the Gran Sasso 3 km underground tunnel of Abruzzes (Aquila) near Roma, Italy (Fig. 11.5). The OPERA detector consists of about 13 million lead plates, which are arranged in a sandwich structure with photo emulsion sheets in between. The use of nuclear photo emulsion is dictated by the fact that the τ has only a short lifetime and thus generates a track as short as 100 micron.

11.3 Neutrinos from Cosmic Rays

We have different sources of natural neutrinos from cosmic rays because we already know that we are incessantly bombarded by these particles without knowing it. These neutrinos can be from different origins.

♣ Primordial Neutrinos

They are relics of the Big Bang. Like in the case of the photons, one may also imagine that neutrinos decouple from matter when the temperature decreases at about 3000°C for about 2 seconds after the Big Bang and can constitute the cosmic background or fossile radiation. One may expect about 100 million of these neutrinos per m^3 of the Universe, though small compared to 10^{23} protons in 1 cm^3 of water, this number is huge compared to the few 1–2 protons inside 1 m^3 of the Universe. However, they have small energy of about 10^{-3} eV making their detection difficult.

◇ Solar and Supernova Neutrinos

Neutrinos can be also formed via nucleosynthesis happening in the Sun and stars where the liberated proton interacts with the e^- to produce a neutron and an electron neutrino: $p + e^- \rightarrow n + \nu_e$. These neutrinos escape from the core of the Sun or of the exploding star and travel to the Earth, which, in addition to the emitted photons, are *a good signature for identifying the supernovae.* In 1987, the discovery of the supernova from the explosion of the star labelled SN1987A located at 160 000 light years has increased by a factor 2 the number of expected neutrinos in the detector during 10 seconds, 2 hours after the detection of emitted light (photons). This result also indicates that *the mass of the emitted neutrinos should be small as they arrive at the same time as the photon.*

• *BOREXINO solar neutrino experiment at Gran Sasso*
It includes measuring the ν_e produced by two protons fusion: $p + p \rightarrow d + e^+ + \nu_e$. Initiated by the GALLEX experiment, one uses a gallium target bombarded by neutrino flux which transforms into a germanium. After a complicated chemical reaction, one arrives at the previous $p + p$ process and counts the number of emitted neutrinos. At present, the previous detector is replaced by the BOREXINO detector, with liquid scintillators, which searches for solar neutrinos having energy less than 1 MeV. The background is reduced by the use of liquid in which contamination by products having natural radioactivity is less than 10^{-15} g per g of liquid, which corresponds to 2 emissions of electron per day inside the 300 m^3 of the detector. *GALLEX experiment and SAGE experiment in Russia have detected copious low-energy neutrinos below 400 keV which are the primary components of the solar neutrino flux.*

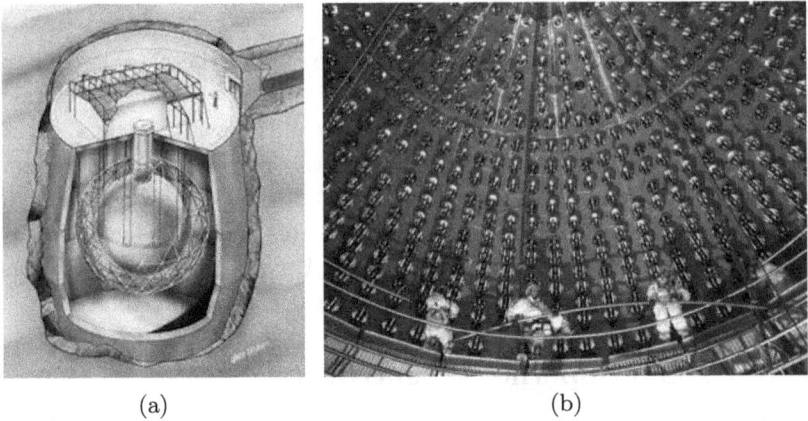

(a)   (b)

Figure 11.6: (a) SNO detector (courtesy of the SNO collaboration), (b) Borexino detector at Gran Sasso.

- *Sudbury Neutrinos Observatory (SNO)*

The SNO detector is installed 2070 m underground in the Creighton mines near Sudbury (Ontario, Canada) (see Fig. 11.6). It is similar to Super Kamiokande where it is surrounded by 7000 tons of ultra-pure ordinary water inside a cavity of 22 m width and 34 m height. However, outside, one has a spherical transparent plastic recipient of 17 m diameter containing 1000 tons of heavy pure water which detects the neutrino for about 1 neutrino per hour. *This experiment has found that the neutrino transforms to another (oscillation) when they travel from the Sun to the Earth.*

♡ Atmospheric Neutrinos

- *Sources*

One can also perform experiments using cosmic rays or aliens that nature gives to us. *Atmospheric neutrinos are produced when cosmic ray particles from outer space collide with the Earth's atmosphere (Fig. 11.7a) at a distance of 30 km from the ground* which is about 3 times the altitude of long distance flights. The rays interact with air and produce charged pions $\pi^{\pm}$. The $\pi^{\pm}$ will decay in turn to a muon which will be accompanied by his invisible wife, the muon neutrino. The muon will then decay to an electron accompanied by his wife, the electron neutrino. But in the latter reaction, the muon will be also accompanied by the image of his wife who is the anti-muon neutrino. By this process, then, we arrived to produce three invisible

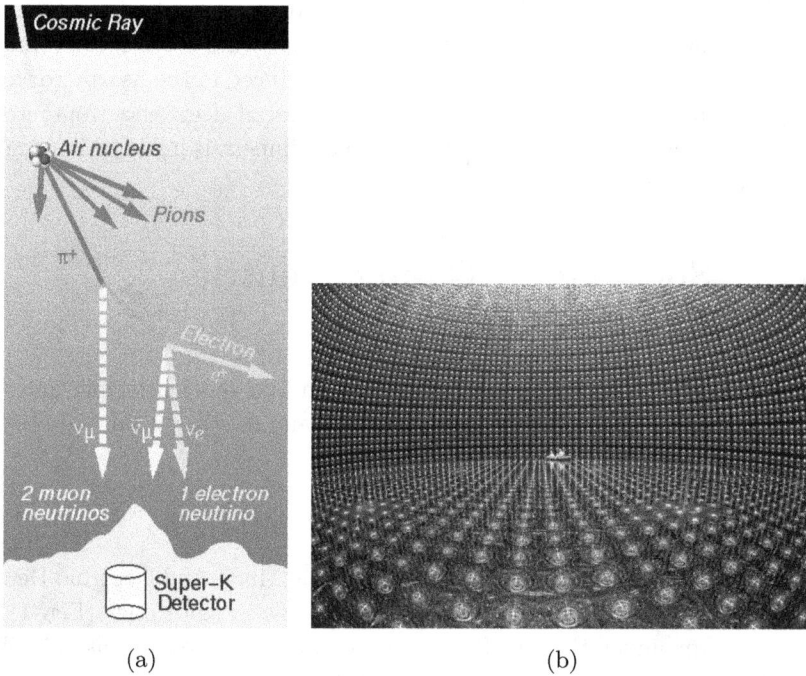

Figure 11.7: (a) X-ray cosmic; (b) Interior of the Super-K detector.

particles.[1] To retrieve them, one can make an underground experience by stopping the parasite charged particles.

• *Super-Kamiokande*

The Super-K detector (see Fig. 11.7b) is located 1500 m underground in the Mozumi old gold mine in Hida's Kamioka area. It consists of a cylindrical stainless steel tank 41.4 m high and 39.3 min diameter holding 50 000 tons of ultra-pure water. The tank volume is divided by a stainless steel super-structure into an inner detector region of 33.8 m in diameter and 36.2 m in height and an outer detector which consists of the remaining tank volume. The water in the tank acts as both the target for neutrinos, and the detecting medium for the by-products of neutrino interactions. The inside surface of the tank is lined with 11 146 light collectors (photo-multiplier tubes) having 50 cm diameter each. In addition to the inner detector, which is

---

[1]Actually, it is a mixture of these neutrinos which are produced when they oscillate due to the fact that one of them or all of them are massive.

used for physics studies, the additional outer detector is also instrumented light sensors to detect any charged particles entering the central volume, and to shield it by absorbing any neutrons produced in the nearby rock. In addition to the light collectors and water, a forest of electronics, computers, calibration devices, and water purification equipment is installed in or near the detector cavity.

11.4 Some Other Cosmic Neutrino Experiments

Some other cosmic neutrino experiments on the same principle and for similar goals as previous experiments, which aim to study solar and cosmic neutrinos, are listed as follows:

♣ Ice-Cube

This experiment replaces the previous Antartic Muon And Neutrino Detector (AMANDA) and is located in the south pole in Antarctica (Fig. 11.8) about 2 km under the ice. It consists of spherical optical sensors called Digital Optical Modules (DOM), each with a photomultiplier tube (PMT) and a single board data acquisition computer which sends digital data to the counting house on the surface above the array. *It is designed to look for high-energetic neutrinos in the TeV range. Ice-Cube is more sensitive to neutrinos in the northern hemisphere where the neutrino enters the southern ice from below.* The PMT will track the muon as it travels through the telescope allowing to reconstruct the original neutrino's energy and direction for determining its origin in the galaxy. Detection of 28 very energetic neutrinos likely from solar system has been announced in 2013.

Figure 11.8: Ice-Cube site in the Antarctica south pole.

◇ ANTARES as a Submarine Detector

Another method for fishing the neutrino is the use of a submarine detector. Astronomy with a Neutrino Telescope and Abyss environmental RESearch (ANTARES) 2.5 km deep and 40 kilometers off the coast of Toulon, France. The 150-meter wide, 300-meter high ANTARES need a deep, flat and featureless section of sea floor. ANTARES faces environmental and technological challenges. The telescope will have 900 optical modules tight along 12 electrical cables, all of which must be kept leak-free as a tiny leak would destroy the PMT inside. A small mechanical weakness in the glass optical module could cause it to implode at the depth of the sea, where the pressure is 200 to 240 times of that at the surface. Salt water contains a radioactive isotope of potassium, whose decay produces a speeding electron. That electron can emit blue and UV light, simulating a muon and fooling the closest PMT. Ocean currents as strong as 18 cm per second can move the top of the ANTARES detector strings by as much as 5 or 10 m. Strong ocean currents, seafishes and other organisms that have learned to overcome the lack of natural light by producing their own bioluminescence also stimulate other sources of light, which are backgrounds to the experiment. *Like in the case of Ice-Cube, the neutrinos can then be detected indirectly through detection of Cherenkov light produced by charged particles (muons) emerging from neutrino interactions* in the sea water or sea bed (see Fig. 11.10). *ANTARES complements the Ice-Cube telescope and will survey the larger part of the Galactic disc and center*, which is less visible by Ice-Cube at the south pole. It uses energetic neutrinos to study particle acceleration mechanisms in active galactic nuclei and gamma-ray bursts, which may

Figure 11.9: ANTARES Experiment in the Mediterranean sea.

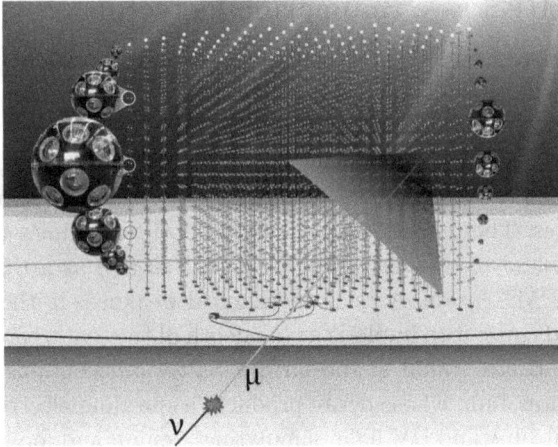

Figure 11.10: Reaction mechanism for detecting neutrino from energetic muons from the Earth for submarine and Ice-Cube experiments.

also shed light on the origin of ultra-high-energy cosmic rays. ANTARES also *studies the oscillation of the neutrino in the Mediterranean sea by using a source of neutrino where one of the entry points is Itampolo in the South of Tuléar (Madagascar).*

♡ Baikal Lake Telescope

Baikal Lake Telescope is similar to the ANTARES experiment. It is located in the Siberian Baikal lake at 1.1 km depth, covers 43 m diameter and 72 m high area of the lake. *It looks through the Earth for neutrinos from the southern hemisphere.* Completed in 1993, it has seen few hundered atmospheric neutrinos.

♠ Anti-neutrino at Super-NEMO

We have mentioned earlier that it is not clear if the neutrino is its own antiparticle. To clarify this problem, one can consider a *double β decay experiment with two neutrons decaying into two protons and two electrons.* If the neutrino is a Dirac particle (neutrino $\neq$ antineutrino) like in the Standard Model, one can, in addition, have an emission of 2 antineutrinos. In the case of Majorana particle (neutrino = antineutrino), the 2 neutrinos annihilate and there is no missing energy. *The experiment consists of the detection of the two produced electrons.* It has been started in the Frejus

Figure 11.11: SuperNEMO detector used to measure rare double β decay at the underground Frejus tunnel in Modane (France).

tunnel under the Alpes (France) in order to protect the detector called NEMO from some parasitic reactions. The detector itself produces a very low radioactive background. However, this kind of process is rare as the average time for a nucleus decay is larger than 10^{25} years, which means that with 100 g of matter which contains about 10^{26} nucleus, one only expects to have about 10 events per year compared with the 10 decays per second of 1 kg of a human body due to natural radioactivity. After 8 years of running the NEMO detector has been replaced by a SuperNEMO detector since 2011 which is installed in the Frejus tunnel in Modane, France. The uranium-238 and thorium-232 activities in the 200 tons of SuperNEMO (Fig. 11.11) recovered by iron, water and woods are less than the radioactivity in human body. *In addition to the neutrino, this experiment also aims to study dark matter and datation.*

Part V
A Brief Cosmos Tour

Chapter 12

General Relativity and Gravitation

12.1 Einstein Field Equations

Introduced by Einstein in 1916, *General Relativity is the generalization of the Special Relativity and Newton's law*, where gravity is described as a space–time geometric property and where the energy and momentum are related to the curvature of space–time independently of the presence of matter and radiation. This theory is governed by the Einstein fields system of partial differential equations or *Einstein field equations* [see Eq. (A.19)] which relate the mass and energy, expressed in terms of the *stress-energy tensor*, to the curvature of space–time (*Ricci–Einstein tensor*) [Gregorio Ricci-Curbastro (1853–1925), Fig. 12.1]. Ricci–Einstein tensor measures the deviation of the geometry determined by a given *Riemann metric* [Bernhard Riemann (1826–1866), Fig. 12.1] which is not constant but varies from point to point of space–time. This Riemann geometry which includes in it the *non-Euclidiean geometry* might differ from that of ordinary Euclidean space. Its predictions for the motions of an object in free fall (liberated from the gravitational force and with a non-negligible velocity compared to the one of light), the propagation of light, the passage of time, the geometry of space differ significantly from those of classical physics. In these cases, the corrections due to the space–time curvature should be considered. However, in the limit where the speed of light is infinite or the object has small velocity (non-relativistic limit), the theory should reproduce the newton law of gravity. In order to appreciate this effect, let us consider the free fall of the Earth considered as a spherical object of mass M and radius R. Its deliberated velocity is given as $2G_N M/R$ where G_N is the gravity constant. This velocity of about 11 km/s is relatively small compared to the one of light such that corrections due to general relativity are negligible though should be taken into account if one wishes to make precision measurements like the GPS synchronization, atomic clock, . . . These effects become sizeable at the scale of stars and galaxies having large masses and densities.

12.2 Einstein Cosmological Constant

The cosmological constant Λ appears in the Einstein equation [Eq. (A.19)]. *Historically, this term has been introduced by Einstein on the basis that the Universe which is a static and homogeneous matter should be unstable and tends to reduce to a point.* The addition of the cosmological constant into the equation cures this problem, which leads Einstein to introduce the *Mach principle* [Ernst Mach (1838–1916), Fig. 12.1] requiring that the gravitational field and the inertia should be determined by the energy content of the Universe. However, according to De Sitter (Fig. 9.13), the presence of this term on the left hand side (LHS) of the equation is inconsistent with the Mach principle [Universe without matter (and then stable) with elliptic geometry but not a flat space]. If put on the right hand side (RHS) of the equation, its origin might be of the microscopic origin but then it

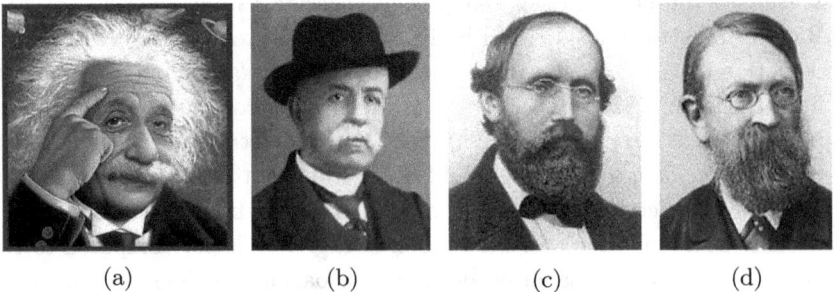

Figure 12.1: (a) A. Einstein, (b) G. Ricci-Curbastro, (c) B. Riemann, (d) E. Mach.

does not explain the repulsive effect between light particles put in a limited part of the space because the cosmological constant tends to separate these particles towards events horizon at finite distance. Einstein abandoned this idea later on after the experimental observation of the temporal evolution of the space–time geometry.

To make us more familiar to the astrophysical or cosmological field of research, let us give some technical word definitions.

12.3 Time Dilation

It states the elapsed time difference between two events measured by observers either moving relatively to each other (relative velocity) or differently situated from gravitational masses (gravity). This is due to the nature of space–time and explains why an atomic clock in the space shuttle runs slightly slower than the one on Earth. An astronaut spending 6 months on the International Space Station (ISR) is younger by 0.007 s than his age on the Earth. The effect would be larger if the astronaut travels near the speed of light of 300 000 km/s (the speed of the ISR is 7 km/s). However, this effect of relative velocity is opposite of the gravity effect where a clock closer to the gravitational mass (the Earth) is expected to go slower than the one away from it. However, these two opposite effects are not equally strong such that the ISR astronauts end up with slower time.

12.4 Gravitational Redshift of Light

The redshift is the reduction of an electromagnetic light or other radiation frequency. Due to the wave and quantum nature of light, it is equivalent to the reduction of photon energy or equivalently to the increase of the light wavelength. The gravitational redshift occurs in a weaker gravitational field due to time dilation where away from the emitting source, the rate at which the time passes relatively increases. On the contrary, *a blueshift is observed when an electromagnetic radiation moves towards a gravitational field.* Another type of redshift is the *cosmological redshift* due to the expansion of the Universe (see Chapter 13) where a few million light-years distant light sources show redshift due the distance from Earth.

12.5 Gravitational Lensing

Predicted by Einstein in 1926, *gravitational lensing refers to the phenomenon of matter (stars, galaxy or cluster of galaxies) bending (distortion by gravitation) the light from a source when it travels to the observer.* Deflected from above and from below by the massive object if it lines up perfectly with the light emitting source, the light seen by the observers appears to form a ring around the object (Einstein's ring). If the lensing object is not lined up perfectly with the source, one observes instead a distorted image of the source. As the deflection angle due to the gravitational effects is proportional to $\sqrt{M/D}$ where M is the mass of the object and D its distance from us (equal to the one of the lens to the source), astronomers using an optical or infrared telescope having a performance of 0.000 28° cannot see the ring of about 0.000 002 5° from a star distant of 100 parsecs (microlensing) (1 parsec = 3.085 677 58 × 10^{13} km = 3.261 633 44 light-years: see units in Table 14.1) but can observe the one of 0.0025° from a galaxy situated at 100 Mpc (weak lensing) or better 0.025° from a cluster of galaxies at 1000 Mpc (strong lensing).

12.6 Gravitational Time Delay

Observed by Irwin Shapiro (1929–, Fig. 1.27) in 1964, the *gravitational time delay effect states that radar signals passing near a massive object take a slightly longer time to reach a target and to return than they would if the massive object were not present.* This is due to the fact that the gravitational potential slows the propagation of light. Shapiro tests this theory by measuring the time delay of about 200 microseconds for a radar signal traveling from the Earth to Venus and back in the presence of the Sun in the favorable case where the Earth, Sun and Venus are aligned.

Chapter 13

The Minimal Big Bang Model

13.1 Friedmann–Lemaître Model

The *Big Bang hypothesis* and the cosmological evolution of the Universe is summarized in Fig. 2.4. *This theory claims that the current Universe we live in is the result of combinations of the infinitely small particles produced from the violent explosion 13.8 billion years ago of a single point in space where the entire Universe was contained.* According to such a theory, one would see 1 second after the Big Bang a 10 billion kelvin sea of neutrons, protons, electrons, anti-electrons (positrons), photons and neutrinos. Then, as time went on, we would see the Universe cool down. The neutrons either decay into protons and electrons or combine with protons to form deuterium (an isotope of hydrogen), other molecules, matters, and eventually planets and other large-scale structure.

The Big Bang hypothesis says that the Universe is expanding and was denser and hotter in the past. That comes from Alexandre Friedmann (1888–1925) in 1922 and independently from Georges Lemaître (1894–1966) in 1927 and later by Georges Gamow (1904–1968) in 1948 (Fig. 13.1).

♣ Friedmann–Lemaître Equations

The Model assumes a homogeneous and isotropic Universe with a positive cosmological constant and expansion parameter $a(t) = R/R_0$ which relates

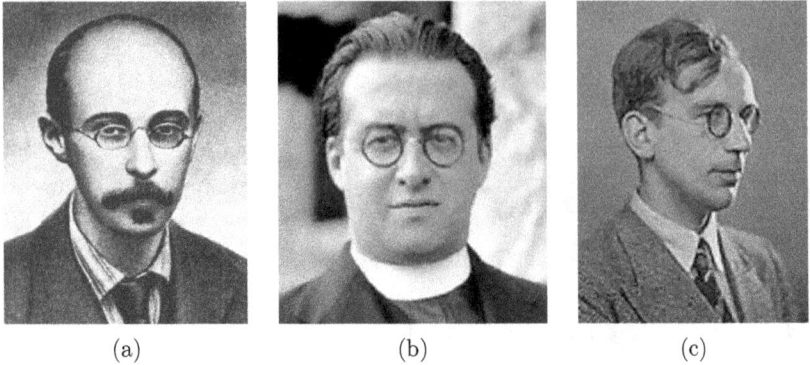

Figure 13.1: (a) A. Friedman, (b) G. Lemaître, (c) G. Gamow.

Figure 13.2: (a) E. Hubble, (b) A. Penzias, (c) R. Wilson.

the size R of the Universe at a time t to the one R_0 at time t_0. Within such assumptions, have been derived from the Einstein equation in Eq. (A.19), the set of Friedmann–Lemaître equations in Eqs. (A.20) and (A.21). Equation (A.21) defines the *critical density*: $\rho_c \equiv 3H^2/(8\pi G_N)$ which is about five atoms of hydrogen per cubic meter.

These concepts have been developed and highlighted by Edwin Powell Hubble (1889–1953) in 1929 (Fig. 13.2). Observing a shift to the red (*redshift*) of the spectra of several galaxies, they show that they are moving away from another with a speed linearly proportional to their distance (*Hubble's law*) via the so-called *Hubble's constant* having the value: $50 \leq H_0 \leq 85$ km s^{-1}Mpc^{-1} as given in Table 14.1.

◇ The Cosmological Constant

• *Zero value*

In the simplest Big Bang model, this constant is put to be zero. Then, in the case of flat space $(c^2/R^2 = 0)$, the Friedmann–Lemaître equation [Eq. (A.20)] implies that the total density of matter is equal to the critical density, while the sum of the kinetic energy is equal to the sum of gravitational energy.

• *Dark energy*

By making the replacement: $\rho \to \rho + \Lambda c^2/(8\pi G_N)$ and $p \to p - \Lambda c^4/(8\pi G_N)$ into the Lemaître–Friedmann equations [Eqs. (A.20) and (A.21)], where ρ and p are the matter density and pressure, one can omit the cosmological constant. In this way, the cosmological constant can be interpreted as arising from a form of energy which has negative pressure, equal in magnitude to its (positive) energy density: $p = -\rho c^2$. Such form of energy, which is a generalization of the notion of a cosmological constant, is called *dark energy*.

13.2 Evolution of the Universe

The evolution of the Universe can be divided into different phases:

♣ The Very Early Universe

This period is poorly known as there are no accelerator experiments which can probe this region of energy. It is itself subdivided into:

• *Planck epoch*

This happened from 0 to 10^{-44} s after the explosion where the temperature was so high such that *the four fundamental forces of nature (gravitational, strong, electromagnetic and weak forces) which we see around us merge (see Fig. 2.3) into one unique force.*

• *Grand Unification epoch*

It happened between 10^{-44} to 10^{-37} s after the Big Bang. As the Universe expanded and cooled, it crossed a transition temperature where the *gravitational force (macroscopic) separates from the three microscopic (strong, electromagnetic and weak) forces.*

• *Electroweak epoch and baryon asymmetry*

— It happened between 10^{-37} to 10^{-10} s after the Big Bang where the temperature of the Universe is low enough ($\leq 10^{28}$ K) such that *the strong* *force splits from the electroweak (electromagnetic $\oplus$ weak) forces and where first particles (more baryonic than antibaryonic matters)[1] have been formed asymmetrically (baryogenesis).* The conditions requested for such an asymmetry to occur in the Standard Model of electroweak interactions (see Chapter 7) do not induce strong enough effects to explain the present baryon asymmetry in the Universe.

— Notice that in an *inflationary cosmology*, the electroweak epoch ends when the *inflationary epoch* begins at around 10^{-32} s. This inflationary epoch or *cosmic inflation* is supposed to be produced by a scalar field called *inflaton* (very similar to the Higgs boson of the electroweak theory: see Chapter 7 and Part III). The cosmic inflation stopped when the inflaton decays into ordinary particles. However, the duration of this decay is unknown at present.

♢ The Early Universe

After cosmic inflation ends, the Universe is filled with a quark–gluon plasma. At this stage, the physics of the early Universe is better understood and less speculative.

♡ Electroweak Symmetry Breaking and the Quark Epoch

It is between 10^{-12} to 10^{-6} s after the Big Bang. After a critical temperature, *it is expected that the Higgs field spontaneously acquires a vacuum expectation value. The particles and gauge bosons which interact with the Higgs boson acquire masses via the Higgs mechanism.* At the end of this epoch, the fundamental gravitational, strong, electromagnetic and weak forces have taken their present forms, and the elementary particles are massive but the temperature of the Universe is still too high to allow quarks to bind for forming hadrons.

• *Hadron epoch*

Between 10^{-6} to 1 s, after the Big Bang, *the quark–gluon plasma that composes the Universe cools to form hadrons* including baryons which we have studied in quantum chromodynamics (QCD) (Chapter 5).

• *Cosmic neutrino background*

At approximately 1 s after the Big Bang, neutrinos decouple and begin traveling freely through space. This cosmic neutrino background, while unlikely

---

[1] A proton and neutron are examples of baryons (see Chapter 5).

to ever be observed in detail since the neutrino energies are very low, is analogous to the cosmic microwave background that was emitted much later.

• *Lepton epoch*

Between 1 and 10 s, *at the end of the hadron epoch, most of hadrons and anti-hadrons annihilate each other, leaving leptons and anti-leptons dominating the mass of the Universe.* Approximately 10 s after the Big Bang, the temperature of the Universe falls to the point at which new lepton/anti-lepton pairs are no longer created and most leptons and antileptons are eliminated in annihilation reactions, leaving a small residue of leptons.

• *Photon epoch*

It is about 10 s to 380 000 years after the Big Bang where *the energy of the Universe is dominated by the photons* which interact frequently with protons and electrons.

— *Nucleosynthesis*

It happened between 3 and 20 minutes after the Big Bang. *Protons and neutrons combine in the process of nuclear fusion to form atomic nuclei* (deuterium) *which rapidly fuses into helium-4.* This leaves about three times more hydrogen than helium-4 (by mass) and only trace quantities of other light nuclei.

— *Remainder of the photon epoch*

For the remainder of the photon epoch, the Universe contained a hot dense plasma of nuclei, electrons and photons. About 380 000 years after the Big Bang, *the temperature of the Universe fell to the point where nuclei could combine with electrons to create neutral atoms.* As a result, photons no longer interacted frequently with matter, the Universe became transparent and the *CMB radiation* was created and then *structure formation* took place.

• *CMB radiation*

According to Gamow, *high densities of the primordial Universe establish a thermal equilibrium between the atoms that slow very intense radiation which must still exist today, although less intense.* Its temperature can be calculated from the age of the Universe, the density of matter and the abundance of helium. This radiation is called CMB radiation or radiation fossil and corresponds to that of a black body at the temperature of $-271°C$ near absolute zero $-273°C$ or $0\,K$. This radiation discovered by Arno Allan Penzias and Robert Woodrow Wilson (Fig. 13.2) in 1965 (Nobel Prize of 1978) is a direct witness of the Big Bang. Later, its fluctuations have been

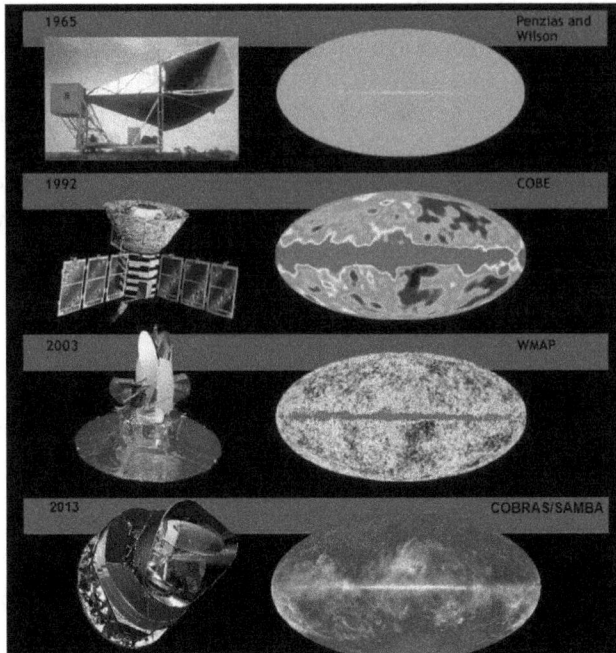

Figure 13.3: CMB radiation. From top to bottom: discovered in 1965, and then study its fluctuations by COBE (1992), WAMP (2003) and Planck (2013).

studied by space probes: COBE in 1992 (Nobel Prize of 2006), WAMP in 2003 and Planck in 2009 (see Fig. 13.3).

♠ Structure Formation

Structure formation in the Big Bang model proceeds hierarchically, with smaller structures forming before larger ones.

• *Quasars, Population III Stars and Reionization*

— 150 millions to 1 billion years after the Big Bang, *the first structures to form from gravitational collapse are quasars*, which are thought to be bright, early active galaxies, and population III stars. The intense radiation they emit reionizes the surrounding Universe. From this point of view, most of the Universe is composed of plasma.

— The first stars, most likely Population III stars, form and start the process of turning the light elements that were formed in the Big Bang (hydrogen, helium and lithium) into heavier elements. However, until now, there have been no observed Population III stars.

• *Galaxies and Population II and I stars*
Large volumes of matter collapse to form a galaxy. Population II stars are formed early on in this process, with Population I stars formed later.

• *Groups, Clusters and Superclusters*
Gravitational attraction pulls galaxies towards each other to form groups, clusters and superclusters.

• *Solar system*
The Solar System began forming about 4.6 billion years ago, or about 9 billion years after the Big Bang. A fragment of a molecular cloud made mostly of hydrogen and traces of other elements began to collapse, forming a large sphere in the center which would become the Sun, as well as a surrounding disk. The surrounding accretion disk would coalesce into a multitude of smaller objects that would become planets, asteroids and comets. *The Sun is a late-generation star, and the Solar System incorporates matter created by previous generations of stars.*

• *Today*
The Big Bang is estimated to have occurred (13.798 ± 0.037) billion years ago. Since the expansion of the Universe appears to be accelerating, its large-scale structure is likely to be the largest structure that will ever form in the Universe.

13.3 Black and Dark Elements of the Universe

♣ Black Holes

A black hole is a region of spacetime in the Universe with a strong gravitational attraction that no particle or electromagnetic radiation (light) can escape from it, hence its name. Then, a black hole cannot be seen but one can see the strong effects of gravity on stars and gas around the black hole. *The theory of general relativity predicts that, in a compact and dense region, the gravitational force is very strong.* The light can be compared with a swimmer pulled by a whirlpool (black hole) in a river or like a car aspirated by a whirlwind.

According to Stephen Hawking (and in a theory of multiverse), a black hole is a door through which one passes from our Universe to another one, i.e. an object/light absorbed by a black hole reappears in another Universe.

A black hole can be as small as an atom but as heavy as few times the mass of the Sun. *Small black holes* formed in the beginning of the Universe formation.

Stellar black holes have masses up to 20 times that of the Sun and can be formed when a very big star collapses and causes a supernova (exploding star).

The very large black holes are called *supermassive black holes* which have masses more than 1 million times that of the Sun. Every large galaxy contains a supermassive black hole at its center. The supermassive black hole at the center of the Milky Way galaxy (Fig. 15.7d) is called Sagittarius A. It has a mass equal to about 4 million times the solar mass and would fit inside a very large ball that could hold a few million Earths.

◇ Dark Matter

During the 1970s and 1980s, various observations showed that there is not sufficient visible matter in the Universe to account for the apparent strength of gravitational forces within and between galaxies. This led to the idea that *up to 96% of the matter in the Universe is dark matter* (called "matière noire" in French, black instead of dark matter!) *and dark energy that does not emit light nor interact with normal baryonic matter.* Therefore, dark matter cannot be seen with telescopes but its existence and properties are deduced from its gravitational effects on visible matter.

Evidence from Big Bang nucleosynthesis, the CMB and structure formation suggests that *about 23% of the mass of the Universe consists of non-baryonic dark matter, whereas only 4% consists of visible, baryonic matter.*

The gravitational effects of dark matter are well understood, as it behaves like a cold, non-radiative fluid that forms haloes around galaxies.

Dark matter has never been detected in the laboratory, and the particle physics nature of dark matter remains completely unknown. As it does not absorb, reflect or emit light, i.e. does not have any electromagnetic interaction, it is extremely hard to detect it. Without observational constraints, there are a number of candidates for dark matter, such as a *stable lightest supersymmetric particle* (LSP), a *weakly interacting massive particle* (WIMP), an *axion* and a *massive compact halo object*. If one of these theories is correct, these invisible particles can be detected indirectly through measurement of missing energy and momentum at the LHC.

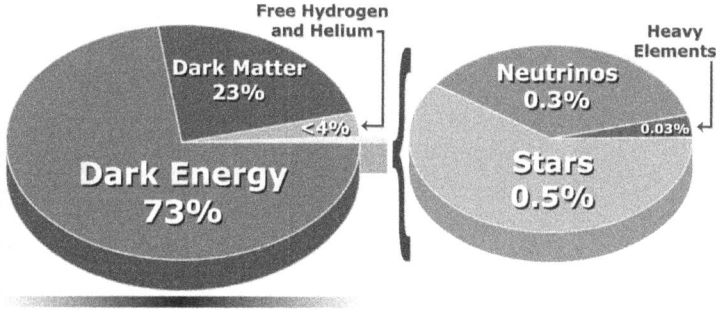

Figure 13.4: Proportion of the different Universe components where about 95% is dark matter ⊕ dark energy.

♡ Dark Energy

Measurements of the redshift magnitude relation for Type Ia Supernovae indicate that the expansion of the Universe has been accelerating since the Universe was about half its present age. However, this observation contradicts the usual picture. It had always been assumed that the matter of the Universe would slow down its rate of expansion. Mass creates gravity, which in turn creates a pull that would slow the expansion. But supernovae observations showed the contrary. *To explain this acceleration, general relativity requires that much of the energy in the Universe consists of a component with a large negative pressure, called "dark energy".* This leads to a repulsive force, which tends to accelerate the expansion of the Universe.

Dark energy is associated with the vacuum in space. It is distributed evenly throughout the Universe, not only in space but also in time such that its effect is not diluted during the expansion of the Universe. The even distribution means that dark energy does not have any local gravitational effects, but rather a global effect on the Universe as a whole.

Measurements of the CMB indicate that the Universe is almost spatially flat, and therefore according to general relativity the Universe must have almost exactly the critical density of mass over energy. However, the mass density of the Universe can be measured from its gravitational clustering, and is found to have only about 27% of the critical density. *Since theory suggests that dark energy does not cluster in the usual way, it is the best candidate for explaining the 73% "missing" energy density of the Universe in addition to the 23% dark matter and 4% baryons* (see Fig. 13.4).

In order not to interfere with Big Bang nucleosynthesis and the CMB, dark energy must not cluster in haloes like baryons and dark matter.

Chapter 14

Beyond the Minimal Big Bang Model

14.1 Success of the Minimal Big Bang Model

Until now, predictions of the Minimal Big Bang Model based on general relativity have been confirmed in all astrophysical experiments. Among these, we quote:

♣ The Recession of Galaxies

This refers to the recession of galaxies between each other with a velocity proportional to the distance which is due the expansion of the Universe.

◇ The Abundance of Light Elements

These include Hydrogen, Helium, Deuterium and Tritium in the interstellar gas where one observes that they are about the same quantity. This feature may eventually indicate that they are of the same origin. As they can only be issued from the proton and neutron above a temperature of about 10^9 K, their presence may indicate that the Universe was in a primordial warm epoch where it was denser. This period is known as a *primordial nucleosynthesis*.

♡ The Cosmic Microwave Background (CMB) Radiation

It was first observed by Penzias and Wilson (Fig. 13.2) in 1964. It was issued from the radiative transition of neutral atoms formed at lower temperature of about 3000 K where the wavelength of the emitted photon 370 000 years after the Big Bang increased with the relative distance in the expanding Universe. Initially observed in the visible and ultraviolet (UV) frequency, it is now observed in the Radio frequencies.

More precise measurements of the CMB radiation are now available. The CMB radiation is a flash of light emitted 3700 years after the Big Bang in a transparent medium. This is because the nucleus and electron plasma transformed into gas of neutral atoms in equilibrium with the radiation which obeys perfectly the well-known Planck law of quantum mechanics with the energy $E = h\nu$ where ν is the light frequency. This frequency can then be related to the temperature of matter and as one observes, at present, it is a cold radiation having a temperature of 2.7 K (see Fig. 2.4). Noting that the wavelength increases linearly with the scale factor which controls the expansion of the Universe and that the wavelength is inversely proportional with the temperature, this fact indicates that the radiation was initially warmer. If this temperature were 1000 times the present levels, it would correspond to the period when atoms were dissociated and where the Universe was opaque. *These precise measurements corresponding to a temperature of 2.726 K show the primordial homogeneity of the Universe, while the apparition of present structure of the Universe may indicate that they originated from the quantum vacuum fluctuations.*

14.2 Minimal Big Bang Model Limitations

♣ Homogeneity of the Universe

The Minimal Big Bang Model cannot explain the initial homogeneity of the Universe because some regions which were not causally related in the past would be related at present. This fact seems to be in contradiction with the Big Bang's assumption of rapid expansion, while here it seems that there is a deceleration.

◇ Curvature of Space–Time

Measurements of the space–time curvature indicates that the Universe is almost flat, while the Minimal Big Bang Model allows different scenarios. In order to have a flat space, one needs to have a fine-tuning of the parameters of the Model which is not quite natural.

14.3 Cosmological Standard Model

♣ Contemporary Big Bang Model does not suffer from the previous problems of the Minimal Big Bang Model though still based on the Friedmann–Lemaître cosmological equations. They are called Cosmological Standard Model or Lambda-CDM (Lambda cold dark matter) model. The model continues to explain the abundances of light elements, the existence of the CMB, the accelerating expansion of the Universe observed in the light from distant galaxies.

◇ However, it should also provide a natural explanation of the primary homogeneity of the Universe and the almost flat space as indicated by different experimental observations.

♡ It should also contain dark matter and dark energy which manifests as:

• *Dark Matter*

It does not or weakly participate(s) in non-gravitational interactions. It is a *cold matter* assumed to be formed by weakly interacting massive (non-relativistic) particles.

• *Dark Energy and Cosmological Constant*

The density of the dark energy arising from a form of a positive energy density ρ, which has negative pressure $p = -\rho c^2$ where c is the light velocity, compensates the one of the vacuum induced by the cosmological constant which is responsible for the acceleration of the Universe. This assumes an important role for the cosmological constant

14.4 Inflationary Cosmology

Initiated by Alan Guth (1947–) and Andrei Linde in 1980 (Fig. 14.1), *cosmic inflation* assumes that between 10^{-36} to 10^{-32} s after the Big Bang, there has been an exponential expansion of space (about 26 order of magnitude) in the early Universe (see Fig. 2.4). The quantum fluctuations in the microscopic inflationary region, magnified to cosmic size, became the seeds

(a)   (b)

Figure 14.1: (a) A. Guth, (b) A. Linde.

for the growth of structure in the Universe. *After this period, the Universe continues to expand but with less acceleration. The inflation mechanism is assumed to be due to a scalar field called inflaton* (very similar to the Higgs boson in particle physics) which is expected to explain the origin of the:

- large-scale structure of the Universe,
- isotropic structure (same in all directions) of the Universe,
- homogeneity of the CMB radiation,
- flatness of space,
- non-observation of magnetic monopoles.

Chapter 15

Excursion to the Infinitely Large

Let us make a short excursion towards the domain of infinitely large says in experimental astrophysics for measuring the properties of planets, galaxies, ...

15.1 Some Astrophysical Constants and Units

We have emphasized for particle physics that the units of measurement are important. We list in Table 15.1 some astrophysical constants and the definitions and equivalence of units in this field of research.

15.2 Radiation Spectra and Experiments

Our knowledge of the Universe is mostly based on the observation of radiations from stars and some other planets. We are all impressed by observing with our naked eyes the stars in the sky during a clear night. However, stars are only part of large spectra of radiations given in Fig. 15.1, where the frequency ranges from the radio wave of 408 megahertz (MHz) to the gamma rays having TeV energy. The observation of these different radiation spectra

Particles and the Universe

are enhanced by various large telescope and/or satellite experiments. Some examples are given as follows:

Table 15.1 Some astrophysical constants.

Observable	Symbol	Value
Newton gravitation constant	G_N	$6.672\ 59(85) \times 10^{-11}$ m^{3}kg^{-1}s^{-2}
Astronomical unit	AU	$1.495\ 978\ 706\ 6(2) \times 10^{11}$ m
Tropical year (equinox to equinox)	yr	$31\ 556\ 925.2$ s
Age of the Universe	t_0	$15(5)$ Gyr
Planck mass	$\sqrt{\frac{\hbar c}{G_N}}$	$1.221\ 047(79) \times 10^{19}$ GeV/c^2
parsec (1 AU/1 arcsec)	pc	$3.085\ 677\ 580\ 7(4) \times 10^{16}$ m = $3.261\ 633\ 44$ light-years (ly)
light-year	ly	$0.3066\ldots$ pc $= 0.9461\ldots \times 10^{16}$ m
Solar mass	$M_\odot$	$1.968\ 92(25) \times 10^{30}$ kg
Solar luminosity	$L_\odot$	3.846×10^{26} W
Solar equatorial radius	$R_\odot$	76.96×10^8 m
Earth's mass	$M_\oplus$	$5.973\ 70(76) \times 10^{24}$ kg
Earth's equatorial radius	$R_\oplus$	$6.378\ 140 \times 10^6$ m
Hubble constant	H_0	$100\ h_0$ km s^{-1}Mpc^{-1} = $h_0 \times (9.778\ 13$ Gyr$)^{-1}$
Normalized Hubble constant	h_0	$0.5 \leq h_0 \leq 0.85$
Critical density of the Universe	$\rho_c = \frac{3H_0^2}{8\pi G_N}$	$2.775\ 366\ 27 \times 10^{11} h_0^2 M_\odot$ Mpc^{-3}
Local halo density	ρ_{halo}	$(2{-}13)10^{-25}$ g cm^{-3} $\approx$ $(0.1{-}0.7)$ GeV/c^2 cm^{-3}
Scaled cosmological constant	$\lambda_0 = \frac{\Lambda c^2}{3H_0^2}$	$-1 < \lambda_0 < 2$
Scale factor for cosmological constant	$\frac{c^2}{3H_0^2}$	$2.853 \times 10^{51} h_0^2$ m^2

♣ Fermi Gamma-Ray Space Telescope (FGST)

This telescope explores the cosmos in the region of 10 keV to 300 GeV with the aim to understand the origin of detected cosmic rays, the mechanism of dark matter and dark energy which constitute 96% of the Universe, the acceleration of jets matter near the black hole, ... This experiment is done on the NASA satellite in Fig. 15.2.

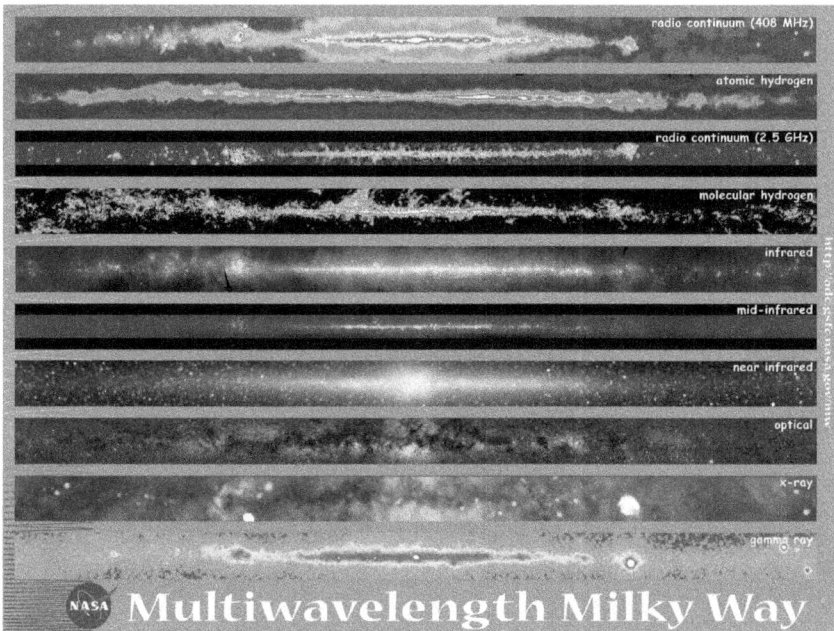

Figure 15.1: Multiwavelength images of Milky Way showing how they appear different frequency bands. For visible light, the center of our galaxy is hidden by gas clouds. Both infrared radiation and gamma rays penetrate these clouds and provide a view of the Galactic center. Infrared observations have revealed the existence of a large black hole at the core of the Galaxy, with a mass corresponding to a million solar masses (courtesy of NASA).

◇ High Energy Stereoscopic System (HESS)

This experiment aims to detect very energetic cosmic gamma rays until 1 TeV energy which may have originated after unusual stellar explosions (supernovae) or in the vicinity of giant black holes suspected to be at the cores of the so-called active galaxies. The telescope site is in Namibia, Africa, southern hemisphere on the capricorn tropics where the sky is known to be clean which facilitates the observation. It consists of 4 telescopes

Figure 15.2: Cosmos exploration with Fermi telescope (FGST) (courtesy of NASA).

placed in the corners of a region of 120 m^2 area and of 250 m diameter (Fig. 15.3) covering the Cherenkov light on Fig. 15.4. The 960 pixels CCD camera covers an area of 1.4 m of diameter which is equivalent to a field angular view of 5° in the sky corresponding to 10 times the diameter of the moon. The gamma (γ)-rays are produced through the cascade reactions at 10 km above the ground: photon (γ) to electron–antielectron (positron) pairs (e^+e^-) and then $e^+ \rightarrow e^+\gamma$ or $e^- \rightarrow e^-\gamma$. They are the messengers of planets at 15 light-years away where we show some examples in Fig. 15.5. Supernova Cassioppia exploded in 1680 BC and sends a shock wave into space which now has expanded to 15 light-years. Crab Nebula is the remainder of a stellar explosion in the year 1054 BC and was the first source of very energetic Cherenkov light. The active galaxy Cygnus A — the small white spot at the center — was discovered in 1989 by the American Whipple. It sends beams of matter across many hundred gamma rays, of thousands of light-years, generating turbulent plumes when they are finally stopped.

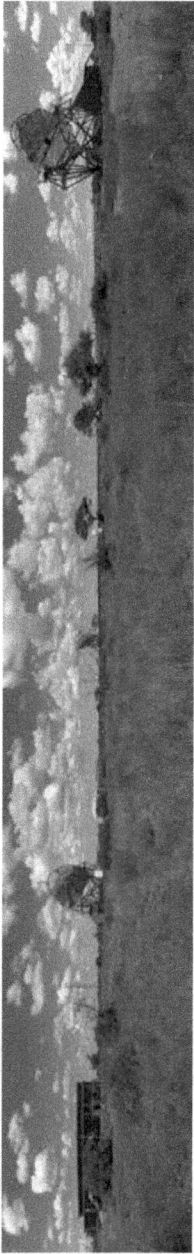

(a)

(c)

(b)

Figure 15.3: HESS telescope: (a) 4 telescopes in the Namibia site, (b) Zoom of one telescope with its mirror, the digital camera is in front, (c) Zoom of the 960 pixels of the CCD camera.

(a)   (b)

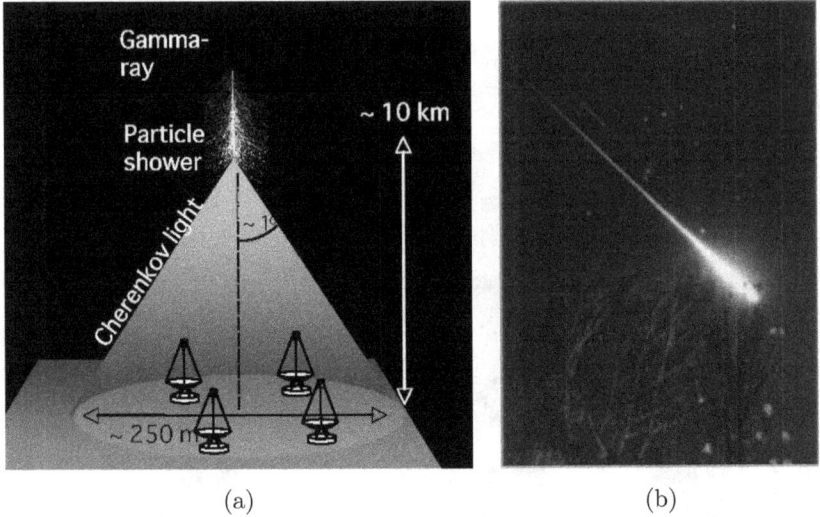

Figure 15.4: Gamma rays stopped by the atmosphere which ressembles a meteorite. The produced Cherenkov light is analyzed by the HESS detectors.

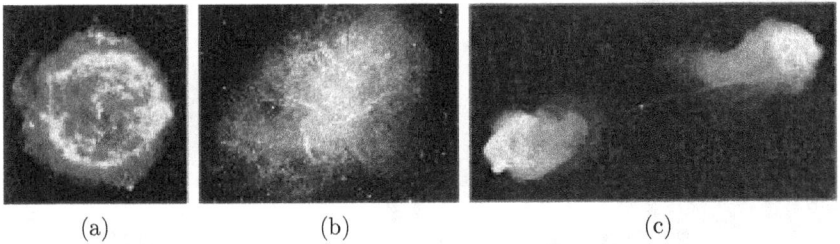

(a)   (b)   (c)

Figure 15.5: (a) Supernova Cassiopeia, (b) Crab Nebula, (c) Cygnus A.

♡ European Southern Observatory (ESO)

The telescope site is in Paranal, desert of Chili (see Fig. 15.6). *It aims to study amas of galaxies at 50 million light-years.* We show in Fig. 15.7 some examples: (a) Spiral galaxy NGC 1232 observed in 1998 at 200 south of the celestial equator (Eridanus constellation) at 100 million light-years; (b) Spiral galaxy NGC 4565 discovered in 2005 at 30 million light-years (Coma Berenice constellation); (c) The bird (angel?) which is composed of two massive spiral galaxies and a third irregular galaxy discovered in 2007

(a)   (b)

Figure 15.6: (a) ESO site and its 4 detectors in the desert of Paranal, Chile, (b) Zoom of the detector.

(a)   (b)   (c)

(d)   (e)   (f)

Figure 15.7: (a) Spiral galaxy NGC 1232, (b) NGC 4565 The Needle Galaxy (courtesy of Ken Crawford), (c) The bird, (d) Central part of the galaxy, (e) Horsehead nebulae, (f) Stellar nursery IC 2944 [(a), (c), (d), (e), (f) are courtesy of ESO].

at 650 million light-years; (d) The central part of Milky Way (our galaxy) observed in 2008 in the near-infrared at 25 000 light-years where the mass of a supermassive blackhole has been measured; (e) Horsehead nebulae at 1300 light-years discovered in 2002 (Orion constellation); (f) Stellar nursery IC 2944 seen in 2013 at 6500 light-years (Centaurus constellation).

15.3 The Space Conquest Era

Since the attempts of Galileo many centuries ago (Fig. 1.2), astronomers have been curious to know the properties of planets and galaxies surrounding our Earth. In fact, *understanding and conquering space has been a goal of mankind for a long time.* One can think of the great discoveries and innovations, explorations of unknown lands, and, in the 20th century, the beginnings of space exploration. We have given previously some examples of the galaxies discovered from modern telescopes and satellites.

Missions for exploring the properties of planets in the solar system dawned with the Sputnik artificial Earth satellite launched by the Soviet Union in 1957 and which was followed by Vostok 1 mission in 12 April 1961 where the *first cosmonaut, Yuri Gagarin (1934–1968) (Fig.* 15.8) *left for Baïkonur cosmodrome in Kazakhstan.* He went around the Earth in 2 hours at an altitude of about 250 km.

Gagarin's success has provoked a huge space conquest competition between the Soviet Union and the United States. In 1965, NASA under the advice and support of President John Fitzgerald Kennedy started the Gemini 6 program of manned flight followed by Gemini 7 and 8. On 21 July 1969 in the Apollo 11 Moon mission, *Neil Armstrong (1930–2012) and Buzz Aldrin (1930–) became the first humans to set foot on the Moon* (Fig. 15.8). The success of this mission has provided more information about the Moon's properties and has stimulated other missions both to the Moon and to some further planets of the solar system (Mars, ...).

15.4 NASA's Searches for Exoplanets

On the other hand, *we are also curious to know if there is life outside of Earth or if there are other planets similar to the Earth.* This motivates the launch of the NASA program of the spacecraft Kepler.

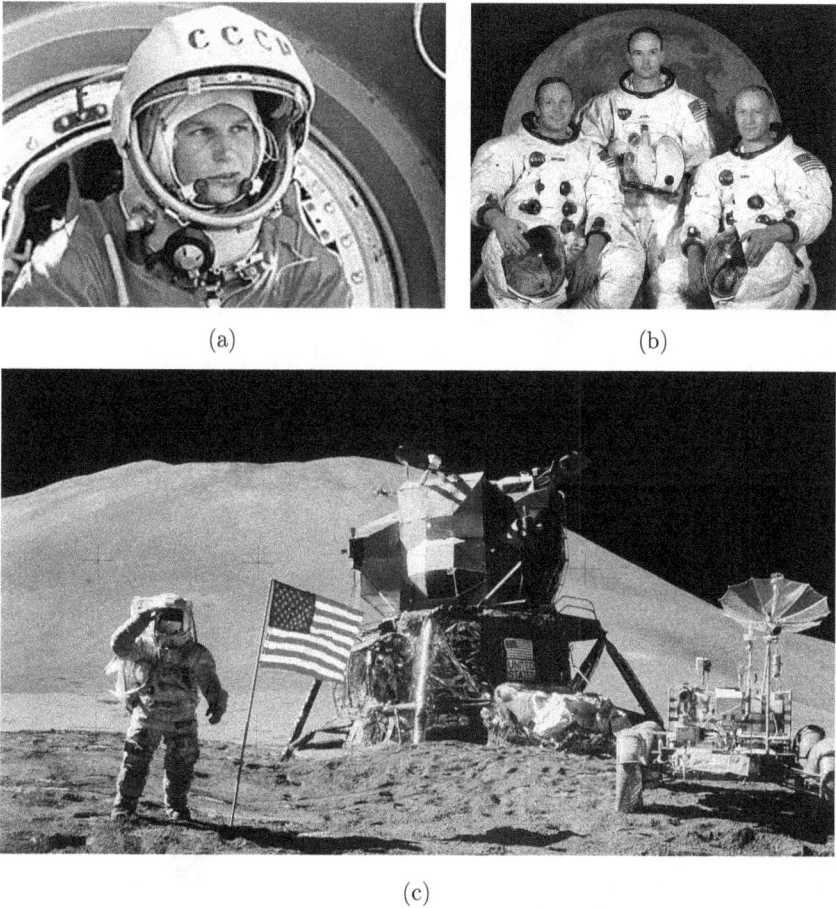

Figure 15.8: (a) Yuri Gagarin, (b) Apollo 11 crew: (from left to right) N. Armstrong, M. Collins, B. Aldrin, (c) First walk of Neil Armstrong on the Moon.

♣ Kepler

Kepler [named after the astronomer Johannes Kepler (1571–1630) (Fig. 15.9)] is part of NASA's Explorer Program of relatively low-cost, focused missions. *It is a space observatory launched by NASA to discover Earth-like planets orbiting other stars.*

Designed to survey a portion of our region of the Milky Way to discover up to dozens of Earth-size extrasolar planets in or near the habitable zone

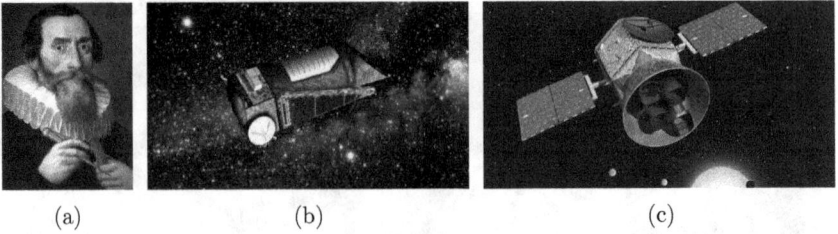

(a)   (b)   (c)

Figure 15.9: (a) Johannes Kepler, (b) The spacecraft Kepler, (c) Transiting Exoplanet Survey Satellite (TESS) (courtesy of NASA).

around a star and estimate how many of the billions of stars in our galaxy have such planets, Kepler is a photometer that continually monitors the brightness of over 145 000 main sequence stars in a fixed field of view.

Four of the newly confirmed exoplanets were found to orbit within habitable zones of their corresponding stars: three of the four, Kepler-438b, Kepler-442b and Kepler-452b, are near-Earth-size and likely rocky; the fourth, Kepler-440b, is a super-Earth.

◇ Kepler-452b

Newly discovered in July 2015, Kepler-452b (Fig. 15.10) is the smallest planet to date discovered orbiting in the habitable zone of a G2-type star like our Sun where one can have water. It is 1.5 billion years older than Earth.

Kepler-452b is 60% larger in diameter than Earth and previous research suggests that it has a good chance of being rocky. While Kepler-452b is larger than Earth, *its 385-day orbit is only 5 percent longer.*

The planet is 5 percent further from its parent star Kepler-452 than Earth is from the Sun. The Kepler-452 System is 1.5 million years older than the solar system. Kepler-452 has about the same temperature as the Sun, and is 20 percent brighter and has a diameter 10 percent larger. Kepler-425b is about 5 times as heavy as the Earth and the surrounding atmosphere is denser. One may expect that the analysis of Kepler-452b can give an idea on how our planet will evolve in the future.

The Kepler-452 System is located 1400 light-years away in the constellation Cygnus, that is about 14 millions of billions of kilometers from the Earth. Even though it is obviously hopeless to go physically there, if extraterrestrial/alien life exists on Kepler-452b, one may hope to exchange some messages with its population. However, assuming that the velocity of

(a)

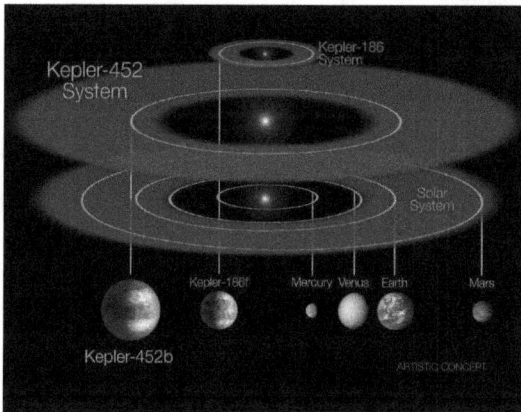

(b)

Figure 15.10: (a) Kepler-452b compared to Earth, (b) Kepler-452 System compared with the Kepler-186 System and the solar system. Kepler-186 is a miniature solar system that would fit entirely inside the orbit of Mercury (courtesy of NASA).

the message does not exceed the one of light (maximal velocity), one needs to wait, at least, about 1400 years to receive it, which is beyond our lifetime! The only hope is that some extraterrestrial messages have already been sent some 1400 years ago and will be detected soon!

♡ The Transiting Exoplanet Survey Satellite (TESS)

TESS is a planned space telescope (Fig. 15.9) for NASA's Explorers Program, designed to search for exoplanets using the transit method and to continue and enlarge the Kepler program.

The primary mission objective for TESS is to survey the brightest stars near the Earth for transiting exoplanets over a two-year period. It will focus on G-type stars (often called a yellow dwarf or G dwarf, which has about 0.8 to 1.2 solar masses and surface temperature of between 5300 and 6000 K; the Sun is the best known (and most visible) example of a G-type main sequence star) and K-type stars (also referred to as orange dwarf or K dwarf, they have masses between 0.45 and 0.8 times the mass of the Sun and surface temperatures between 3900 and 5200 K; better-known examples include Alpha Centauri B and Epsilon Indi) .

These stars are of particular interest in the search for extraterrestrial life because they are stable on the main sequence for a very long time (15 to 30 billion years, compared to 10 billion for the Sun) and have apparent magnitudes brighter than magnitude 12. Approximately 500 000 stars will be studied, including the 1000 closest red dwarfs, across an area of the sky 400 times larger than that covered by Kepler. TESS is expected to discover more than 3000 transiting exoplanet candidates, including those which are Earth-size or larger.

Part VI
Fundamental Researches and Society

Chapter 16

International Research Centers

16.1 International Research Centers

The European Center of Nuclear Research (CERN) in Geneva, Switzerland and the International Center for Theoretical Physics (ICTP) or Abdus Salam Center (the name of its founder) in Trieste, Italy are good examples of international collaborations and international research centers showing that science is universal and has no borders. These collaborations and centers put together persons of different ethnics and racial origins, who do not have the same culture and the same tradition, who have different beliefs and who come from different countries which do not have the same social and political organizations. One can have in the same group an Israeli and an Iranian, a Russian and an American, an Indian and a Pakistani, ... This fact also indicates the peaceful nature of research in particle physics.

♣ CERN, Geneva

It was founded in 1954 (its 60th birthday was celebrated in 2014) by 12 European countries (Germany, Belgium, Denmark, France, Greece, Italy, Norway, United Kingdom, Sweden, Switzerland, Netherlands, Yugoslavia). *It was created well before the European Commission Policy* which was created in 1957. It is often remembered on different occasions that

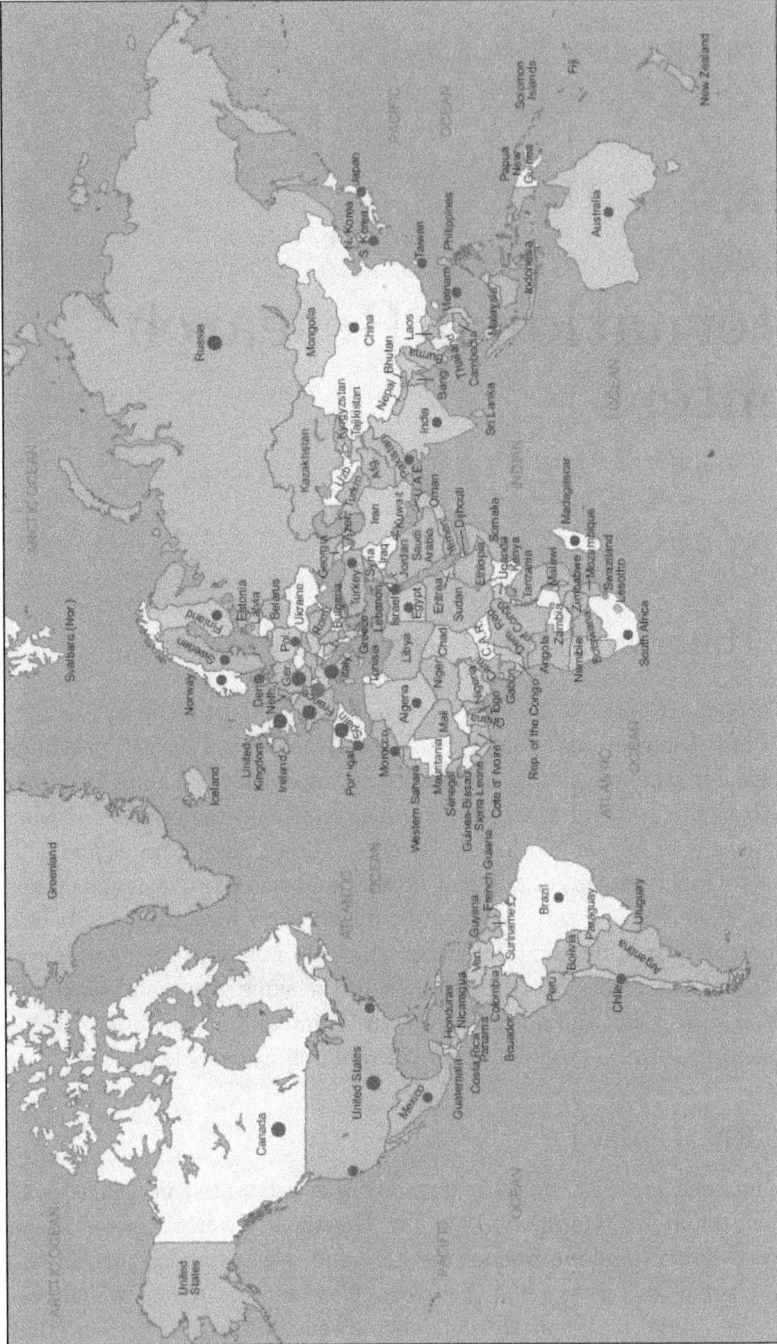

Figure 16.1: Countries involved (blue circle) in the experiments of the LHC in Geneva (brown circle). Madagascar partly participates by sending student trainees at the LHCb.

"CERN brings people closer". *CERN also provides an opening for emerging countries* that are not able to develop the physics of particles, either by special contracts for the training of young people through more or less stays, by summer schools that allow young people (like students from Madagascar) to spend 2 months at CERN by taking courses in particle physics and performing some experimental works inside a specialized group.

◇ ICTP, Trieste

The ICTP in Trieste, Italy *created by Prof. Abdus Salam in 1964* and supported by UNESCO, the United Nations and the Italian State *aimed to provide a workplace for researchers from developing countries.* Just like CERN, ICTP helps these researchers and students to participate in summer school programs or spend a few months in Trieste.

16.2 LHC as an International Collaboration

LHC is an example of a big international collaboration, that enables the design and construction of the accelerator and detectors for the analysis of the data. A detector such as ATLAS involves 174 teams from 38 countries with 3000 physicists, technicians and engineers including 800 students.

♣ Structure of the collaboration

It is like a *Small State*. It includes a *Board* involving representatives of each laboratory involved in the project. The *Council* elects a *Spokesperson* for a term of a few years who runs the executive part of the collaboration. He names in turn the members of the management team that will daily follow the project and take urgent decisions. *Committees* are formed to manage the Group's publications and to select speakers representing the collaboration in different conferences. All subjects are discussed at the level of physicists including doctoral candidates at a plenary meeting which many, who may not be on-site, can follow in front of their computers through a videoconferencing screen via the internet at staggered hours. The decision is made after a consensus. Data analyses are done by small groups that integrate physicists from different laboratories regardless of their country of origin.

◇ Doctoral Students

They participate fully in data analyses both at the individual and team level. This allows them to interact with experts but also with doctoral students from other countries with different educational systems. This wealth of contacts plays a positive role in the training of the students and their participation in a research at the frontiers of knowledge stimulates them. However, they still separate themselves into small groups for their daily work and the relationship between professor and students has not changed.

♡ Costs and Social Implications of LHC

We also saw that the construction of the LHC was very expensive but this is not quite a negative aspect because it has allowed to develop companies that are subcontracted in this project. The LHC is also expensive from the viewpoint of the salaries of staff. It helps feed families of technicians, engineers and physicists who are involved in this project. *All these show the indirect contribution of fundamental research to the development of our society.*

16.3   High-Technology Impacts of HEP

♣ LHC

We saw in the example of the LHC (previous section) that *our thirst for understanding the Universe indirectly obliges us to design new forms of high technology*:
- *Cryogenics*
One reaches a sidereal temperature of $-271°C$ where the magnet is bathed in superfluid helium in order to obtain a stable superconducting state. This temperature corresponds to the most cooled temperature of our Universe in the fossil state.
- *The ultra vacuum*
It corresponds to 10^{-13} times the atmospheric pressure to avoid collisions between protons and molecules in the vacuum chamber such that one loses as little as possible protons in the beam packages.
- *Detectors*
One uses the design of the individual detectors based on G. Charpak's wire sensors. These are used in our daily lives to control containers in ports such as le Havre (France), Toamasina (Madagascar), ...

• *High-technology computer*

— Broadband 10 gigabytes compared to ADSL of tens of megabytes.
— Creation of software to select the 600 millions per second shock of the two protons.
— Concept of grid computing inherited from Tim Lee-Berners web that allows complicated calculations on a network series-connected computers in the world. This concept is currently used by professionals and Microsoft under the terminology of *Cloud*.

◇ Former Technological Developments

Since long, physics has always been the percussor of technological developments. We have:

• Domesticated light, thanks to our understanding of the photon by the equations of Maxwell in 1865.

• Built our old-fashioned TV through our understanding of the property of the electron which has enabled us to design the electron gun and to curve their trajectory by magnetic fields. It is still with these techniques that we build a particle accelerator which has energies 5 million times more powerful (LEP) than our TV.

• Differentiated between radio waves and waves of our mobile phone by the understanding of their wavelengths.

• Established a medical imaging because we have understood the wavelengths of X-rays and magnetic resonance imaging (MRI or IRM in French) with high-resolution coming from quantum properties of the nucleus (spin magnetic moments).

• Treated cancer with beams of protons (proton or/and hadron therapy like, for instance, the Orsay Center) that bombard the ill cells without affecting healthy cells. This is a great advantage over chemotherapy where radioactive products spread throughout the body and often result in irreversible side effects.

Chapter 17

Popularizing High-Energy Physics

17.1 Transmission of Knowledge

In addition to his specialized research work, the mission of a researcher is the transmission of knowledge by training, communication and popularization:

♣ Training

It is often done by:
- imparting at higher levels,
- supervision of the PhD students.

◇ Communication

It is done:
- through the media (written press, television, interviews),
- by organizing conferences accessible to a wide untrained audience,
- through the organization of high-level international conferences.

♡ Popularization

It is done either by:
- communication as mentioned previously, or
- giving qualitative seminars for college and high-school students.

17.2 QCD Series of International Conferences

Created in Montpellier in 1985 as Conference on Non-Perturbative Methods, the QCD–Montpellier Series of International Conferences is organized regularly every 2 years in Montpellier and celebrates its 30th anniversary in 2015. The QCD–Montpellier conference series have become *Classics QCD Conferences* with a good reputation. This conference has many original aspects:

♣ Goal and Motivation

The aim is to organize regularly a conference which puts together different experts in QCD and PhD students/Post-Docs working in this field in order to make a status of the research done during the last two years and to present recent results both in theory and in experiments. The new experimental results are often presented there before their presentation in the classical large scale European Physical Society (EPS) and International Conference in High-Energy Physics (ICHEP).

♢ Size and Participants

It involves experts in QCD, PhD students and Post-Docs who often present for the first time their talks at the international level. The conference is a medium size about 100–120 participants of comprising equal number of theorists and experimentalists. The idea was to have compromise between small sized specialized conferences/workshops and the previous large scale conferences.

♡ Spirit

The goal of the conference is two-fold. First is to have advanced talks presented by the experts in the field, while the second is to give an opportunity to young physicists to present their works at the international level. The medium size of the conferences and the good organization of the social events create a family relaxing atmosphere during the conference which facilitates the communications among the different participants.

♠ Quality and Sponsors

The QCD–Montpellier conferences have a good reputation for the high quality of the different scientific contributions and for the well-organized social events. They were named as the Euroconferences from 1994 to 2000 with funds attributed by the European Commission in Brussels. They also get regular (symbolic) support from the French National Center of Scientific Research (CNRS), the University of Montpellier 2, the Mayor of Montpellier and the Region of Languedoc–Roussillon.

♣ Committee Members and Prestigious Participants

The committee members of the QCD conference are worldwide experts in the field (see the conference site: *http://www.lupm.univ-montp2.fr/users/qcd/* and the sample of photos in Figs. 17.1 to 17.3). The good reputation of the conference has always attracted the participation of prestigious physicists. Notable among them are the three Nobel Prize winners: Prof. J. Steinberger (Fig. 17.4) in 1997, Prof. D. Gross (Fig. 17.4) in 1998 and Prof. G. 't Hooft (Fig. 17.5) in 1998 and 2002.

Figure 17.1: Committee members of QCD08 at the CNRS, Montpellier. From left to right: (Front) J.-M. Richard and the author; (Back) K. Zalewski, H. Leutwyler, E. de Rafael and A. Di Giacomo.

Figure 17.2: Committee members of QCD10 at the CNRS, Montpellier. From left to right: (Front) V. Zakharov, A. Di Giacomo, the author, K. Zalewski and H. Fritzsch; (Back) S. Bethke, J.-M. Richard, E. de Rafael, U. Gastaldi and H.G. Dosch.

Figure 17.3: Committee members of QCD12 at the CNRS, Montpellier. From left to right: E. de Rafael, the author, H.G. Dosch and S. Bethke.

(a)   (b)

Figure 17.4: (a) J. Steinberger (Nobel Prize of 1988), (b) D. Gross (Nobel Prize of 2004).

Figure 17.5: From left to right: The author (chairman), G. 't Hooft (Nobel Prize of 1999), Max Levita (Vice-Mayor of Montpellier) at the QCD02 conference (July 2002 at Le Corum Congress Centre, Montpellier).

Figure 17.6: QCD02 conference participants including G. 't Hooft (Nobel Prize of 1999) and his wife (center) at the Corum, Montpellier (July 2002).

17.3 Developing HEP in Madagascar

One of my priorities during these last 10 years has been to promote and to introduce particle or high-energy physics in Madagascar.

♣ Developing Countries

Remember that this discipline has always been reserved for the major economic power countries with the exception of India and has never been taught traditionally in emerging countries, in particular, sub-saharian Africa. It has been introduced recently in the countries of the Maghreb, South Africa and the countries of Latin America.

◇ Introduction of HEP in Madagascar

In 2001, I started the 1st international conference in High-Energy Physics, the HEPMAD series (see Figs. 17.7 and 17.8) which was to be held alternately with the QCD–Montpellier conferences. When I introduced this field of research in Madagascar in 2001, I had to start from scratch because even field theory was not taught at the university. However, I was spoilt by the thirst and the curiosity of young students to learn this field.

Figure 17.7: Inauguration of the 1st HEPMAD conference (2001, Madagascar). From left to right: the former Dean of the University R. Rakotomahanina, John Ellis of CERN, Geneva, the author, the Dean of the University P. Rakotobe, the Director of the Nuclear Research Institute Raoelina Andriambololona, the Ministry of Research.

Figure 17.8: Participants at the fifth international conference HEPMAD 2009 held at the Foreign Ministery of Antananarivo (21–30 August 2009).

Figure 17.9: iHEPMAD Master and PhD students (class of 2014–2015).

♡ Creation of the iHEPMAD Institute

This enthusiasm of the young students led us to create the research institute of High-Energy Physics in Madagascar (iHEPMAD) attached to the University of Antananarivo which serves as an official platform for the organization of the series of HEPMAD conferences and especially to train young Malagasy students in this new discipline.

♠ IN2P3/CNRS Support

To achieve this, we have benefitted from the support of the IN2P3/CNRS and of the laboratory in Montpellier (France) for the organization of the HEPMAD conferences.

♣ CERN Schools

Later, with the help of John Ellis of CERN, Geneva (Switzerland), we have been able to send since 2002 young PhD students regularly to CERN summer schools and, after, to receive training at the LHCb group of

Figure 17.10: Conference at the ENS, Antananrivo (29 October 2014).

Figure 17.11: Conference at the ASJA University, Antsirabe (25 November 2014).

Figure 17.12: Conference at the Lycee Ampefiloha, Antananarivo (31 October 2014).

Figure 17.13: Conference at the Lycee Antsirabe (25 November 2014).

CERN thanks to the signed agreement of common interest between CERN, the University and the iHEPMAD in Antananarivo. Some students have benefited from thesis scholarships to continue working inside the LHCb group and Fermilab.

◇ ICTP Schools

Some other students were sent to theoretical schools such as the ICTP, Trieste (Italy) to study string theory.

♡ African Schools

More recently, the students doing their Masters degree were sent to the new African School of Physics (South Africa in 2010, Ghana in 2012 and Senegal in 2014) where they have given a good impression.

♠ PhD Students of the iHEPMAD

After these various international training, some students continued to prepare their theses at the iHEPMAD of Antananarivo either in theory or

AFRICA
Developing high-energy physics in Madagascar

The 6th High-Energy Physics International Conference in Madagascar, HEP-MAD 13, took place on 4–10 September at the Ministry of Foreign Affairs in Antananarivo. This series of conferences – initiated in 2001 by Stephan Narison of the Laboratoire Univers et Particules in Montpellier and formerly supported by IN2P3/CNRS – alternates with the series of QCD conferences that Narison started in Montpellier in 1985. There were around 50 participants, including 12 from other countries.

New results from the ATLAS, CMS and LHCb experiments were presented, as well as from NA48 at CERN, Belle at KEK and Babar at SLAC. There were also theoretical contributions on Higgs-like models and QCD non-perturbative approaches such as QCD spectral sum-rules. In addition to high-energy physics, this conference series includes contributions from national researchers on other branches of physics such as climatology, nuclear physics and the environment. This allows researchers in Madagascar to have international visibility and publication of their research in the SLAC eConf online proceedings.

The HEP-MAD conferences are part of a programme for promoting high-energy physics in Madagascar and, more generally,

developing countries, according to the ideas expressed by the late Abdus Salam while Narison was a postdoc at the International Centre for Theoretical Physics (ICTP) in Trieste in 1979. (He continues to be supported by ICTP as a consultant.)

In addition to the HEP-MAD conferences, two other main activities have been established. The creation of a high-energy-physics research institute in Madagascar in 2004 provides a platform for training PhD students, some of whom have gone to CERN as summer students and worked for up to two months with LHCb, thanks to the effort of John Ellis, who was one of CERN's advisers for non-member states, and the LHCb collaboration. Some of the students also participated in the

Participants at HEP-MAD 13, with Stephan Narison at front, centre right. (Image credit: HEP-MAD.)

African School of Physics (*CERN Courier* November 2012 p36). A second activity centred on the popularization of high-energy physics at high schools and the general public in different regions of Madagascar. An elementary introduction to the field – the book *Particle Physics: From the Ionian School to the Higgs Boson* – is under preparation.

To manage these various activities and other developments, Narison created the Association Gasy Miara-Mandroso (AGMM) – Malagasy growing and advancing together – in 2009. More recently, the University of Antananarivo offered land inside the campus for construction of a high-energy-physics research institute.

In recognition of these activities and for developing science in Madagascar, Narison was nominated Grand Officier de l'Ordre National Malgache in January 2012 and Associate Member of the Malagasy National Academy in February 2013.
● For more about HEP-MAD 13, see www.lupm.univ-montp2.fr/users/qcd/econf13/index13.htm.

Figure 17.14: CERN Courier, Jan/Feb 2014 edition (page 43).

experiment where they collaborated with some other PhD students or/and physicists from abroad.

♣ Particle Physics in High Schools

The positive assessment of these actions has also stimulated the organization of conferences at high schools to give the students an idea about particle physics to enable them to make informed choices regarding their future research direction. This initiative was instructive because there has been an increasing number of young students who wish to study this field.

These different developments of activities in Madagascar are well summarized in the CERN Courier article in Fig. 17.14.

Part VII
Epilogue

Summary

Throughout this book, we have seen the evolution of science from the antique Greek era until the discovery of the Higgs boson in July 2012. We have also seen the rapid progress in physics researches, since the discovery of the electron in 1897 by J. Thompson, which is due to the complementary theoretical and experimental researches. Since few years, we have also observed complementary developments in particle physics (infinitely small scale) and in astrophysics and cosmology (infinitely large scale) for accelerating our understanding of the origin of the Universe. Since the reaching of electroweak scale of few 100 GeV by accelerators, we have seen the success of QCD, QED and Standard Model (SM) gauge theories in explaining the three microscopic strong, electromagnetic and weak forces.

1. QED

The QED program of regularization and renormalization of Dyson, Feynman, Schwinger, and Tomonaga (1930–1940) describes the electromagnetic phenomena with impressive precisions. The classic example is the agreement between theory and experiment on the anomalous magnetic moment $(g-2)$ of the electron, as well as the precise evaluation of the one of the muon which seems to reveal new physics beyond the SM.

2. QCD

It perfectly describes the strong interaction. The hadronic properties at low energy are explained by non-perturbative approaches such as the spectral sum rules, the lattice simulation, the effective theories and the potential models. The jets of hadrons at high energies are well described by perturbative calculations. Several processes give a very precise value of the QCD coupling constant α_s. However, it has not yet found any convincing analytical solution to the problem of the confinement of quarks and gluons.

3. Electroweak Standard Model

All the particles of the SM were discovered from 1894 to 2012. However, although the SM is adequate to explain most of the phenomena observed up to now, it is unable to explain the following questions:

♣ Why is the up quark very light (2.8 MeV)?
◊ Why is the top quark very heavy (173 GeV)?
♡ Why are neutrinos massive?
♠ Why does the Universe contains a high proportion of dark matter?
♣ Why is there essentially matter but not antimatter?
◊ ...

4. Beyond the Standard Model

Though quite successful at present energies, the previous unanswered questions lead us to think that our understanding of the origin of the Universe is not complete enough. Thus, we should go beyond these Standard Models (SM) of strong and electroweak interactions, if we want to build a more ambitious Theory of Everything (ToE). Unfortunately, there is no clear research direction to follow and numerous models beyond the SM exist in the literature. Most probably, the future LHC results will clarify the situation. We have to wait and see.

5. Concluding Remarks

Though the different facts discussed previously explain the progress of researches in particle physics. We have also mentioned briefly recent progresses in astrophysics and shown the complementarity of these two fields of research. We wish that the future will surprise us, which will make these fields more exciting and more attractive. There will still be much to do before arriving at the time $t = 0$, the origin of the Big Bang, though we are already very close (10^{-35} s) to the explosion (if this assumption is correct)!

Message to the Readers

♣ In writing this book, I have tried to present an overview of the evolution of research in particle physics, from the thoughts of the Greek philosophers 610 BC until the recent discovery of the Higgs boson 4 July 2012 at the LHC of CERN, Geneva (Switzerland), using simple language without too many technical difficulties and in a concise manuscript. However, I have provided, in a special Appendix (Part VIII), some necessary and useful tools for those who want to work in this field.

♢ The aim is to share and transmit the passion and the knowledge gained from this fundamental research, which is "particle physics or high energy physics", to a wide audience who does not necessarily have a technical basis in physics.

♡ I have shown the usefulness of this research for the understanding of the origin of our Universe and of the different laws that govern it.

♠ I have also discussed its indirect long term consequences on the different technologies that we already use everyday as well as the new advanced technologies that will help later and which already exist at the prototype stage when one designed and built machines dedicated to the large scale experiments such as the LHC.

♣ I hope that the initial goal of accessibility of this book to a large audience is achieved and that the reader with a sound basis in physics is able to acquire, in this book, new knowledge on particle and astrophysics.

Thanks for your patience!

Merci pour votre patience!

Misaotra amin'ny faharetana!

Acknowledgements

The writing of this book is a long term project initiated in 2013 and motivated by different public seminars and lectures delivered at the QCD-Montpellier and HEPMAD conference series, at different high schools in Montpellier and in Madagascar, at the National Academy, universities and "Alliance Française" in Madagascar. I wish to thank the LUP (Montpellier), ICTP (Trieste) and the Association Gasy Miara-Mandroso (AGMM) who have supported in different ways my various missions in Madagascar leading to such an achievement. The efficient and pleasant collaboration with Kah-Fee and his team from WSPC is appreciated. Last but not least are the valuable advice and comments of my family when I first started writing this book.

Part VIII
Appendix: Useful Notes for Advanced Readers

Appendix A

Special and General Relativity

A.1 Special Relativity

♣ Four-vector

In 4-dimension, the covariant four-vector is defined as:

$$x_\mu = (ct, -x, -y, -z) \equiv (x_0, x_1, x_2, x_3), \tag{A.1}$$

in the metric signature $(+, -, -, -)$.[1] The scalar product is:

$$x^2 = x_\mu x^\mu = c^2 t^2 - \vec{x}^2. \tag{A.2}$$

The four-vector momentum is defined by:

$$p_\mu = i\hbar \partial_\mu \equiv i\hbar \frac{\partial}{\partial x_\mu} \equiv i\hbar \left(\frac{1}{c} \frac{\partial}{\partial t}, -\vec{\nabla} \right), \tag{A.3}$$

which is the derivative operator in space–time where $\mu = 0, 1, 2, 3$; $\vec{\nabla}$ (pronounced as nabla) is the *Laplacian operator* defined in Eq. (D.7), which is the sum of the partial derivatives with respect to spatial coordinates x_1, x_2, x_3. The *D'Alembertian operator* reads:

$$\Box = \partial_\mu \partial^\mu \equiv \nabla^2 = (1/c^2)(\partial^2/\partial t^2) - \vec{\nabla}^2. \tag{A.4}$$

---

[1] Along all discussions in this book, this metric will be adopted.

◇ Lorentz Transformations

• *The Lorentz factor*
It is defined as:

$$\gamma = \frac{1}{\sqrt{1 - \beta^2}} = \frac{dt}{d\tau},$$ (A.5)

where: $\beta = v/c$ is the ratio of the relative velocity v between two inertial frames over the speed of light c; t is the time coordinate and τ the *proper time* which measures the time intervals in the observer's own frame:

$$d\tau^2 = dt^2 - \frac{1}{c^2} d\vec{x}^2.$$ (A.6)

• *The Lorentz transformation*
From a system of coordinates $K(x, y, z, t)$ to another frame $K'(x', y', z', t')$ moving with a relative speed v on the x axis (K measures K' to move with a velocity v), it reads:

$$x' = \gamma(x - vt), \quad y' = y, \quad z' = z, \quad t' = \gamma\left[t - (v/c^2)x\right].$$ (A.7)

— The Lorentz transform of a velocity V into another speed V' can be written:

$$V' = (V - v) \Big/ \left(1 - \frac{vV}{c^2}\right).$$ (A.8)

— It is easy to check that if $V = c$ then V' is also equal to c, i.e. the light velocity is invariant under Lorentz transformations.

♡ The Energy–Momentum

• *Invariant mass and Einstein relation*
In 4-dimension space–time, the energy is a four-vector called energy–momentum. For a particle of a mass m_0 (invariant mass or rest mass or proper mass) and momentum $\vec{p}$ (vector of a 3-dimension space), *the invariance of the energy reads*:

$$E^2 - p^2 c^2 = m_0^2 c^4.$$ (A.9)

m_0 corresponds to the mass in the rest frame of the particle ($p = v = 0$) in Eq. (A.9) and leads to the famous Einstein relation:

$$E_0 = m_0 c^2.$$ (A.10)

• *Lorentz transform*

In a reference frame having a speed v, the energy can be written as:

$$E = m_0 c^2 \gamma = \frac{m_0 c^2}{\sqrt{1 - v^2/c^2}}, \tag{A.11}$$

and the momentum as:

$$\vec{p} = m_0 \vec{v} \gamma = \frac{m_0 \vec{v}}{\sqrt{1 - v^2/c^2}}. \tag{A.12}$$

These relations show that *the mass m of a moving object depends on γ*:

$$m = m_0 \gamma, \tag{A.13}$$

with which, the *relativistic momentum* can be written as:

$$\vec{p} = m\vec{v}, \tag{A.14}$$

like for classical mechanics, while the *relativistic kinetic energy* reads:

$$E_c = E - E_0 = (\gamma - 1) m_0 c^2. \tag{A.15}$$

Taking the ratio of Eqs. (A.12) and (A.11), one can deduce the velocity v:

$$\vec{v} = \frac{\vec{p} c^2}{E}. \tag{A.16}$$

• *Zero mass and speed of light limits*

For $m = 0$ (photon without mass) one has $v = c$, which implies, in this case, that $\gamma = 1/0$ from its definition in Eq. (A.5). Therefore, E and p given by Eqs. (A.11) and (A.12) have indeterminate values (0/0). However, Eq. (A.16) leads to: $E = pc$ for $v = c$ which is the ratio of Eq. (A.12) over Eq. (A.11) and which is well-defined. This fact indicates the peculiar feature of the Lorentz transformation!

• *Classical mechanics limit*

For v much smaller than c ($v \ll c$) which implies $\beta^2 \ll 1$, one can make the Taylor expansion of the square root:

$$\gamma \simeq 1 + \frac{\beta^2}{2} + \frac{1}{2} \cdot \frac{3}{4} \beta^4, \tag{A.17}$$

indicating that to a first approximation:

$$\vec{p} \simeq m\vec{v}, \quad \text{and} \quad E_c = E - E_0 = \frac{1}{2} m v^2. \tag{A.18}$$

One recovers the result of classical mechanics for the quantity of movement (impulsion) and kinetic energy.

A.2 General Relativity and Cosmology

♣ Einstein Field Equations

They are used in general relativity. *They describe how mass and energy represented by the stress–energy tensor are related to the curvature of space-time.* They read in a covariant form:

$$G_{\mu\nu} + \Lambda g_{\mu\nu} = \frac{8\pi G_N}{c^4} T_{\mu\nu}, \qquad (A.19)$$

where c is the light speed, Λ is the cosmological constant, G_N is the gravitational constant, $G_{\mu\nu}$ is the Ricci–Einstein tensor and $T_{\mu\nu}$ is the stress–energy tensor.

◇ Friedmann–Lemaitre Equations

Assuming an homogeneous and isotropic Universe with positive cosmological constant and expansion parameter $a(t) = R/R_0$ which relates the size R of the universe at a time t to the one R_0 at time t_0, the solution to the previous Einstein equations lead to the Friedmann–Lemaitre equations:

$$\frac{\dot{a}^2}{a^2} = \frac{8\pi G_N}{3} \rho_{\text{tot}} - \frac{c^2}{R^2 a^2} + \frac{\Lambda}{3}, \qquad (A.20)$$

and

$$\frac{d\rho}{dt} + 3\rho_{\text{tot}} \frac{\dot{a}}{a} = 0, \qquad (A.21)$$

where $\rho_{\text{tot}} \equiv \rho + 3P/c^2$ is the total density which is the sum of the mass density ρ and the one due to the pressure P; c is the speed of light and Λ is the cosmological constant. *The second term in the 1st equation c^2/R^2 characterizes the curvature of the space*: it is zero for a flat space which has zero curvature, negative for a negative curvature (hyperbolic space) and positive for a positive curvature (spheric space).

Appendix B

Quantum Mechanics — Field Theory

B.1 Creation and Annihilation Operators

♣ Definition

The *creation operator* is denoted by $a_n^\dagger$ and the *annihilation operator* by a_n. The annihilation operator a_n decreases the state $|n\rangle$ by one unit quantum:

$$a_n|n\rangle = \sqrt{n}|n-1\rangle, \tag{B.1}$$

while the operator of creation $a_n^\dagger$ makes it grow in one unit:

$$a_n^\dagger|n\rangle = \sqrt{n+1}|n+1\rangle. \tag{B.2}$$

- *The vacuum $|0\rangle$ or ground state*
  It is defined as the state cancelled by the operator of annihilation $a_n|0\rangle = 0$.

◇ Hamiltonian $\mathcal{H}$

- *Harmonic oscillator*
The total energy of the harmonic oscillator which is the eigenvalue of the hamiltonian is:

$$E_n = \epsilon a^\dagger a = \hbar\omega\left(n + \frac{1}{2}\right), \tag{B.3}$$

showing that the energy of the ground state $n = 0$ is not zero. ω is the frequency of the oscillator.

• *Bosons*

In terms of creation $(a_B^\dagger)$ and annihilation (a_B) operators, the Hamiltonian for bosons is written:

$$\mathcal{H}_B = \sum_B \epsilon(a_B^\dagger a_B + b_B^\dagger b_B), \tag{B.4}$$

where ϵ is the energy of a particle, $b_B^\dagger$ and b_B are creation and annihilation operators of anti-particles. Taking account that *the couples* $(a_B, a_B^\dagger)$ *and* $(b_B, b_B^\dagger)$ *commute*, the corresponding eigenvalue is the total energy:

$$E = \sum_B \epsilon(N_B + \bar{N}_B), \tag{B.5}$$

where the additive constant $E_0 = \sum \epsilon$ is removed. This energy E is positive definite.

• *Fermions*

For fermions, the Hamiltonian operator is:

$$\mathcal{H}_F = \sum_F \epsilon(a_F^\dagger a_F - b_F^\dagger b_F), \tag{B.6}$$

but taking into account that *the couples* $(a_F, a_F^\dagger)$ *and* $(b_F, b_F^\dagger)$ *anti-commute*, i.e. $b_F^\dagger b_F = -b_F b_F^\dagger + 1$, one can deduce the total energy which is positive definite:

$$E_F = \sum_F \epsilon(N_F + \bar{N}_F), \tag{B.7}$$

where the additive constant $E_0 = \sum \epsilon$ has been removed.

B.2 Schrödinger Equation

The Schrödinger equation reads:

$$\mathcal{H}|\psi(x,t)\rangle = i\hbar \frac{\partial}{\partial t}|\psi(x,t)\rangle, \tag{B.8}$$

where $\hbar = h/2\pi$ is the reduced Planck constant, i is a pure imaginary (complex) number where the square is $i^2 = -1$. $\mathcal{H}$ is the Hamiltonian which is the total energy: sum of the kinetic energy $E_c = mv^2/2 = p^2/2m$ and potential energy V in classical mechanics (v is the velocity and p the impulsion or momentum). *For a stationary object*, the Schrödinger equation simplifies to:

$$\mathcal{H}|\psi(x,t)\rangle = E|\psi(x,t)\rangle, \tag{B.9}$$

where E is its proper value.

B.3 Klein–Gordon Equation for Scalar Boson

♣ Wave Function

In terms of the annihilation (a_n) and creation (a_n^+) operators for a particle and (b_n, b_n^+) for an anti-particle, *the wave function of a scalar particle* is (in general) written as:

$$\phi(x) = \frac{1}{\sqrt{2\epsilon}} \sum_{n,p} a_n e^{-ipx} + b_n^\dagger e^{+ipx}, \tag{B.10}$$

and its complex conjugate:

$$\phi^\dagger(x) = \frac{1}{\sqrt{2\epsilon}} \sum_{n,p} a_n^\dagger e^{+ipx} + b_n e^{-ipx}, \tag{B.11}$$

which is valid for a scalar (spin 0) and vector (spin 1) boson like the photon. For the photon, one can also note that the particle is equal to its anti-particle such that $(a_n, a_n^\dagger) = (b_n, b_n^\dagger)$.

◇ Plane Wave

For a plane wave, the energy ϵ satisfies the condition $\epsilon^2 = p^2 + m^2$ for a given momentum p^2 in 3-dimension, i.e. it can have two values:

$$\epsilon = \pm\sqrt{p^2 + m^2}, \tag{B.12}$$

while only the positive solution has a physical meaning.

♡ Klein–Gordon Equation

The Klein–Gordon equation for a particle of momentum p and mass m reads:

$$(p^2 - m^2)\phi(t,r) = 0. \tag{B.13}$$

In a developed form, it can be written as:

$$\left(-\frac{1}{c^2}\frac{\partial^2}{\partial t^2} + \nabla^2\right)\phi(t,r) = \frac{m^2}{\hbar^2}\phi(t,r), \tag{B.14}$$

which can be re-written in a more mathematical form as:

$$\left(\Box + \frac{m^2}{\hbar^2}\right)\phi(t, r) = 0,\qquad\text{(B.15)}$$

with: $\Box = (1/c^2)(\partial^2/\partial t^2) - \vec{\nabla}^2$ is called the *D'Alembertian*.

B.4 Second Quantization of Dirac for Fermion

♣ Fermion Wave Function

The wave function of a fermion is written:

$$\psi(x) = \frac{1}{\sqrt{2\epsilon}}\sum_p a_p u_p e^{-ipx} + b_p^\dagger v_p e^{+ipx},\qquad\text{(B.16)}$$

and its conjugate:

$$\bar{\psi}(x) \equiv \psi^+\gamma_0(x) = \frac{1}{\sqrt{2\epsilon}}\sum_p \bar{u}_p a_p^\dagger e^{+ipx} + b_p \bar{v}_p e^{-ipx},\qquad\text{(B.17)}$$

where γ_0 is a 4×4 gamma matrix (see below), u_p and v_p are Dirac spinors related to the wave functions $\psi_p = u_p e^{-ipx}$ for the particle and $\psi_{-p} = v_p e^{+ipx}$ for anti-particle. The sum runs over the impulsions p and spin states $\pm 1/2$. The pairs of annihilation and creation operators $(a_p, a_p^\dagger)$ of a particle anti-commute $(a_p^\dagger = -a_p)$ in the case of fermions (Fermi–Dirac statistics). The pair $(b_p, b_p^\dagger)$ for anti-particle obeys similar properties. In the case of bosons (Bose–Einstein statistics), one just takes $u = v = 1$ and considers that operators commute $(b_p^\dagger = b_p)$. The fermion fields are normalized as:

$$\bar{u}(p)u(p) = 2m, \quad \bar{v}(p)v(p) = -2m,\qquad\text{(B.18)}$$

where: $\bar{u} = u^\dagger\gamma_0$ is the hermitian conjugate of the fermion field and $\bar{v} = v^\dagger\gamma_0$ is the hermitian conjugate of the anti-fermion field.

◇ Dirac Equation

The Dirac equation for a fermion reads in momentum space:

$$(p^\mu\gamma_\mu - m)u(p) = 0 = \bar{u}(p)(p^\mu\gamma_\mu - m),\qquad\text{(B.19)}$$

where γ_μ are 4×4 matrices (tables with 4 lines and 4 columns) called Dirac matrices which have some properties discussed in Appendix E [not to be confused with γ of the Lorentz transformation in Eq. (A.5)]. The *Dirac equation for the anti-fermion* (positron in the case of electron) corresponding to the solution of negative energy in Eq. (B.12) is:

$$(p^\mu \gamma_\mu + m)v(p) = 0 = \bar{v}(p)(p^\mu \gamma_\mu + m). \qquad (B.20)$$

Here and in the following, *we shall use the Landau's book notation* [3]:

$$\hat{p} \equiv \gamma_\mu p^\mu, \qquad (B.21)$$

which is equivalent to: $\not{p}$ in the Bjorken–Drell book notation [3].

B.5 Ordered Products

♣ Normal Ordered Products or Normal Ordering

All dynamical variables depending quadratically on field operators taken at the same space time like the Lagrangian, currents,... are *normal ordered products* in the sense that all annihilation operators are to the right of creation operators which would commute (anti-commute) for bosons (fermions). This is denoted by:

$$\text{normal ordered of } \bar{\psi}(x)\psi(x) \equiv \quad : \bar{\psi}(x)\psi(x): \qquad (B.22)$$

• *Bosons*
Let $b^\dagger$ and b be the boson's creation and annihilation operators. Then, they *satisfy the commutation rules*:

$$[b^\dagger, b^\dagger] = [b, b] = 0, \qquad (B.23)$$

$$[b, b^\dagger] = 1. \qquad (B.24)$$

Then, one obtains *the normal ordered products*:

$$: bb^\dagger := b^\dagger b =: b^\dagger b : \qquad (B.25)$$

Combined with the commutation relation in Eq. (B.24), it gives:

$$bb^\dagger = b^\dagger b + 1 =: bb^\dagger : +1 \quad \text{or} \quad bb^\dagger - : bb^\dagger := 1. \qquad (B.26)$$

For multiple operators:

$$: b^\dagger bb^\dagger bbb^\dagger := b^\dagger b^\dagger b^\dagger bbb = \left(b^\dagger\right)^3 b^3. \qquad (B.27)$$

For two different bosons, we have:

$$: b_1^\dagger b_2 := b_1^\dagger b_2, \qquad : b_2 b_1^\dagger := b_1^\dagger b_2. \qquad (B.28)$$

• *Fermions*

Let $f^\dagger$ and f be the fermion's creation and annihilation operators. Then, they satisfy the *anti-commutation rules*:

$$\{f^\dagger, f^\dagger\} = \{f, f\} = 0, \tag{B.29}$$

$$\{f, f^\dagger\} \equiv f f^\dagger + f^\dagger f = 1. \tag{B.30}$$

The *normal ordering for a single fermion* is:

$$: f^\dagger f := f^\dagger f, \qquad : f f^\dagger := -f^\dagger f. \tag{B.31}$$

Combined with the previous anti-commutation relations, it gives:

$$f f^\dagger = 1 - f^\dagger f = 1 + : f f^\dagger : . \tag{B.32}$$

One can also easily show that:

$$: f f^\dagger f f^\dagger := f^\dagger f^\dagger f f = 0. \tag{B.33}$$

For two different fermions, one has:

$$: f_1^\dagger f_2 := f_1^\dagger f_2, \qquad : f_2 f_1^\dagger := -f_1^\dagger f_2. \tag{B.34}$$

◇ Time Ordered Product

In quantum field theory (QFT), the time-ordered product or usually called $\mathcal{T}$ *product* of two operators $\mathcal{O}_1(x)$ and $\mathcal{O}_2(y)$ is defined as:

$$\mathcal{T} \mathcal{O}_1(x) \mathcal{O}_2(y) = \theta(x_0 - y_0) \mathcal{O}_1(x) \mathcal{O}_2(y) \pm \theta(y_0 - x_0) \mathcal{O}_2(y) \mathcal{O}_1(x), \tag{B.35}$$

with $\theta(x)$ the Heaviside step function which is equal to 1 if $x \geq 0$ and zero otherwise; the sign $\pm$ depends if the operators are bosonic or fermionic. If the operators are bosonic, the sign is always positive. If the operators are fermionic, the sign depends on the number of operator interchanges necessary to achieve the proper time ordering. It is customary to range the operators from the largest time at the left to the smallest time at the right:

$$\mathcal{T} \{ \mathcal{O}_1(t_1) \mathcal{O}_2(t_2) \ldots \mathcal{O}_n(t_n) \} = \sum_p \theta \left(t_{p_1} > t_{p_1} \ldots t_{p_n} \right) \epsilon(p)$$

$$\times \mathcal{O}_{p_1}(t_{p_1}) \mathcal{O}_{p_2}(t_{p_2}) \ldots \mathcal{O}_{p_n}(t_{p_n}), \tag{B.36}$$

where the sum runs over p and over n possible permutations. $\epsilon(p) = 1$ for bosonic operators and have the sign of the number of permutations for fermionic operators.

♡ Free-Field Propagators or Chronological Contractions

In the case of QED, we have two propagators:
- *Fermion (electron) propagator*

It corresponds to the quantity:

$$iS(x) = \langle 0|T\psi(x)\bar{\psi}(0)|0\rangle \equiv \overline{\psi(x)\bar{\psi}(0)}, \qquad (B.37)$$

which is the Green's function solution of the Dirac equation of motion:

$$\left(i\gamma^\mu \frac{\partial}{\partial x_\mu} - m\right)\overline{\psi(x)\bar{\psi}(0)} = i\delta^{(4)}(x), \qquad (B.38)$$

where ψ and m are the free-electron field and mass. It follows from the definition of the T product:

$$T\psi(x)\bar{\psi}(0) = \theta(x)\psi(x)\bar{\psi}(0) - \theta(-x)\psi(0)\bar{\psi}(x), \qquad (B.39)$$

where the minus sign comes from Fermi–Dirac statistics. $\theta(x)$ is the Heaviside step function which is equal to 1 if $x \geq 0$ and zero otherwise. Using the Feynman prescription for the contour of the pole $m \rightarrow m - i\epsilon$, one obtains:

$$\overline{\psi(x)\bar{\psi}(0)} = \int \frac{d^4p}{(2\pi)^4} e^{-ipx} \frac{i(\hat{p}+m)}{p^2 - m^2 + i\epsilon}, \qquad (B.40)$$

and its Fourier transform:

$$iS(p) = \int \frac{d^4p}{(2\pi)^4} e^{ipx} \langle 0|T\psi(x)\bar{\psi}(0)|0\rangle$$
$$= \frac{i(\hat{p}+m)}{p^2 - m^2 + i\epsilon}. \qquad (B.41)$$

In terms of the Feynman diagram, the electron (fermion) propagator can be represented by the straight line in Appendix E where p is its momentum.

• *Photon propagator*
It corresponds to the quantity:

$$iD_{\mu\nu}(x) = \langle 0|T A_\mu(x) A_\nu(0)|0\rangle \equiv \overline{A_\mu(x) A_\nu}(0), \qquad \text{(B.42)}$$

which is the Green's function solution of the equation of motion:

$$\left\{ \left(\Box_x - \lambda^2\right) g_\mu^\rho + (1 - 1/\alpha_G) \frac{\partial}{\partial_\mu} \frac{\partial}{\partial^\rho} \right\} \overline{A_\rho(x) A_\nu}(0) = -i g_{\mu\nu} \delta^{(4)}(x), \quad \text{(B.43)}$$

which follows from the definition of the chronological product:

$$T A_\mu(x) A_\nu(0) = \theta(x) A_\mu(x) A_\nu(0) + \theta(-x) A_\mu(0) A_\nu(x), \qquad \text{(B.44)}$$

with the solution:

$$iD_{\mu\nu}(x) = \int \frac{d^4 k}{(2\pi)^4} e^{-ikx} \left[(-i) \frac{g_{\mu\nu} - (1 - \alpha_G)\frac{k_\mu k_\nu}{k^2 - \lambda^2 + i\epsilon}}{k^2 - \lambda^2 + i\epsilon} \right], \qquad \text{(B.45)}$$

where λ is the photon mass and α_G is the covariant gauge parameter. Its Fourier transform is:

$$iD_{\mu\nu}(k) = \int \frac{d^4 k}{(2\pi)^4} e^{+ikx} \langle 0|T A_\mu(x) A_\nu(0)|0\rangle$$

$$= (-i) \frac{g_{\mu\nu} - (1 - \alpha_G)\frac{k_\mu k_\nu}{k^2 - \lambda^2 + i\epsilon}}{k^2 - \lambda^2 + i\epsilon}. \qquad \text{(B.46)}$$

In terms of the Feynman diagram, the photon propagator can be represented by the wavy line in Appendix E by replacing the gluon by a photon.

B.6   Wick's Theorem

Introduced by Gian Carlo Wick (1909–1992, Fig. B.1), *the theorem states that the T product of n linear operators is equal to the sum of the corresponding normal ordered products : : with all possible chronological*

Figure B.1: G.C. Wick.

contractions:

$$T\{\mathcal{O}_1(t_1)\mathcal{O}_2(t_2)\ldots\mathcal{O}_n(t_n)\} = : \mathcal{O}_1(t_1)\mathcal{O}_2(t_2)\ldots\mathcal{O}_n(t_n) :$$
$$+ : \overline{\mathcal{O}_1(t_1)\mathcal{O}_2}(t_2)\ldots\mathcal{O}_n(t_n) :$$
$$+ \cdots$$
$$+ : \overline{\mathcal{O}_1(t_1)\ldots\mathcal{O}_{n-1}}(t_{n-1})\mathcal{O}_n(t_n) :$$
$$+ : \overline{\mathcal{O}_1(t_1)\ldots\mathcal{O}_n}(t_n) :$$
$$+ \cdots$$
$$+ : \overline{\mathcal{O}_1(t_1)\mathcal{O}_2}(t_2)\overline{\mathcal{O}_3(t_3)\mathcal{O}_4}(t_4)\ldots\mathcal{O}_n(t_n) :$$
$$+ \cdots \tag{B.47}$$

As an application, one can consider the second order S-matrix amplitude:

$$\mathcal{A}^{(2)} = \frac{(-ie)^2}{2!}\int d^4x d^4x' T \mathcal{L}_I^{(0)}(x)\mathcal{L}_I^{(0)}(x'), \tag{B.48}$$

where: $\mathcal{L}_I^0(x) = -e : \bar{\psi}\gamma_\mu\psi : A^\mu$ is the interaction Lagrangian. Therefore:

$$\mathcal{A}^{(2)} = \frac{(-ie)^2}{2!}\int d^4x d^4x' T\left(: \bar{\psi}(x)\gamma^\mu\psi(x) : A_\mu(x)\right)$$
$$\left(: \bar{\psi}(x')\gamma^\nu\psi(x') : A_\nu(x')\right). \tag{B.49}$$

The T-*product of the photon field* reads:

$$T A_\mu(x)A_\nu(x') = : A_\mu(x)A_\nu(x') : + \overline{A_\mu(x)A_\nu}(x'). \tag{B.50}$$

The $\mathcal{T}$-*product of the fermion current* reads:

$$\mathcal{T}\left(: \bar{\psi}(x)\gamma^\mu\psi(x) :: \bar{\psi}(x')\gamma^\nu\psi(x') :\right) = : \bar{\psi}(x)\gamma^\mu\psi(x)\bar{\psi}(x')\gamma^\nu\psi(x') :$$
$$+ : \overline{\bar{\psi}(x)\gamma^\mu\psi(x)\bar{\psi}}(x')\gamma^\nu\psi(x') :$$
$$+ : \bar{\psi}(x)\gamma^\mu\overline{\psi(x)\bar{\psi}}(x')\gamma^\nu\psi'(x) :$$
$$+ : \bar{\psi}(x)\gamma^\mu\overline{\psi(x)\bar{\psi}}(x')\gamma^\nu\psi(x') :,$$

$$(\text{B.51})$$

where one should understand that the contraction terms correspond to the propagators. Combining these different terms in Eqs. (B.50) and (B.51), one has eight contributions to Eq. (B.49) that can be represented by eight Feynman diagrams which the readers can easily draw as an exercise.

B.7 Scattering Matrices

The scattering matrix or S-matrix $S_{fi} \equiv \langle f|S|i\rangle$ which quantifies the transition amplitude from an initial state $|i\rangle$ to a final state $\langle f|$ (bra and ket notation of Dirac) can be related to the *scattering amplitude* $\mathcal{T}_{fi}$ by the relation:

$$S_{fi} = \delta_{if} + i(2\pi)^4\delta^{(4)}(P_i - P_f)\mathcal{T}_{fi}. \qquad (\text{B.52})$$

δ_{if} is the delta function which is equal to 1 for $i = f$ and 0 for $i \neq f$, while $\delta^{(4)}$ is the the space–time delta function which is 1 if $P_i = P_f$ (condition of momentum conservation) and 0 if $P_i \neq P_f$. i is the pure imaginary number with $i^2 = -1$. The "reduced" invariant amplitude is:

$$\mathcal{M}_{fi} = \mathcal{T}_{fi}(2\epsilon_1 V \dots 2\epsilon_i V)^{1/2}(2\epsilon'_1 V \dots 2\epsilon'_f V)^{1/2}, \qquad (\text{B.53})$$

where $\epsilon_1, \dots, \epsilon_i$ are the energies of the initial states, $\epsilon'_1, \dots, \epsilon'_f$ that of final states. V is the normalization volume where stands a particle.

B.8 Decay Rate or Width of a Particle

The partial rate of a particle of mass M decaying into n particles in the rest frame is:

$$d\Gamma = \frac{(2\pi)^4}{2M}|\mathcal{M}|^2 \mathcal{N} d\Phi_n, \qquad (\text{B.54})$$

where:

$$d\Phi_n(P, p_1, \ldots, p_n) = \delta^{(4)}\left(P - \sum_{i=1}^{n} p_i\right) \prod_{i=1}^{n} \frac{d^3 p_i}{(2\pi)^3 2\epsilon_i}, \tag{B.55}$$

is the element of the n-body phase space and $\mathcal{N} = 1/N!$ in case where there are N identical final produced particles.

In the particular case of *two-body decays of a particle* (P, M) into two particles (p_1, m_1) and (p_2, m_2), the kinematic is simple in the rest frame:

$$E_1 = (M^2 - m_2^2 + m_1^2)/2M,$$

$$|\vec{p}_1| = |\vec{p}_2| \equiv \frac{1}{2M}\lambda^{1/2}(M^2, m_1^2, m_2^2), \tag{B.56}$$

where: $\lambda(a^2, b^2, c^2) = [a^2 - (b+c)^2][a^2 - (b-c)^2]$ is the phase space factor. In terms of these quantities, the decay rate reads:

$$d\Gamma = \frac{1}{32\pi^2}|\mathcal{M}|^2 \frac{|\vec{p}_1|}{M^2} d\Omega, \tag{B.57}$$

where $d\Omega = d\phi_1 d(\cos\theta_1)$ is the solid angle of the produced particle 1. Properties of the δ-function have been used when integrating over the momentum $d^3 p_1$ and the energy $d(\epsilon_1 + \epsilon_2)$.

B.9 Cross-Section

The collision of two particles of four momenta $P_1 \equiv (\epsilon_1, \vec{P}_1)$ and $P_2 \equiv (\epsilon_2, \vec{P}_2)$ and of masses M_1 and M_2 which produce n particles $p'_n \equiv (\epsilon'_n, \vec{p'_n})$ is:

$$d\sigma = \frac{(2\pi)^4}{4I}|\mathcal{M}|^2 \mathcal{N} d\Phi'_n, \tag{B.58}$$

where the phase space factor is:

$$d\Phi'_n(P_1 + P_2, p'_1, \ldots, p'_n) = \delta^{(4)}\left(P_1 + P_2 - \sum_{a=1}^{n} p'_a\right) \prod_{a=1}^{n} \frac{d^3 p'_a}{(2\pi)^3 2\epsilon'_a}. \tag{B.59}$$

The square of the particle flux is:

$$I^2 = (P_1 P_2)^2 - M_1^2 M_2^2 = \frac{1}{4}\lambda(s, M_1^2, M_2^2), \tag{B.60}$$

where $s = (P_1 + P_2)^2$ is a Lorentz invariant quantity. In the case of two produced particles p'_1 and p'_2, one obtains in the center of mass of the system: $\epsilon = \epsilon_1 + \epsilon_2 = \epsilon'_1 + \epsilon'_2$ and $\vec{P}_1 = -\vec{P}_2 \equiv \vec{P}$ and $\vec{p}'_1 = -\vec{p}'_2 \equiv \vec{p}'$. Integrating over $d^3 p'_1$ and $d(\epsilon'_1 + \epsilon'_2)$ eliminates the δ function and leads to the differential cross-section:

$$d\sigma = \frac{1}{64\pi^2} |\mathcal{M}|^2 \frac{|\vec{p}'|}{|\vec{P}|} \frac{1}{\epsilon^2} d\Omega', \tag{B.61}$$

where Ω' is the solid angle of $\vec{p}'_1$ relative to $\vec{P}_1$. In the particular case of elastic scattering, one has $|\vec{P}| = |\vec{p}'|$. One can introduce the invariant $t = (P_1 - p'_1)^2$. In the center of mass energy frame ($\vec{P}_1 = -\vec{P}_2 \equiv \vec{P}$) and for a given constant energy ϵ, its derivative reads: $dt = 2|\vec{P}||\vec{p}'| d(\cos\theta)$, while $I = |\vec{P}|\epsilon$. Then, a trivial substitution leads to:

$$d\sigma = \frac{1}{64\pi} |\mathcal{M}|^2 \frac{dt}{I^2} \frac{d\phi}{2\pi}. \tag{B.62}$$

Using the fact that the cross-section, I and the azimut angle ϕ are invariant under the Lorentz transformation, one can integrate the angle $d\phi$ leading to a simple expression for the cross-section of $1 + 2$ to $3 + 4$ particles.

Appendix C

$SU(N)$ Lie Algebra

The following discussions and notations come from the books and reviews in Refs. [30–32].

C.1 Definition

The $SU(N)$ Lie group [Sophus Lie (1842–1899) (Fig. C.1)] is the special unitary group of degree N of $N \times N$ unitary matrices with determinant 1. Its dimension is $N^2 - 1$. The $N^2 - 1$ generators T_a of the Lie algebra obey the commutation relation:

$$[T_a, T_b] = i f_{abc} T_c, \tag{C.1}$$

and the trace and hermitian properties:

$$\text{Tr}\, T_a = 0, \qquad T_a = T_a^\dagger. \tag{C.2}$$

f_{abc} are called *structure constants* of the $SU(N)$ group which are *real* and totally antisymmetric. $T_a^\dagger$ is the hermitian conjugate. They are normalized as:

$$f_{abc} f_{dbc} = N \delta_{ad}. \tag{C.3}$$

C.2 Gluons in the Adjoint Representation

In this so-called $(N^2 - 1)$ or adjoint representation, the generators are represented by $(N^2-1) \times (N^2-1)$ matrices. (N^2-1) of them are represented by the structure constants:

$$(T_a)_{bc} = -i f_{abc}, \tag{C.4}$$

Figure C.1: Sophus Lie.

with the properties:

$$f_{abe}f_{cde} = \frac{2}{N}\left[\delta_{ac}\delta_{bd} - \delta_{ad}\delta_{bc}\right] + d_{ace}d_{dbe} - d_{ade}d_{bce},$$

$$f_{abe}d_{cde} + f_{ace}d_{dbe} + f_{ade}d_{bce} = 0, \qquad (C.5)$$

where d_{abc} is a real and totally symmetric tensor:

$$d_{abb} = 0,$$

$$d_{abc}d_{dbc} = (N - 4/N)\,\delta_{ad}. \qquad (C.6)$$

In this representation, the trace properties are:

$$\mathrm{Tr}\, T_a T_b = N\delta_{ab},$$

$$\mathrm{Tr}\, T_a T_b T_c = \frac{i}{2}N f_{abc},$$

$$\mathrm{Tr}\, T_a T_b T_c T_d = \delta_{ab}\delta_{cd} + \delta_{ad}\delta_{bc} + \frac{N}{4}\left(d_{abe}d_{cde} - d_{ace}d_{dbe} + d_{ade}d_{bce}\right).$$

$$(C.7)$$

We also have the useful property:

$$(T_a)_{bc}(T_a)_{cd} = (C_2(G) \equiv N)\,\delta_{bd}. \qquad (C.8)$$

C.3 Quarks in a Fundamental Representation

In this so-called N or fundamental representation, the generators are represented by $N \times N$ unitary matrices:

$$T_a = \frac{1}{2}\lambda_a, \qquad (C.9)$$

with the properties:

$$[\lambda_a, \lambda_b] = 2if_{abc}\lambda_c,$$

$$\{\lambda_a, \lambda_b\} = \frac{4}{N}\delta_{ab} + 2d_{abc}\lambda_c, \tag{C.10}$$

The trace properties are:

$$\mathrm{Tr}\,\lambda_a = 0$$

$$\mathrm{Tr}\,\lambda_a\lambda_b = 2\delta_{ab}$$

$$\mathrm{Tr}\,\lambda_a\lambda_b\lambda_c = 2(d_{abc} + if_{abc})$$

$$\mathrm{Tr}\,\lambda_a\lambda_b\lambda_c\lambda_d = \frac{4}{N}\left(\delta_{ab}\delta_{cd} - \delta_{ac}\delta_{bd} + \delta_{ad}\delta_{bc}\right) +$$
$$2(d_{abe}d_{cde} - d_{ace}d_{abe} + d_{ade}d_{bce}) +$$
$$2i(d_{abe}f_{cde} - d_{ace}f_{abe} + d_{ade}f_{bce}). \tag{C.11}$$

Some other useful relations are:

$$(\lambda_a)_{\alpha\beta}(\lambda_a)_{\beta\gamma} = 4\left(C_2(R) \equiv \frac{N^2-1}{2N}\right)\delta_{\alpha\gamma},$$

$$(\lambda_a)_{\alpha\beta}(\lambda_a)_{\gamma\delta} = 2\left(\delta_{\alpha\delta}\delta_{\beta\gamma} - \frac{1}{N}\delta_{\alpha\beta}\delta_{\gamma\delta}\right)$$

$$= \frac{2(N^2-1)}{N^2}\delta_{\alpha\delta}\delta_{\beta\gamma} - \frac{1}{N}(\lambda_a)_{\alpha\beta}(\lambda_a)_{\gamma\delta},$$

$$(\lambda_b\lambda_a\lambda_b)_{\alpha\beta} = -\frac{2}{N}(\lambda_a)_{\alpha\beta},$$

$$(\lambda_a\lambda_b)_{\alpha\beta}(T_b)_{ca} = N(\lambda_c)_{\alpha\beta}. \tag{C.12}$$

C.4 The *SU*(2) Group

The $SU(2)$ Lie algebra is generated by $2^2 - 1$ 2×2 Pauli matrices multiplied by the imaginary number i ($i^2 = -1$) obeying the previous general properties. The Pauli matrices read:

$$\sigma_1 = \begin{pmatrix} 0 & 1 \\ 1 & 0 \end{pmatrix} \quad \sigma_2 = \begin{pmatrix} 0 & -i \\ i & 0 \end{pmatrix} \quad \sigma_3 = \begin{pmatrix} 1 & 0 \\ 0 & -1 \end{pmatrix}. \tag{C.13}$$

C.5 The $SU(3)$ Group

In the case of $SU(3)$, the Lie algebra is generated by $3^2 - 1$ 3×3 Gell-Mann matrices which are generalizations of the Pauli matrices for $SU(2)$. They read explicitly:

$$\lambda_1 = \begin{pmatrix} 0 & 1 & 0 \\ 1 & 0 & 0 \\ 0 & 0 & 0 \end{pmatrix} \qquad \lambda_2 = \begin{pmatrix} 0 & -i & 0 \\ i & 0 & 0 \\ 0 & 0 & 0 \end{pmatrix} \qquad \lambda_3 = \begin{pmatrix} 1 & 0 & 0 \\ 0 & -1 & 0 \\ 0 & 0 & 0 \end{pmatrix}$$

$$\lambda_4 = \begin{pmatrix} 0 & 0 & 1 \\ 0 & 0 & 0 \\ 1 & 0 & 0 \end{pmatrix} \qquad \lambda_5 = \begin{pmatrix} 0 & 0 & -i \\ 0 & 0 & 0 \\ i & 0 & 0 \end{pmatrix} \qquad \lambda_6 = \begin{pmatrix} 0 & 0 & 0 \\ 0 & 0 & 1 \\ 0 & 1 & 0 \end{pmatrix}$$

$$\lambda_7 = \begin{pmatrix} 0 & 0 & 0 \\ 0 & 0 & -i \\ 0 & i & 0 \end{pmatrix} \qquad \lambda_8 = \frac{1}{\sqrt{3}} \begin{pmatrix} 1 & 0 & 0 \\ 0 & 1 & 0 \\ 0 & 0 & -2 \end{pmatrix}.$$

$$\text{(C.14)}$$

Therefore:

$$f_{123} = +1$$

$$f_{147} = f_{156} = f_{246} = f_{257} = f_{345} = -f_{367} = \frac{1}{2} \,,$$

$$f_{458} = f_{678} = \frac{\sqrt{3}}{2}, \qquad \text{(C.15)}$$

and:

$$d_{118} = d_{228} = d_{338} = d_{888} = \frac{1}{\sqrt{3}},$$

$$d_{146} = d_{157} = -d_{247} = d_{256} = d_{344} = d_{355} = -d_{366} = -d_{377} = \frac{1}{2},$$

$$d_{448} = d_{558} = d_{668} = d_{778} = -\frac{1}{2\sqrt{3}}. \qquad \text{(C.16)}$$

The other components which cannot be obtained by permutation of indices of the earlier ones are zero.

Appendix D

Gauge Theory

The following discussions and notations come from the books and reviews in Refs. [30–32].

D.1 Path Integral

♣ Principle of Least Action

Feynman (Fig. 4.1) pointed out that if light is a wave and hence a field, principle of Fermat (1603/1608–1665) (the light path is the minimum optical path) can be deduced from the more general principle of Huygens (Fig. 1.5), which stipulates that the amplitude of a field is the sum of the contributions of all the field waves that depart from a source S and arrive at the observer $\mathcal{O}$ after reflection on the mirror. Waves will then browse multiple paths and the light will choose one which minimizes its time of travel. In the case of very low wavelengths (geometrical optics), this minimum path is that of the distance of Fermat. By applying this summation to the motion of the particles considered as quanta of the field, Feynman gets in the classical limit, where the reduced Planck constant $\hbar = 0$, the principle of least action corresponding to Fermat's principle in geometrical optics. *He thus managed to unify the wave and the particle.*

◇ The Lagrangian

In classical mechanics, *the Lagrangian L is the difference between the kinetic E_c and potential V energies*:

$$L = E_c - V, \qquad (D.1)$$

while *the hamiltonian $\mathcal{H}$ corresponds to their sum* like we have seen in Chapter 1 and Appendix B.

An amplitude can be expressed as a sum of the action $\mathcal{S}$ when one moves from an initial point A at a time t_1 to another point B at a time t_2, which can be symbolically written as a functional integral or path integral:

$$\mathcal{A} = \int_{A(t_1)}^{B(t_2)} e^{-\mathcal{S}}. \qquad (D.2)$$

The action $\mathcal{S}$ can be expressed as the sum (integral) of the Lagrangian L over an infinitesimal time interval dt:

$$\mathcal{S} = \frac{i}{\hbar} \int dt \ L(x,t) = \frac{i}{\hbar} \int d^4x \ \mathcal{L}(x,t), \qquad (D.3)$$

where one has introduced, for convenience, the *Lagrangian density $\mathcal{L}$* with L is its sum (integral) over an infinitesimal volume d^3x of 3-dimensions:

$$L = \int d^3x \ \mathcal{L}. \qquad (D.4)$$

The Lagrangian density can be expressed in terms of the particle fields and characterizes a specific gauge theory.

♡ Euler–Lagrange Equation

It is a differential equation whose solutions are the functions for which a given functional is stationary. In Lagrangian mechanics, because of Hamilton's principle of stationary action, the evolution of a physical system is described by the solutions to the Euler–Lagrange equation for the action of the system. If the Lagrangian L is described by the functional variable $q(t)$ and of its derivative $\dot{q}(t)$ where t is the time variable, the equation reads:

$$\frac{\partial L}{\partial q}((t, q(t), \dot{q}(t)) - \frac{d}{dt} \frac{\partial L}{\partial \dot{q}}((t, q(t), \dot{q}(t)) = 0. \qquad (D.5)$$

♠ Amplitudes (S-Matrix)

In the approach of least action, to calculate the transition amplitude (or S-matrix) to go from initial point A at the time t_i to a final point B at the time t_f, must be considered the sum on all paths that meet the initial and final conditions. Each path is characterized by its weight that is the exponential of the classical action $\mathcal{S}$ with respect to the variation of the Lagrangian L (difference between kinetic E_c and potential energy V) in 3-dimensional space multiplied by $(i/\hbar)$. In terms of an equation [see Eq. (D.3)], it means that: $\mathcal{S} = (i/\hbar) \int dt\, L$.

The complete path or functional integral corresponds to the summation over an infinite number of complex weights (infinite number of variables of integration). Then, it permits to calculate the transition amplitude.

D.2   Maxwell's Equations

Maxwell law is the unification of electric and magnetic forces known later as the electromagnetic forces.

♣ Classical Physics Notations

At the microscopic level, the Maxwell's differential equations read:

Law's name	Maxwell's equations	
Gauss's law for electricity	$\nabla \cdot E = \frac{\rho}{\epsilon_0}$,	
Gauss's law for magnetism	$\nabla \cdot B = 0$	(D.6)
Faraday's law of induction	$\nabla \times E = -\partial B / \partial t$	
Ampère's circuital law	$\nabla \times B = \mu_0 \left(J + \epsilon_0 \frac{\partial E}{\partial t} \right)$	

where E and B are vector electric and magnetic fields which have a time-dependence; ρ and J are charge and current densities; ϵ_0 and μ_0 are the permittivity and permeability of a free space;

$$\nabla = \frac{\partial}{\partial x_1} + \frac{\partial}{\partial x_2} + \frac{\partial}{\partial x_3} \qquad (D.7)$$

is the *Laplacian operator* in a space of 3-dimensions. Then, $\nabla\cdot$ is the *divergence operator* and $\nabla\times$ is the *curl operator* (*vector product*) which describes the infinitesimal rotation of a vector field.

◇ Covariant Notation and Gauge Invariance

In order to illustrate the advantage of the covariant notation, let us consider the *Lorentz force equation* which describes the motion of a particle having a charge e in the presence of an electric $\vec{E}$ and magnetic $\vec{B}$ fields:

$$\frac{d\vec{p}(\tau)}{d\tau} = e\left(\vec{E}\frac{cdt}{d\tau} + \frac{d\vec{x}(\tau)}{d\tau}\times\vec{B}\right), \qquad (D.8)$$

where $x(\tau)$ is the particle trajectory and τ the proper time:

$$c^2 d\tau^2 = dx_\mu x^\mu = dx_0^2 - d\vec{x}^2 = c^2 dt^2 - d\vec{x}^2. \qquad (D.9)$$

Introducing the *anti-symmetric electromagnetic field strength tensor* $F_{\mu\nu}(x)$ defined as:

$$F_{0i} = -E_i, \qquad i = 1, 2, 3,$$

$$F_{12} = -B_3, \quad F_{23} = -B_1, \quad F_{31} = -B_2, \qquad (D.10)$$

the previous Lorentz force equation can be written in the covariant form:

$$\frac{dp_\mu}{d\tau} = -eF_{\mu\nu}\frac{dx^\nu}{d\tau}. \qquad (D.11)$$

One can associate to the electromagnetic field tensor $F_{\mu\nu}(x)$ its *dual*:

$$\tilde{F}_{\mu\nu}(x) = \frac{1}{2}\epsilon_{\mu\nu\rho\sigma}F^{\rho\sigma}(x), \qquad (D.12)$$

Then, in a *covariant notation*, the Maxwell's equations read:

$$\partial^\nu\tilde{F}_{\mu\nu}(x) = 0 \quad \text{and} \quad \partial^\nu F_{\mu\nu}(x) = j_\mu(x), \qquad (D.13)$$

where: $j_\mu(x)$ is the current source density: $j_0(x) \equiv \rho(x)$ is the charge density and $\vec{j}(x)$ is the electric current density. The first set of Maxwell equations in Eq. (D.13) is automatically satisfied if $F_{\mu\nu}(x)$ is expressed in terms of the four-vector potential $A_\mu(x)$:

$$F_{\mu\nu} = \partial_\mu A_\nu - \partial_\nu A_\mu, \qquad (D.14)$$

where ∂_μ is the partial derivative with respect to the position x_μ in 4-dimension space–time. Conversely, this equation implies the existence of a vector potential $A_\mu^\theta(x)$ defined as:

$$A_\mu^\theta(x) = A_\mu(x) - \partial_\mu\theta(x), \qquad (D.15)$$

where $\theta(x)$ is an arbitrary scalar field. The transformation in Eq. (D.15) is called gauge transformation and the vector potential $A_\mu(x)$ is called *gauge field*. The Maxwell equations in Eq. (D.13) are invariant under these gauge transformations.

D.3 QED Lagrangian

♣ Electromagnetic Field Sector

One can check that the *Maxwell's equations in Eq.* (D.13) *without radiations* $(j_\mu(x) = 0)$ are the Euler–Lagrange equations [see Eq. (D.5)] corresponding to the *Lagrange density*:

$$\mathcal{L}_A = -\frac{1}{4}F_{\mu\nu}F^{\mu\nu}, \tag{D.16}$$

which is invariant under the gauge transformation in Eq. (D.15).

◇ Fermionic Sector

The *Dirac equation*:

$$(i\gamma^\mu\partial_\mu - m)\,\psi(x) = 0, \tag{D.17}$$

which governs the free motion of an electron is the equation of motion *associated to the Lagrange density*:

$$\mathcal{L}_\psi = \frac{i}{2}\bar{\psi}\gamma^\mu\overleftrightarrow{\partial_\mu}\psi - m\bar{\psi}\psi, \tag{D.18}$$

where γ_μ is the Dirac matrix with the properties in Appendix E; $\psi(x)$ is the electron field; $\bar{\psi}(x) \equiv \psi^\dagger(x)\gamma_0$. We have introduced the notations:

$$\bar{\psi}\overleftrightarrow{\partial_\mu}\psi \equiv \bar{\psi}(\partial_\mu\psi) - (\partial_\mu\bar{\psi})\psi. \tag{D.19}$$

♡ First Kind Gauge Invariance and Noether Current

We consider the *abelian $U(1)$ group set of transformations*:

$$\psi(x) \to e^{ie\theta}\psi(x), \tag{D.20}$$

where θ *is a real number*. As the fermion fields appear only in the combination $\bar{\psi}\psi$, it is clear that the previous Lagrangian is invariant under these transformations:

$$\delta\mathcal{L} = 0 \equiv \frac{\partial\mathcal{L}}{\partial\psi}\delta\psi + \frac{\partial\mathcal{L}}{\partial(\partial_\mu\psi)}\delta\partial_\mu\psi. \tag{D.21}$$

Using the Euler–Lagrange equation of motion:

$$\frac{\partial\mathcal{L}}{\partial\psi} - \partial_\mu\frac{\partial\mathcal{L}}{\partial(\partial_\mu\psi)} = 0, \tag{D.22}$$

one can write $\delta\mathcal{L}$ as a total divergence:

$$\delta\mathcal{L} = \partial_\mu \left(\frac{\partial\mathcal{L}}{\partial(\partial_\mu\psi)} \delta(\partial_\mu\psi)\delta\psi \right) \quad : \quad \delta\psi = ie\theta\psi(x). \tag{D.23}$$

It means that, to the set of transformations in Eq. (D.20), is associated a conserved *Noether current*:

$$J_\mu(x) \equiv ie\frac{\partial\mathcal{L}}{\partial(\partial_\mu\psi)}\psi = -e\bar{\psi}(x)\gamma_\mu\psi(x) \quad : \quad \partial^\mu J_\mu(x) = 0, \tag{D.24}$$

to which is associated the *charge*:

$$Q = \int d^3\vec{x} J_0(\vec{x}, t), \tag{D.25}$$

which is the electric charge which is conserved, i.e. Q is a constant of motion:

$$\frac{dQ}{dt} = 0. \tag{D.26}$$

Therefore a possible form of the *Lagrangian interaction* is:

$$\mathcal{L}_I = J_\mu(x)A^\mu(x) = -e\bar{\psi}(x)\gamma_\mu\psi(x)A^\mu(x). \tag{D.27}$$

♠ Second Kind or Local Gauge Invariance

As the θ parameter of the first kind gauge invariance is a real number, it does not affect the vector potential part $\mathcal{L}_A$ of the Lagrangian. Let us now consider the case where θ *depends on the space–time* x. It is clear that this change does not affect the mass term $m\bar{\psi}\psi$ of the Lagrangian but induces an extra term in:

$$\partial_\mu\psi(x) \to e^{ie\theta}\partial_\mu\psi(x) + ie\partial_\mu\theta(x)e^{ie\theta}\psi(x). \tag{D.28}$$

For solving this problem, one introduces a generalized derivative called *covariant derivative*:

$$D_\mu = \partial_\mu + ieA_\mu, \tag{D.29}$$

which transforms as:

$$\begin{aligned} D_\mu\psi(x) &\to \left(\partial_\mu + ieA'_\mu\right)\psi'(x) \\ &= e^{ie\theta}\left(\partial_\mu + ieA'(x)\right)\psi(x) + ie\partial_\mu\theta(x)e^{ie\theta}\psi(x). \end{aligned} \tag{D.30}$$

This remaining extra term can now be absorbed by redefining the gauge field:

$$A_\mu(x) \to A'_\mu(x) \equiv A_\mu(x) - \partial_\mu\theta(x), \tag{D.31}$$

which is nothing but the gauge transformation in Eq. (D.15).

♣ Canonical Quantization and Gauge Fixing Term

• *Canonical quantization*

For a given Lagrangian $\mathcal{L}$, canonical quantization *treats a field $\phi(x)$ and its conjugate momentum*:

$$\pi(x) = \frac{\partial \mathcal{L}}{\partial \dot{\phi}}, \tag{D.32}$$

as operators obeying the canonical commutation relations at a time $t = 0$:

$$[\phi(x), \phi(y)] = 0, \quad [\pi(x), \pi(y)] = 0 , \quad [\phi(x), \pi(y)] = -i\hbar\delta(x - y). \tag{D.33}$$

It states that operators built from ϕ and π can be formally defined at other times via the time evolution generated by the Hamiltonian $\mathcal{H}$:

$$\mathcal{O}(t) = e^{it\mathcal{H}}\mathcal{O}e^{-it\mathcal{H}}. \tag{D.34}$$

A straightforward application of this procedure to the previous QED Lagrangian leads to difficulties as the time derivative of the time component $\partial A_0/\partial t$ does not appear in this Lagrangian such that the *conjugate momentum*:

$$\pi_\mu(x) = \frac{\partial \mathcal{L}}{\partial(\partial_0 A_\mu)} = F_{\mu 0}, \tag{D.35}$$

has a zero time component:

$$\pi_0(x) = 0. \tag{D.36}$$

From the previous canonical commutation relations where $\phi(x) \equiv A_\mu(x)$, one can see that $A_0(x)$ commutes with all other operators, i.e. it is a c-number in contrast with the space components $A_i(x)$ ($i = 1, 2, 3$). This is due to the fact that the field $A_\mu(x)$ *is a vector potential having four components while it describes a particle, the photon, which has only two components*.

• *Lorenz condition*

This can be achieved by fixing a non-covariant gauge called *radiation gauge*. Rewriting the Maxwell's equations in Eq. (D.13) for free fields ($j_\mu(x) = 0$) in terms of the Dalembertian opertator $\Box \equiv -\partial_\mu\partial^\mu$:

$$\Box A_\mu(x) + \partial_\mu(\partial_\nu A^\nu)(x) = 0, \tag{D.37}$$

the Ludvig Valentin Lorenz (1829–1891) (Fig. D.1) condition (*often mistakenly called: Lorentz condition*):

$$(\partial_\nu A^\nu)(x) = 0, \tag{D.38}$$

implies the simple form for the equation of motion:

$$\Box A_\mu(x) = 0. \tag{D.39}$$

However, this condition does not entirely fix the gauge field as we still have the *second kind of gauge transformation* in Eq. (D.31) with:

$$\Box\theta(x) = 0. \tag{D.40}$$

- *Radiation or Coulomb gauge*

The gauge field $A_\mu(x)$ can be entirely fixed by choosing $\theta(x)$ in such a way that:

$$A'_\mu(x) = A_\mu(x) - \partial_\mu \int_0^t A_0(t',t)dt', \tag{D.41}$$

i.e. the scalar component $A_0(x)$ vanishes. In this gauge:

$$A_0(x) = 0, \qquad \vec{\nabla}\cdot\vec{A} = 0. \tag{D.42}$$

In the momentum space, at taking the Fourier transform of the field:

$$A_\mu(x) = \frac{1}{(2\pi)^{3/2}} \int d^4k\ \delta(k^2)e^{ikx} A_\mu(k), \tag{D.43}$$

the *radiation gauge condition* implies:

$$\vec{k}\cdot\vec{A}(k)|_{k^2=0} = 0, \tag{D.44}$$

which states that the longitudinal components of the fields have disappeared.

- *Covariant gauge*

Alternatively, by using the method of *Lagrangian multipliers*, one can introduce a new field $B(x)$ in the Lagrangian when the gauge field $A_\mu(x)$ is restricted by the previous Lorenz condition in Eq. (D.38). Then, the unconstrained Lagrangian has the extra term:

$$\mathcal{L}_B = -\partial_\mu A^\mu(x)B(x). \tag{D.45}$$

The equation of motion for $A_\mu(x)$ in the presence of the source $J_\mu(x)$ in Eq. (D.24) becomes:

$$\partial^\nu F_{\mu\nu}(x) = -e\bar{\psi}(x)\gamma_\mu\psi(x) + \partial_\mu B(x), \tag{D.46}$$

which implies:

$$\Box B(x) = 0, \tag{D.47}$$

indicating that B obeys a free field equation, while the $A(x)$ free field equation of motion in Eq. (D.37) becomes:

$$\Box A_\mu(x) + \partial_\mu(\partial_\nu A^\nu - B(x))(x) = J_\mu(x). \tag{D.48}$$

Then, the new Lagrangian becomes invariant under the *restricted local gauge transformations*:

$$A_\mu(x) \to A_\mu(x) - \partial_\mu\theta(x), \quad \psi(x) \to e^{ie\theta}\psi(x), \quad B(x) \to B(x), \tag{D.49}$$

with the restriction:

$$\Box \theta(x) = 0. \tag{D.50}$$

In this case, one can use the canonical quantization without lost of covariance because:

$$\pi_\mu(x) = \frac{\partial \mathcal{L}}{\partial(\partial_0 A_\mu)} = F_{\mu 0} - g_{\mu 0}B(x), \tag{D.51}$$

such that $\pi_0(x) \neq 0$.

• *Stueckelberg Lagrangian and massive photon*

In the case of massive photon, the Lagrangian possesses the quadratic term:

$$\mathcal{L}_\lambda = \frac{\lambda^2}{2} A^\mu(x)A_\mu(x), \tag{D.52}$$

where λ is the photon mass. In practice, this term serves for regulating the infrared divergence encountered in perturbative calculations. However, this term violates gauge invariance:

$$\frac{\lambda^2}{2}A^\mu(x)A_\mu(x) \to \frac{\lambda^2}{2}A^\mu(x)A_\mu(x) - \lambda^2(\partial_\mu\theta(x))A^\mu(x). \tag{D.53}$$

It can be reabsorbed by the introduction of a new field $S(x)$ coupled to $A_\mu(x)$ via a derivative coupling:

$$\mathcal{L}_S = A^\mu(x)\partial_\mu S(x), \tag{D.54}$$

where $S(x)$ transforms as:

$$S(x) \to S(x) + \lambda^2\theta(x). \tag{D.55}$$

By identifying $S(x)$ with $B(x)$, and *adding a new quadratic term*:

$$\mathcal{L}_\alpha = \frac{\alpha_G}{2} B(x)B(x), \tag{D.56}$$

one obtains the *Stueckelberg Lagrangian*:

$$\mathcal{L}_{SG} = \mathcal{L}_A + \mathcal{L}_\psi + \mathcal{L}_I + \mathcal{L}_\lambda + \mathcal{L}_S + \mathcal{L}_\alpha. \tag{D.57}$$

One can notice that $\mathcal{L}_B$ in Eq. (D.45) and $\mathcal{L}_S$ in Eq. (D.54) are equivalent as they differ only by a total derivative $\partial_\mu(A^\mu(x)B(x))$ and then lead to the same equation of motion. This Lagrangian is invariant under the *restricted gauge transformations* defined in Eq. (D.49) but with the restriction:

$$(\Box - \alpha_G \lambda^2)\,\theta(x) = 0 \ . \tag{D.58}$$

The corresponding equations of motion are:

$$\partial^\nu F_{\mu\nu}(x) = -e\bar{\psi}(x)\gamma_\mu \psi(x) + \partial_\mu B(x) + \lambda^2 A_\mu(x), \tag{D.59}$$

the Dirac equation in the presence of interaction:

$$(i\gamma^\mu \partial_\mu - m)\,\psi(x) = e\bar{\psi}(x)\gamma_\mu \psi(x) A^\mu(x), \tag{D.60}$$

and:

$$\partial_\mu A^\mu(x) = \alpha_G B(x). \tag{D.61}$$

From Eqs. (D.59) and (D.61), one finds that $B(x)$ obeys the free field equation:

$$(\Box - \alpha_G \lambda^2)B(x) = 0, \tag{D.62}$$

which is the same as the restriction on the set of gauge transformations in Eq. (D.58).

◇ Complete QED Lagrangian

Eliminating $B(x)$ in the expression of the Lagrangian through the relation in Eq. (D.61), one arrives at the final QED Lagrangian with a gauge fixing term and for a massive photon:

$$\mathcal{L}_{QED} = \mathcal{L}_A \equiv -\frac{1}{4}F_{\mu\nu}F^{\mu\nu} +$$

$$\mathcal{L}_\psi \equiv \frac{i}{2}\bar{\psi}\gamma^\mu \overleftrightarrow{\partial_\mu}\psi - m\bar{\psi}\psi +$$

$$\mathcal{L}_\lambda \equiv \frac{\lambda^2}{2}A^\mu(x)A_\mu(x) +$$

$$\mathcal{L}_\alpha \equiv -\frac{1}{2\alpha_G^2}\left(\partial_\mu A^\mu(x)\right)^2 +$$

$$\mathcal{L}_I \equiv -e\bar{\psi}(x)\gamma_\mu \psi(x) A^\mu(x). \tag{D.63}$$

$\mathcal{L}_\alpha$ is the *gauge fixing term. The gauge parameter α_G is 0 in the Lev Landau (1908–1968, Nobel Prize of 1962) (Fig. D.1) gauge and 1 in the 't Hooft–Feynman gauge.*

Figure D.1: (a) L.V. Lorenz, (b) L. Landau.

Figure D.2: (a) J.C. Ward, (b) Y. Takahashi, (c) A.A. Slavnov, (d) J.C. Taylor.

The gauge field $A_\mu(x)$ can be decomposed into a transverse vector field $A_\mu^T(x)$ $(\partial_\mu A_\mu^T(x) = 0)$ and a free scalar field $1/\alpha_G \partial_\mu A^\mu(x)$ components:

$$A_\mu^T(x) = A_\mu^T(x) + \frac{1}{\alpha_G \lambda^2} \partial_\mu \left(\partial_\mu A^\mu(x)\right). \tag{D.64}$$

They obey the equations of motion:

$$\left(\Box - \lambda^2\right) A_\mu^T(x) = J_\mu(x), \qquad \left(\Box - \alpha_G \lambda^2\right) \partial_\mu A^\mu(x) = 0, \tag{D.65}$$

which are respectively the equation of motion of a massive vector boson coupled to a conserved vector current and of a free massive scalar field.

♡ Ward–Takahashi identity

John Clive Ward (1924–2000) and Yasushi Takahashi (1924–2013) (Fig. D.2) introduce the Ward–Takahashi identity which is the one between correlation functions that follows from the global or gauged symmetries of a theory, and which remains valid after renormalization. It was originally used to relate the wave function renormalization of the electron to its vertex renormalization form factor, guaranteeing the cancellation of the ultraviolet divergence to all orders of perturbation theory. Later uses include the extension of the proof of Goldstone's theorem to all orders of perturbation theory.

More explicitly, let us consider the full fermion–photon–fermion vertex interaction:

$$V_\mu(p, p+q) = \int d^4x d^4y \; e^{i(px+qy)} \langle 0|T\psi(x)A_\mu(y)\bar{\psi}(0)|0\rangle, \qquad (D.66)$$

where p and $p+q$ are the momenta of the ingoing and outgoing fermions and q the momentum of the incident photon (see Fig. 4.6a). Expressing it in terms of the *amputated vertex* $\Gamma_\mu(p, p+q)$ and of the external propagators, it reads:

$$V_\nu(p, p+q) = -(ie)iS(p+q)\Gamma^\mu(p, p+q)iS(p)iD_{\mu\nu}(q), \qquad (D.67)$$

where S and $D_{\mu\nu}$ are the fermion and photon propagators:

$$S(p) = -i \int d^4x \; e^{ipx} \langle 0|T\psi(x)\bar{\psi}(0)|0\rangle,$$

$$D_{\mu\nu} = -i \int d^4x \; e^{ipx} \langle 0|T A_\mu(x)A_\mu(0)|0\rangle, \qquad (D.68)$$

where T indicates *the time-ordered product* or T product as defined in Eq. (B.36). *The Takahashi (Fig. D.2) identity states that:*

$$q_\mu\Gamma^\mu(p, p+q) = S^{-1}(p+q) - S^{-1}(p), \qquad (D.69)$$

while the limit of this relation for $q \rightarrow 0$ *has been obtained by Ward (Fig. D.2):*

$$\Gamma^\mu(p, p+q) = \frac{\partial S^{-1}(p+q)}{\partial q_\mu}\bigg|_{q=0}. \qquad (D.70)$$

D.4   Standard Model Lagrangian

Analogous expression can be obtained for the Standard Model based on the group $SU(2)_L \times U(1)$. However, *we have here two groups and then two families of gauge fields that we call* A_μ^a *and* B_μ *associated respectively to the* $SU(2)_L$ *and* $U(1)$ *gauge groups.* The former ressembles the gluons of

QCD but instead of $8 = 3^2 - 1$ one has $3 = 2^2 - 1$ gauge fields, while B_μ ressembles the photon of QED. Another main difference is that the *fermion enters with left chirality*. Then, one has to introduce the chiral fermion field: $\psi_{L(R)} = (1 \mp \gamma_5)\psi/2$.

♣ Gauge Fields Sector

The gauge field part of the Lagrangian reads:

$$\mathcal{L}^A_{WS} = -\frac{1}{4}F_{\mu\nu}F^{\mu\nu} - \frac{1}{4}G^a_{\mu\nu}G^{\mu\nu}_a, \tag{D.71}$$

where:

$$F^{\mu\nu} = \partial^\mu B^\nu - \partial^\nu B^\mu, \qquad G^{\mu\nu}_a = \partial^\mu A^\nu_a - \partial^\nu A^\mu_a + g_2 f_{abc}A^\mu_b A^\nu_c, \tag{D.72}$$

where $F^{\mu\nu}$ and $G^{\mu\nu}_a$ are the field strength tensors associated to the fields B^μ and A^μ_a of the $U(1)$ and $SU(2)_L$ groups. g_2 is the coupling of the $SU(2)_L$ group and the index a runs from 1 to 3.

◇ Fermionic Sector

The corresponding Lagrangian reads for the leptonic sector:

$$\mathcal{L}^\psi_{WS} = \bar{\psi}_L\gamma_\mu D^\mu_L\psi_L + \bar{\psi}_R\gamma_\mu D^\mu_R\psi_R, \tag{D.73}$$

where the covariant derivatives are:

$$D^\mu_L = \partial^\mu + \frac{i}{2}g_1 B^\mu - \frac{i}{2}g_2\lambda^a A^\mu_a, \quad D^\mu_R = \partial^\mu + ig_1 B^\mu, \tag{D.74}$$

where g_1 and g_2 are the couplings of the $U(1)$ and $SU(2)_L$ groups. The lepton fields are:

$$\psi_L \equiv \frac{1}{2}(1 - \gamma_5)\begin{pmatrix} \nu_l \\ l \end{pmatrix}, \qquad \psi_R \equiv \frac{1}{2}(1 + \gamma_5)\,l, \tag{D.75}$$

which have respectively hypercharges $Y_L = -1$ and $Y_R = -2$; ν and l are respectively the neutrino and associated charged lepton fields.

♡ Higgs Sector

Unlike QED and QCD, *the Standard Model Lagrangian possesses a part coming from the scalar Higgs field*. The self-interaction term reads:

$$\mathcal{L}^\phi_{WS} = (D^\mu\phi)^\dagger(D^\mu\phi), \tag{D.76}$$

where:

$$D^\mu = \partial^\mu - \frac{i}{2}g_1 B^\mu - \frac{i}{2}g_2\lambda^a A^\mu_a \tag{D.77}$$

is the covariant derivative associated to the Higgs field which reads:

$$\phi \equiv \begin{pmatrix} \phi^+ \\ (\phi_1 + i\phi_2)/\sqrt{2} \end{pmatrix}. \tag{D.78}$$

It has a hypercharge Y= +1. ϕ^+ is the charged and $\phi_{1,2}$ the neutral scalar fields. $\phi^\dagger$ is the hermitian conjugate of ϕ. The Lagrangian contains a term corresponding to *the coupling of the scalar field to fermions called Yukawa coupling*:

$$\mathcal{L}^f_{WS} = g_f \bar{\psi}_R \phi \psi_L + \text{h.c.} \tag{D.79}$$

This term gives masses to fermions. The *Higgs potential* reads:

$$V(\phi^\dagger \phi) = \mu^2 \phi^\dagger \phi + \lambda \left(\phi^\dagger \phi \right)^2, \tag{D.80}$$

where λ is positive which corresponds to the lowest energy which is the vacuum. μ^2 is taken to be negative in order to break the symmetry with the *Higgs expectation value*:

$$\langle 0|\phi_1|0 \rangle = \sqrt{\frac{-\mu^2}{2\lambda}} \equiv \nu_\phi/\sqrt{2}. \tag{D.81}$$

In this way, *the potential takes the form of the mexican cap* in Fig. 7.6. The charged lepton, $W^\pm$ and Z^0 bosons acquire masses:

$$m_l = g_\psi \left(\frac{\nu_\phi}{\sqrt{2}} \right), \qquad M_{W^\pm} = g_2 \left(\frac{\nu_\phi}{2} \right), \qquad M_Z = \sqrt{g_1^2 + g_2^2} \left(\frac{\nu_\phi}{2} \right), \tag{D.82}$$

while the photon $\mathcal{A}$ remains massless. The *physical boson fields* are combinations of the A_μ and B_μ fields:

$$\mathcal{A}_\mu = \frac{g_2 B_\mu - g_1 A_\mu^{(3)}}{\sqrt{g_1^2 + g_2^2}} \qquad \text{photon}$$

$$Z_\mu = \frac{g_1 B_\mu + g_2 A_\mu^{(3)}}{\sqrt{g_1^2 + g_2^2}} \qquad Z^0 \text{ boson}$$

$$W_\mu^\pm = \frac{A_\mu^{(1)} \mp A_\mu^{(2)}}{\sqrt{2}} \qquad W^\pm \text{ boson.} \tag{D.83}$$

In terms of the physical fields, the *lepton–boson interaction Lagrangian* reads:

$$\mathcal{L}_{int} = \frac{g_2}{\sqrt{2}} \left(J_c^\mu W_\mu^+ + \text{h.c} \right) + e \mathcal{A}_\mu J_{em}^\mu + \frac{g_2}{\cos \theta_W} \left(J_3^\mu - \sin^2 \theta_W J_{em}^\mu \right) Z_\mu, \tag{D.84}$$

where the currents are:

$$J_c^\mu = \bar\nu_L \gamma^\mu l_L, \quad J_{em}^\mu = \bar l \gamma^\mu l, \quad J_3^\mu = \frac{1}{2} \left(\bar\nu_{lL} \gamma^\mu \nu_{lL} - \bar l_L \gamma^\mu l_L \right), \qquad (D.85)$$

with:

$$\nu_{iL} = \frac{1}{2} \left(1 - \gamma_5 \right) \nu_l, \quad l_L = \frac{1}{2} \left(1 - \gamma_5 \right) l_L, \qquad (D.86)$$

and:

$$\frac{G_F}{\sqrt{2}} = \frac{g_2^2}{8 M_W^2} = \frac{1}{2\nu_\phi^2}; \qquad g_2 \equiv -\frac{e}{\sin\theta_W}, \qquad g_1 \equiv -\frac{e}{\cos\theta_W}. \qquad (D.87)$$

G_F is the Fermi coupling, θ_W is the *Glashow–Weinberg–Salam angle* with an experimental value:

$$\sin^2 \theta_W = 0.23116(12). \qquad (D.88)$$

For experimental purpose, one also defines the ratio called ρ *parameter*:

$$\rho = \left(\frac{M_W}{M_Z \cos\theta_W} \right)^2, \qquad (D.89)$$

which quantifies the ratio of the neutral to the charged current cross-section.

D.5 QCD Lagrangian

The difference with QED is that, *due to the non-abelian nature of the $SU(3)_c$ gauge group, one can have self interactions among the eight colored gluons* as explained in Chapter 6. The gauge field strength tensor then becomes:

$$G_a^{\mu\nu} = \partial^\mu A_a^\nu - \partial^\nu A_a^\mu + g f_{abc} A_b^\mu A_c^\nu, \qquad (D.90)$$

while the covariant derivative is:

$$D_\mu = \partial_\mu - ig(\lambda_a/2) A_\mu^a, \qquad (D.91)$$

where a new self interaction term appears. $a = 1, \ldots, 8$ is the number of colored gluons; f_{abc} is called structure constant which characterizes the group $SU(3)_c$ and λ^a is a 3×3 color matrix with the property:

$$[\lambda_a, \lambda_a] = 2i f_{abc} \lambda^c. \qquad (D.92)$$

g is the $SU(3)_c$ gauge coupling analogue of the electric charge e in QED.

D.6 BRST Transformation for QCD

♣ Faddeev–Popov–DeWitt Ghosts

A new scalar field introduced by Ludvig Dmitrievich Faddeev (1934–) and Victor Nikolaevich Popov (1937–1997) and (within the context of gravity) by Bryce Seligman DeWitt (1923–2004) (Fig. D.3) called *Faddeev–Popov ghost (scalar field φ_a with a negative norm) is introduced in the QCD Lagrangian*:

$$\mathcal{L}_{FP} = -\partial_\mu \bar{\varphi}_a D^\mu \varphi^a, \tag{D.93}$$

which is necessary for eliminating unphysical particles from the theory and for maintaining the consistency of the path integral formulation. For instance, if one considers the process $\bar{q}q \to gg \to \bar{q}q$, one can notice that due to the propagation of the longitudinal and scalar gluon polarizations along the internal gluon lines, the contributions of higher order diagrams violate the unitarity of the S-matrix (see the Feynman diagram representation of the process in Fig. D.4). This problem can be avoided by the introduction of the ghost fields as shown in Fig. D.5 which cancel such contributions. In QED, the analogous process $e^+e^- \to \gamma\gamma \to e^+e^-$ does not have these drawbacks as unphysical contributions from the longitudinal and scalar components of the photons vanish due to gauge invariance and to the conservation of the electromagnetic current.

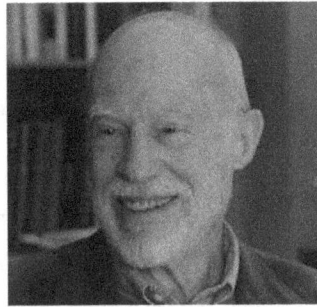

(a)   (b)

Figure D.3: (a) L.D. Faddeev, (b) B.S. Dewitt.

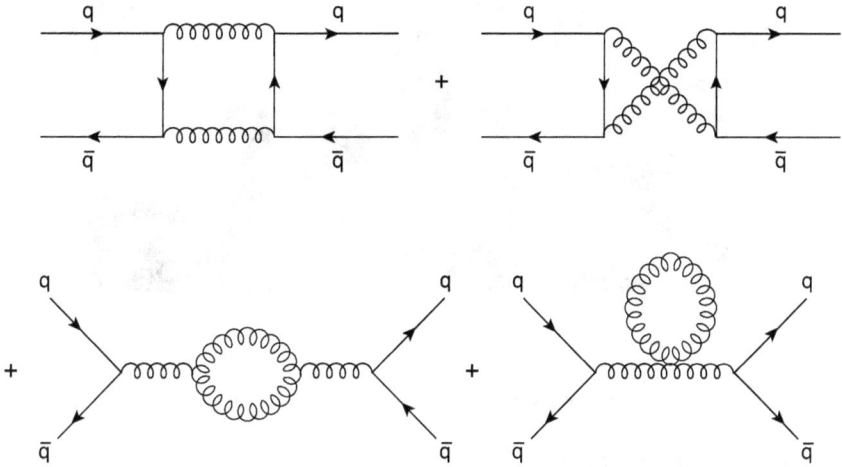

Figure D.4: Gluon contributions to the $\bar{q}q \to \bar{q}q$ process.

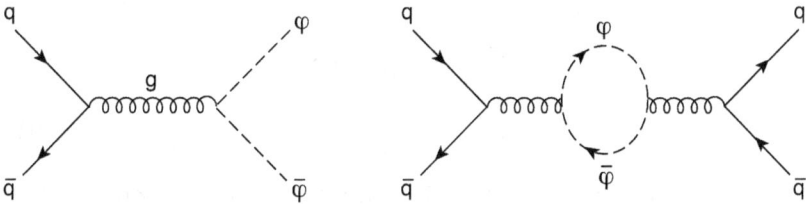

Figure D.5: Ghost contributions to the $\bar{q}q \to \bar{q}q$ process.

◇ Slavnov–Taylor Identities

The Slavnov–Taylor identities, introduced by Andrei Alexeevich Slavnov (1939–) and J.C. Taylor (1930–) (Fig. D.2), is a *generalization of the Ward–Takahashi identities of QED to non-abelian gauge theory*. Like the latter, it is an identity between correlation functions that follows from the global or local gauge symmetry of a theory, and which is not affected by renormalization, i.e. it is valid for all orders of the perturbation theory.

♡ BRST Transformation

The QCD Lagrangian is locally invariant under the BRST [Carlo Becchi (1939–), Alain Rouet (1947–), Raymond Stora (1930–2015) and Igor

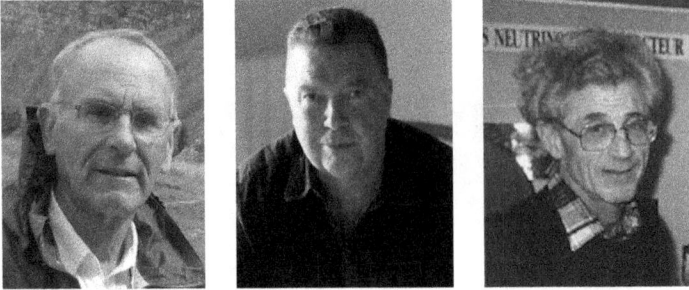

(a)   (b)   (c)

Figure D.6: (a) C. Becchi, (b) A. Rouet, (c) R. Stora.

Viktorovich Tyutin (1940–)] (*Fig. D.6*) *transformation:*

$$A_\mu(x) \to A_\mu(x) + \omega D_\mu \, \varphi,$$

$$\psi_i(x) \to \exp\left(-ig\omega \vec{T}.\vec{\varphi}\right) \psi_i,$$

$$\bar{\varphi} \to \bar{\varphi} + \frac{\omega}{\alpha_G} \partial_\mu A^\mu,$$

$$\varphi \to \varphi - \frac{1}{2} g\omega \vec{\varphi} \times \vec{\varphi}, \tag{D.94}$$

where $\vec{T}$ is the color matrix; $\omega(x)$ is an arbitrary parameter; ψ_i, A_μ and φ are respectively the quark, gluon and ghost fields. *The BRST transformation is a method for quantization of fields which generates the Slavnov–Taylor–Ward–Takahashi identities.* One can, for instance, prove that order by order in perturbation theory, the non-transverse part of the gluon propagator:

$$iD_{\mu\nu}^{ab}(k) = \int d^4x \, e^{ikx} \langle 0|\mathcal{T}A_\mu^a(x)A_\nu^b(0)|0\rangle. \tag{D.95}$$

remains the same as for the free propagator:

$$k^\mu k^\nu iD_{\mu\nu}^{ab}(k) = -i\alpha_G \delta^{ab}. \tag{D.96}$$

D.7 Dimensional Renormalization

♣ The MS and $\overline{MS}$-Schemes

In QED, it is natural to use the *on-shell renormalization scheme*:

$$\Gamma_R(q^2) = \Gamma_B(q^2) - \Gamma_B(q^2 = 0), \tag{D.97}$$

for defining a renormalized Green's function Γ_R having a momentum q^2 (the sub-index B indicates the bare/naked Green's function). This is natural as the photon and electron are observed in QED, and then are on their mass-shells. For an electron self-energy diagram, e.g. one can subtract at $p^2 = m_e^2$, which is not the case of QCD, as quarks are off-shell due to confinement. Therefore, there is freedom for choosing the renormalization schemes in QCD. In the so-called:

't Hooft Minimal Subtraction (MS) − scheme (Fig. 6.2),

one can first calculate the Green's function in n-dimension space–time (dimensional regularization) where the ultraviolet (UV) and infrared (IR) divergences are transformed into poles in $\epsilon \equiv n - 4 \to 0$ originating from the properties of the Γ-function (see Appendix E). More generally, these poles are of the form:

$$\Delta = \sum_{p=1} \frac{Z^{(p)}}{\epsilon^p}. \tag{D.98}$$

In this scheme, one only has to eliminate the $1/\epsilon$ poles of the Green's functions using the renormalization constants. One should notice that, for renormalizable theories like QCD, the $Z^{(p)}$ are local i.e. constants and will appear as counterterms in the initial Lagrangian constrained by the Slavnov–Taylor identities (Fig. D.2). Later on, it has been remarked by William A. Bardeen, Andrzej J. Buras, (Figs. 9.2 and 9.5) that the combination in Eq. (E.52) appears always in the stage of the calculation. Therefore, these authors found that it is natural to also subtract the constant terms $\ln 4\pi - \gamma_E$ together with the ϵ-pole:

$$\frac{2}{\epsilon} \to \frac{2}{\hat{\epsilon}} \equiv \frac{2}{\epsilon} + \ln 4\pi - \gamma_E. \tag{D.99}$$

This is the modified (improved) version of the MS-scheme, and called:

$$\overline{MS}-scheme,$$

which is most commonly used in the literature and will be used in the forthcoming discussions of this book.

◇ Renormalization Constants

Taking into account the dimension obtained in the $4 - \epsilon$ world (see Table D.1) via the mass scale ν, *one has relations between renormalized*

Table D.1 Dimensions of the couplings and fields in n-dimensions.

Name	Notation	Dimension
gauge coupling	g	$\frac{1}{2}(4-n)$
quark mass	m_i	1
covariant gauge parameter	α_G	0
fermion field	$\psi_j(x)$	$\frac{1}{2}(n-1)$
gluon field	$A^a_\mu(x)$	$\frac{1}{2}(n-2)$
Faddeev–Popov field	$\varphi^a(x)$	$\frac{1}{2}(n-2)$

and bare parameters:

$$g^R = \nu^{-\epsilon/2}\, g^B\, Z_\alpha^{-1/2} \ :\ g^2/4\pi \equiv \alpha_s,$$
$$m_j^R = m_j^B Z_m^{-1},$$
$$\alpha_G^R = \alpha_G^B Z_G^{-1},$$
$$\left(\psi_j^\alpha\right)^R = \nu^{\epsilon/2}\left(\psi_j^\alpha\right)^B (Z_{2F})^{-1/2},$$
$$\left(A^a_\mu\right)_R = \nu^{\epsilon/2}\left(A^a_\mu\right)_B (Z_{3YM})^{-1/2},$$
$$(\varphi^a)_R = \nu^{\epsilon/2}(\varphi^a)_B \left(\tilde{Z}_3\right)^{-1/2}, \tag{D.100}$$

where the *renormalization constants* are:

$$Z_\Gamma \equiv 1 + \Delta_\Gamma = 1 + \sum_{p=1} \frac{Z_\Gamma^{(p)}}{\epsilon^p}. \tag{D.101}$$

One can introduce the renormalization constant for the *quark–gluon–quark vertex* as:

$$\left(g\bar{\psi}A\psi\right)_R = \left(g_B\bar{\psi}_B A_B \psi_B\right)\nu^\epsilon Z_{1F}^{-1}, \tag{D.102}$$

which corresponds to the Feynman diagrams:

to which one can define the renormalized coupling;

$$g_B^F = Z_{1F} \, Z_{3YM}^{-1/2} \, Z_{2F}^{-1} \, g_R. \tag{D.103}$$

Analogously, one can introduce the *three-gluon renormalization constant* (Z_{1YM}) corresponding to the vertex:

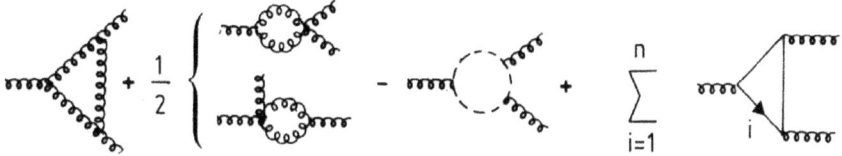

which reads:

$$(g^{YM} AAA)_R = (g_B^{YM} A_B^3) Z_{1YM}^{-1}, \tag{D.104}$$

to which one can define the renormalized coupling:

$$g_B^{YM} = Z_{1YM} \, Z_{3YM}^{-3/2} \, g_R. \tag{D.105}$$

Analogous relations can be obtained by introducing the *ghost–gluon–ghost vertex renormalization constant* $\left(\tilde{Z}_1\right)$ from the vertex diagram:

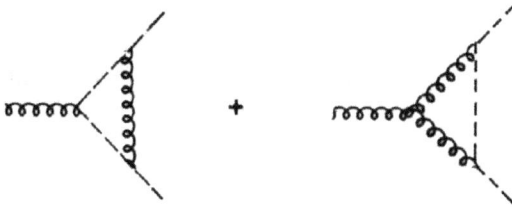

leading to:

$$\tilde{g}_B = \tilde{Z}_1^{-1} \, Z_{3YM}^{-1/2} \, g_R, \tag{D.106}$$

and the renormalization constant (Z_5) from the four-gluon vertices leading to:

$$\left(g_B^{(5)}\right)^2 = Z_5 \, Z_{3YM}^{-2} \, g_R^2. \tag{D.107}$$

All the previous bare couplings are not arbitrary but constrained by the BRST invariance:

$$g_B^{YM} = \ldots\ldots = g_{(5)}^B, \tag{D.108}$$

leading to *the Slavnov–Taylor identities*:

$$Z_{3YM}/Z_{1YM} = \tilde{Z}_3/\tilde{Z}_1 = Z_{2F}/Z_{1F},$$
$$Z_5 = Z_{1YM}^2/Z_{3YM}. \tag{D.109}$$

This is the analogue of the QED relation:

$$Z_{1F} = Z_2. \tag{D.110}$$

The mass renormalization constant is:

$$m_B = \left(Z_m \equiv Z_4\, Z_{2F}^{-1}\right) m_R, \tag{D.111}$$

and the gauge one is:

$$\alpha_G^B = \alpha_G^R\, Z_G^{-1}\, Z_{3YM}. \tag{D.112}$$

Z_{3YM} comes from the evaluation of the gluon propagator:

Z_{2F} and Z_m come from the quark self-energy diagram, which can be parametrized as:

$$\Sigma = m_B \Sigma_1 + (\hat{p} - m_B)\Sigma_2, \tag{D.113}$$

and leads to:

$$Z_{2F} \equiv \frac{1}{1 - \Sigma_2|_{pole}}, \qquad Z_m = 1 - \Sigma_1|_{pole}, \tag{D.114}$$

More generally, for a Green's function with N_G, N_{FP} *and* N_F *external gluons, ghost and fermion fields, one can associate the renormalization constants*:

$$Z_\Gamma = \left(Z_{3YM}^{1/2}\right)^{-N_G} \left(Z_3^{1/2}\right)^{-N_{FP}} \left(Z_{2F}^{1/2}\right)^{-N_F}. \tag{D.115}$$

Expressions of these renormalization constants are known from standard diagram techniques which we shall discuss in the next Appendix E.

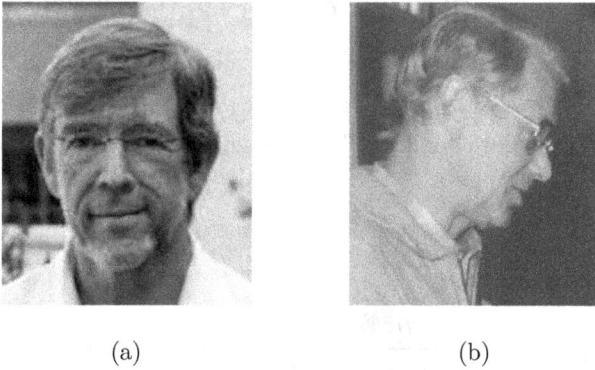

(a) (b)

Figure D.7: (a) C. Callan, (b) K. Symanzik.

D.8   The Renormalization Group

Renormalization invariance states that physical observables must be independent of the renormalization scheme chosen in their theoretical evaluation. The differential approach to renormalization invariance was pioneered by Stueckelberg–Peterman (Fig. 4.4) and by Gell–Mann–Low (Figs. 4.4 and 5.1), where it has been pointed out that the QED coupling constant is momentum dependent due to the definition of the renormalized charge. Such a consideration led to write a differential equation for the photon propagator. Later on in 1970, the study of the scaling behavior in field theory (experimental observation of the Bjorken scaling in deep inelastic scattering, Figs. 5.8 and 5.11) gave rise to the Curtis Callan (1942–), Kurt Symanzik (1923–1983) (Figs. D.7) Equation (CSE), which is a very powerful technique for expressing the renormalization invariance constraints on the short distance behavior of the Green functions. The CSE takes into account the fact that scaling cannot be strictly implemented because of the necessity of a mass scale in the theory. In the dimensional regularization, such a mass scale renders the coupling constants dimensionless (see Table 2.1). A generalization of the uses of the CSE to arbitrary Green functions has been proposed by G. 't Hooft and S. Weinberg in 1973. The central idea was to treat g, m_i, α_G as coupling constants of various interaction terms in the Lagrangian. *One can use the invariance of physical quantities under the renormalization group in order to study the asymptotic behavior of the Green's functions.* This can be done as follows using the renormalization group equation (RGE).

♣ The Renormalization Group Equation (RGE)

Let us consider the renormalized Γ_R ϵ-regularized Green function[1]:

$$\Gamma^R(\nu, p_1, \ldots, p_N; g, \alpha_G, m_i) = Z_\Gamma \Gamma^B(\nu, p_1, \ldots, p_N; g, \alpha_G, m_i), \quad \text{(D.116)}$$

where: Γ_B is the bare quantity. The ν-independence of Γ_B implies the zero of the total derivative:

$$\nu \frac{d\Gamma_B}{d\nu} = 0, \quad \text{(D.117)}$$

which is equivalent to:

$$\left\{ \nu \frac{\partial}{\partial \nu} + \nu \frac{d\alpha_s}{d\nu} \frac{\partial}{\partial \alpha_s} + \sum_j \frac{\nu}{m_j} \frac{dm_j}{d\nu} m_j \frac{\partial}{\partial m_j} + \nu \frac{d\alpha_G}{d\nu} \frac{\partial}{\partial \alpha_G} \right.$$

$$\left. - \frac{1}{Z_\Gamma} \nu \frac{dZ_\Gamma}{d\nu} \right\} \Gamma^R = 0, \quad \text{(D.118)}$$

where standard notations have been used. By introducing *the universal β function and anomalous dimensions γ_i*:

$$\alpha_s \beta(\alpha_s) = \nu \frac{d\alpha_s}{d\nu} \big|_{g_B, m_B \ fixed},$$

$$\gamma_m = -\frac{\nu}{m_i^R} \frac{dm_i^R}{d\nu} \big|_{g_B, m_B \ fixed},$$

$$\gamma_i = \frac{\nu}{Z_i} \frac{dZ_i}{d\nu} \big|_{g_B, m_B \ fixed}, \quad \text{(D.119)}$$

one can transform Eq. (D.118) into the RGE:

$$\left\{ \nu \frac{\partial}{\partial \nu} + \beta(\alpha_s)\alpha_s \frac{\partial}{\partial \alpha_s} - \sum_j \gamma_m(\alpha_s)m_j \frac{\partial}{\partial m_j} + \beta_G \frac{\partial}{\partial \alpha_G} - \gamma_\Gamma \right\} \Gamma^R = 0.$$

$$\text{(D.120)}$$

For N_G, N_{NP} and N_F external gluon, ghost and fermion lines:

$$\gamma_\Gamma = -\frac{1}{2} \left[N_G \gamma_{3YM} + N_F \gamma_{2F} + N_{FP} \tilde{\gamma}_3 \right]. \quad \text{(D.121)}$$

Using the mass-independence property of the β-function in dimensional regularization, one can show [33] that it is the coefficient of the $1/\hat{\epsilon}$ pole of the renormalization constant Z_α defined in Eq. (D.100) while the mass anomalous dimension γ_m is the opposite of the one of Z_m. Their expressions to higher order in α_s are known in the literature [33]. They read to lowest

---

[1]The following discussions come from the review in Refs. [30–32].

order:

$$\beta(\alpha_s) = \left[\beta_1 = -\frac{1}{2}\left(11 - \frac{2n_f}{3}\right)\right]\left(\frac{\alpha_s}{\pi}\right),$$

$$\gamma_m(\alpha_s) = [\gamma_1 = 2]\left(\frac{\alpha_s}{\pi}\right), \qquad (\text{D.122})$$

for three colors and n_f quark flavors. It is important to notice, as already mentioned earlier, that *an asymptotic free theory needs a negative value of the β-function which is achieved in QCD for $n_f \leq 16$* which is the case as only six quark flavors have been observed.

◇ Solution of the RGE

If D is the dimension of Γ in units of mass and if one scales the momenta $p_1, \ldots, p_N$ by a dimensionless factor λ, the *Euler theorem* on homogeneous function gives:

$$\left\{\lambda\frac{\partial}{\partial\lambda} + \sum_j m_j\frac{\partial}{\partial m_j} + \nu\frac{\partial}{\partial\nu} - D\right\}\Gamma^R(\lambda p_1, \ldots, \lambda p_N; \alpha_s, \alpha_G, m_j, \nu) = 0.$$

$$(\text{D.123})$$

Introducing for convenience the dimensionless variables:

$$t \equiv \ln\lambda \qquad x_j \equiv m_j/\nu, \qquad (\text{D.124})$$

one arrives at *the desired form of the RGE*:

$$\left\{-\frac{\partial}{\partial t} + \beta(\alpha_s)\alpha_s\frac{\partial}{\partial\alpha_s} - \sum_j(1 + \gamma_m(\alpha_s))x_j\frac{\partial}{\partial x_j} + \beta_G\frac{\partial}{\partial\alpha_G} + D - \gamma_\Gamma\right\}$$

$$\times \Gamma^R(e^t p_1, \ldots, e^t p_N; \ \alpha_s, \ \alpha_G, \ x_j, \ \nu) = 0, \qquad (\text{D.125})$$

with the solution:

$$\Gamma^R(e^t p_1, \ldots, e^t p_N; \ \alpha_s, \ \alpha_G, \ x_j, \ \nu)$$

$$= \lambda^D\Gamma^R(p_1, \ldots, p_N; \ \overline{\alpha}_s, \ \overline{\alpha}_G, \ \overline{x}_j, \ t = 0)\exp\left\{-\int_0^t dt'\gamma_\Gamma\left[\overline{\alpha}_s(t', \alpha_s)\right]\right\}.$$

$$(\text{D.126})$$

where $\overline{\alpha}_s$, $\overline{\alpha}_G$ and $\overline{x}_j$ are respectively the *running QCD coupling, gauge and mass*, solutions of the differential equations:

$$\frac{d\overline{\alpha}_s}{dt} = \overline{\alpha}_s\beta(\overline{\alpha}_s) \quad : \quad \overline{\alpha}_s(0, \alpha_s) = \alpha_s^R(\nu),$$

$$\frac{d\alpha_G}{dt} = \beta_G(\overline{\alpha}_s) \quad : \quad \overline{\alpha}_G(0, \alpha_s) = \alpha_G(\nu), \qquad (\text{D.127})$$

and:

$$\frac{d\overline{x}_i}{dt} = -\Big[1 + \gamma_m\,(\overline{\alpha}_s)\Big]\overline{x}_i(t) \quad : \quad \overline{x}_i(0,\alpha_s) = x_i^R(\nu). \tag{D.128}$$

♡ Running Coupling and Quark Masses

Solving the differential equation in Eq. (D.127), the expression of the *running coupling*, to one-loop accuracy is:

$$a_s^{(0)}(t,\alpha_s) = \frac{a_s(\nu)}{1 - \beta_1 a_s(\nu)t} \equiv \frac{1}{-\beta_1}\frac{1}{\frac{1}{2}\ln\left(-q^2/\Lambda^2\right)}, \tag{D.129}$$

where:

$$a_s \equiv \frac{\alpha_s}{\pi}\,, \qquad t \equiv \frac{1}{2}\ln\frac{-q^2}{\nu^2}. \tag{D.130}$$

and $\Lambda \simeq 350$ MeV for the three light quark flavors is the QCD scale which one measures from experiments. In the same way, we can also deduce the *running quark mass* to lowest order from Eq. (D.128):

$$\overline{m}_q(\nu) = \hat{m}_q\,(-\beta_1 a_s(\nu))^{-\gamma_1/\beta_1}\,, \tag{D.131}$$

where the *renormalization group invariant quark mass* $\hat{m}_q$ has been introduced in the similar way as Λ. Values of these running masses, at a given value of the scale $\nu=2$ GeV for the light quarks u, d, s and at the value of the running mass for the heavy quarks, can be found in Table 6.1.

Appendix E

Kits for Feynman Diagrams Calculus

Feynman has intuitively illustrated by diagrams the complex mathematical path integral when one does a perturbative expansion of the action S. We give as follows the Feynman rules for QCD, QED and the Standard Model. These rules together with some other properties (Dirac matrices, $SU(n)$ Lie algebra, momentum integrals,...) are useful for evaluating some amplitudes or some Green's functions entering in the expression of cross-sections and decay rates.[1]

E.1 Feynman Rules for QCD

The following rules for QCD come from the books in [33] and [31].

♣ Factors Induced by Closed Loops

$$\int \frac{d^n p}{(2\pi)^n} \qquad \text{for each loop integration}$$

$$(-1) \qquad \text{for each closed fermion loop}$$

---

[1]The following discussions come from the books and reviews in Ref. [30–32].

Figure E.1: Richard Feynman.

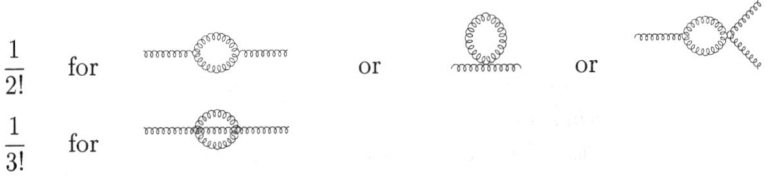

$$\frac{1}{2!} \quad \text{for} \quad \text{(diagram)} \quad \text{or} \quad \text{(diagram)} \quad \text{or} \quad \text{(diagram)}$$

$$\frac{1}{3!} \quad \text{for} \quad \text{(diagram)}$$

◇ Factors Induced by External or Internal Lines

ingoing quark: $\xrightarrow{\quad p \quad}\bullet$ $(2\pi)^{-3/2} u(p,\lambda)$

ingoing antiquark: $\xrightarrow{p}\ \xleftarrow{\ -p\ }\bullet$ $(2\pi)^{-3/2} \bar{v}(p,\lambda)$

outgoing quark: $\bullet\xrightarrow{\quad p \quad}$ $(2\pi)^{-3/2} \bar{u}(p,\lambda)$

outgoing antiquark: $\bullet\xrightarrow{p}\ \xleftarrow{\ -p\ }$ $(2\pi)^{-3/2} v(p,\lambda)$

ingoing gluon: $\xrightarrow{\quad k \quad}\bullet$ $(2\pi)^{-3/2} \epsilon^{\mu}(k,\eta)$

outgoing gluon: $\bullet\xrightarrow{\quad k \quad}$ $(2\pi)^{-3/2} \epsilon_{\mu}^{*}(k,\eta)$

♡ Propagators

quark:

$$\frac{i}{\hat{p}-m+i\epsilon'}\delta_{ij}$$

gluon:

$$(-i)\frac{\delta_{ab}}{k^2+i\epsilon'}\left[g_{\mu\nu}-(1-\alpha_G)\frac{k_\mu k_\nu}{k^2}\right]$$

ghost:

$$(-i)\frac{\delta_{ab}}{k^2+i\epsilon'}$$

♠ Vertices

quark–gluon–quark:

$$(ig)\gamma_\mu T_{ij}^a$$

ghost–gluon–ghost:

$$(-ig)f^{abc}p_\mu$$

3-gluon:

$$(gf^{abc})\left[g_{\mu\nu}(k+q)_\rho - g_{\nu\rho}(q+r)_\mu + g_{\rho\mu}(r-k)_\nu\right]$$

4-gluon:

$$(-ig^2)\left[f^{abe}f^{cde}\left(g^{\mu\sigma}g^{\nu\rho}-g^{\mu\rho}g^{\nu\sigma}\right)+f^{ace}f^{bde}\times\right.$$
$$\left.\left(g^{\mu\nu}g^{\sigma\rho}-g^{\mu\rho}g^{\nu\sigma}\right)+f^{ade}f^{cbe}\left(g^{\mu\sigma}g^{\nu\rho}-g^{\mu\nu}g^{\sigma\rho}\right)\right]$$

E.2   Feynman Rules for QED

The Feynman rules for QED can be deduced from the ones of QCD by replacing the wavy gluon line with a photon line. On the vertices, the QCD coupling g will be replaced by the electric charge $(-e)$ and by taking 1 instead of the color matrix T_{ij}^a. In QED, 3 and 4-photon couplings do not exist as well as a ghost particle because of the $U(1)$ abelian character of the theory.

E.3   Feynman Rules for the Standard Model

The following rules for the Standard Model come from the review article in [32].

♣ Propagators

$W^{\pm}$ boson:

$$(-i)\frac{\delta_{ab}}{k^2 - M_W^2 + i\epsilon}\left\{g^{\mu\nu} - (1-\alpha_G)\frac{k^\mu k^\nu}{k^2 - \alpha_G M_W^2}\right\}$$

Z^0 boson:

$$(-i)\frac{\delta_{ab}}{k^2 - M_Z^2 + i\epsilon}\left\{g^{\mu\nu} - (1-\alpha_G)\frac{k^\mu k^\nu}{k^2 - \alpha_G M_Z^2}\right\}$$

lepton–photon–lepton:

$$(-ie)\bar{\ell}\gamma^\mu\ell$$

scalar–photon–scalar:

$$(-ie)(k + k')_\mu$$

W–photon–W:

$$(+e)[g^{\nu\rho}(2k+q)^\mu - g^{\rho\mu}(k-q)^\nu - g^{\mu\nu}(2q+k)^\rho]$$

scalar–photon–W:

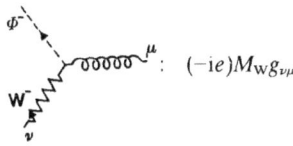

$\qquad : \quad (-\mathrm{i}e)M_{\mathrm{W}}g_{\nu\mu}$

neutrino–lepton–W:

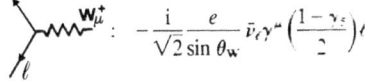

$\qquad : \quad -\dfrac{\mathrm{i}}{\sqrt{2}}\dfrac{e}{\sin\theta_{\mathrm{w}}}\,\bar{\nu}_{\ell}\gamma^{\mu}\left(\dfrac{1-\gamma_{5}}{2}\right)\ell$

◇ Vertex

neutrino–lepton–scalar:

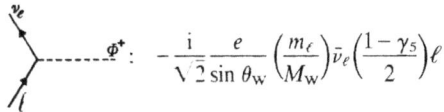

$\qquad : \quad -\dfrac{\mathrm{i}}{\sqrt{2}}\dfrac{e}{\sin\theta_{\mathrm{w}}}\left(\dfrac{m_{\ell}}{M_{\mathrm{W}}}\right)\bar{\nu}_{\ell}\left(\dfrac{1-\gamma_{5}}{2}\right)\ell$

lepton–lepton–Z:

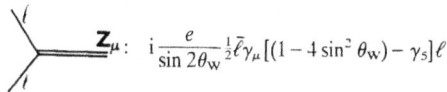

$\qquad : \quad \mathrm{i}\dfrac{e}{\sin 2\theta_{\mathrm{w}}}\tfrac{1}{2}\bar{\ell}\gamma_{\mu}[(1-4\sin^{2}\theta_{\mathrm{w}})-\gamma_{5}]\ell$

neutrino–neutrino–Z:

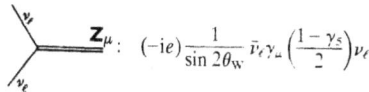

$\qquad : \quad (-\mathrm{i}e)\dfrac{1}{\sin 2\theta_{\mathrm{w}}}\,\bar{\nu}_{\ell}\gamma_{\mu}\left(\dfrac{1-\gamma_{5}}{2}\right)\nu_{\ell}$

quark–quark–photon:

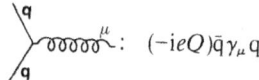

$\qquad : \quad (-\mathrm{i}eQ)\bar{q}\gamma_{\mu}q$

quark–quark–Z:

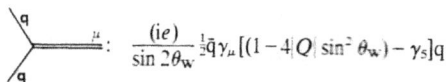

$\qquad : \quad \dfrac{(\mathrm{i}e)}{\sin 2\theta_{\mathrm{w}}}\tfrac{1}{2}\bar{q}\gamma_{\mu}[(1-4|Q|\sin^{2}\theta_{\mathrm{w}})-\gamma_{5}]q$

where Q is the quark charge in units of e.

E.4 Dirac Matrices

♣ Definitions and Properties

These are 4×4 matrices (tables with 4 lines and 4 columns)[2] with the properties:

$$\gamma_{0}^{2}=1\ ,\quad \gamma_{1}^{2}=\gamma_{2}^{2}=\gamma_{3}^{2}=-1,$$

$$\sigma_{\mu\nu}=\frac{\mathrm{i}}{2}[\gamma_{\mu},\gamma_{\nu}]\equiv\frac{\mathrm{i}}{2}(\gamma_{\mu}\gamma_{\nu}-\gamma_{\nu}\gamma_{\mu}),$$

$$\{\gamma_{\mu},\gamma_{\nu}\}\equiv\gamma_{\mu}\gamma_{\nu}+\gamma_{\nu}\gamma_{\mu}=2g_{\mu\nu}, \tag{E.1}$$

---

[2]For more complete discussions, see Refs. [33–35].

where $g_{\mu\nu}$ is the metric tensor with the signature $(1, -1, -1, -1)$ in 4-dimension:

$$g_{\mu\nu} = \begin{pmatrix} 1 & 0 & 0 & 0 \\ 0 & -1 & 0 & 0 \\ 0 & 0 & -1 & 0 \\ 0 & 0 & 0 & -1 \end{pmatrix}. \tag{E.2}$$

One also introduces the γ_5 matrix which quantifies the chirality of the field. Left handed (L) and right handed (R) fields are defined as:

$$\psi_L = \frac{1}{2}(1 - \gamma_5)\psi, \qquad \psi_R = \frac{1}{2}(1 + \gamma_5)\psi. \tag{E.3}$$

They are called *Weyl spinors*. The γ_5 *matrix* is defined as:

$$\gamma_5 \equiv i\gamma_0\gamma_1\gamma_2\gamma_3 \quad \text{or} \quad \gamma_5 \equiv \frac{1}{4!}\epsilon_{\mu\nu\rho\sigma}\gamma^\mu\gamma^\nu\gamma^\rho\gamma^\sigma, \tag{E.4}$$

with the properties:

$$\gamma_5^2 = 1, \qquad \gamma_\mu\gamma_5 = -\gamma_5\gamma_\mu, \qquad [\gamma_5, \sigma_{\mu\nu}] = 0. \tag{E.5}$$

We also have the *hermitian properties*:

$$\gamma_\mu = -\gamma_\mu^\dagger = -\gamma^0\gamma_\mu\gamma^0, \qquad \gamma_0^\dagger = \gamma_0, \qquad \gamma_5 = \gamma_5^\dagger = -\gamma^0\gamma_5\gamma^0, \tag{E.6}$$

for $\mu = 1, 2, 3$, where † means hermitian conjugate. We also have the *charge conjugate properties*:

$$C\gamma_\mu C^{-1} = -\gamma_\mu^T \qquad\qquad C\gamma_5 C^{-1} = \gamma_5^T,$$
$$C\sigma_{\mu\nu}C^{-1} = -\sigma_{\mu\nu}^T \qquad C(\gamma_5\gamma_\mu)C^{-1} = (\gamma_5\gamma_\mu)^T, \tag{E.7}$$

where C is the charge conjugate operator normalized as: $C^2 = -1$ and T means transposed matrix.

◇ Dirac Algebra in n Dimensions

The (anti)-commutation properties of the Dirac matrices in 4-dimensions given in Eq. (E.1) are maintained, but the algebra becomes:

$$\gamma_\mu\gamma^\mu = n\mathbf{1} = g_{\mu\nu}g^{\mu\nu},$$
$$\gamma_\mu\gamma_\alpha\gamma^\mu = (2 - n)\gamma_\alpha,$$
$$\gamma_\mu\gamma_\alpha\gamma_\beta\gamma^\mu = 4g_{\alpha\beta}\mathbf{1} + (n - 4)\gamma_\alpha\gamma_\beta,$$
$$\gamma_\mu\gamma_\alpha\gamma_\beta\gamma_\gamma\gamma^\mu = -2\gamma_\gamma\gamma_\beta\gamma_\alpha - (n - 4)\gamma_\alpha\gamma_\beta\gamma_\gamma. \tag{E.8}$$

♡ Traces of Dirac Matrices

The traces of Dirac matrices in n-dimension can be choosen to be the same as in 4-dimension. The usual properties are:

$$\text{Tr } 1 = 4,$$

$$\text{Tr } \gamma_\mu \gamma_\nu = 4 g_{\mu\nu},$$

$$\text{Tr } \gamma_\mu \gamma_\nu \gamma_\rho \gamma_\sigma = 4 \left(g_{\mu\nu} g_{\rho\sigma} - g_{\mu\rho} g_{\nu\sigma} + g_{\mu\sigma} g_{\nu\rho} \right),$$

$$\cdots,$$

$$\text{Tr } \gamma_{\mu^1} \cdots \gamma_{\mu^m} = 0 \text{ for } m \text{ odd}, \tag{E.9}$$

while:

$$\text{Tr } \gamma_5 = 0,$$

$$\text{Tr } \gamma_5 \gamma_\mu \gamma_\nu = 0,$$

$$\text{Tr } \gamma_5 \gamma_\mu \gamma_\nu \gamma_\rho \gamma_\sigma = 4i\epsilon_{\mu\nu\rho\sigma},$$

$$\cdots,$$

$$\text{Tr } \gamma_5 \gamma_{\mu^1} \cdots \gamma_{\mu^m} = 0 \text{ for } m \text{ odd}, \tag{E.10}$$

where $\epsilon_{\mu\nu\rho\sigma}$ is the *completely antisymmetric tensor*.

E.5 Totally Anti-Symmetric Tensor

It has the same definition in 4- and n-dimension:

$$\epsilon_{\mu\nu\rho\sigma} = \begin{cases} 0, & \text{if two indices are equal} \\ -1, & \text{if } \mu\nu\rho\sigma = 0123 \\ +1, & \text{if } \mu\nu\rho\sigma = 1230, \end{cases} \tag{E.11}$$

while, in n-dimension, one can choose its properties as:

$$\epsilon_{\mu\nu\alpha\beta}\epsilon^{\rho\nu\alpha\beta} = -(n-3)(n-2)(n-1)g_\mu^\rho,$$

$$\epsilon_{\mu\nu\alpha\beta}\epsilon^{\rho\sigma\alpha\beta} = -(n-3)(n-2)\left(g_\mu^\rho g_\nu^\sigma - g_\nu^\rho g_\mu^\sigma \right),$$

$$\epsilon_{\mu\nu\alpha\beta}\epsilon^{\rho\sigma\tau\beta} = -(n-3) \begin{vmatrix} g_\mu^\rho & g_\mu^\sigma & g_\mu^\tau \\ g_\nu^\rho & g_\nu^\sigma & g_\nu^\tau \\ g_\alpha^\rho & g_\alpha^\sigma & g_\alpha^\tau \end{vmatrix}. \tag{E.12}$$

E.6 CPT Transformations

The action of the operators: C ≡ *charge conjugation*, P ≡ *parity transformation*, T ≡ *time reversal*, on the fermion field $\psi(t, \vec{r})$ are:

$$\text{C } \psi(t, \vec{r}) = \gamma_2 \psi^\dagger(t, \vec{r})$$

$$\text{T } \psi(t,\vec{r}) = -i\gamma_1\gamma_3\psi^\dagger(-t,\vec{r})$$
$$\text{PT } \psi(t,\vec{r}) = \gamma_0\gamma_1\gamma_3\psi^\dagger(-t,-\vec{r})$$
$$\text{CPT } \psi(t,\vec{r}) = \gamma_2\gamma_0\gamma_1\gamma_3\psi(-t,-\vec{r}) = i\gamma_5\psi(-t,-\vec{r}), \qquad \text{(E.13)}$$

where: $\psi^\dagger = \bar{\psi}\gamma_0$.

E.7 Polarizations

♣ Unpolarized Cross-Section

In the evaluation of an unpolarized cross-section, one has to sum over polarizations. In the case of fermions, one has:

$$\sum_\lambda u(p,\lambda)\bar{u}(p,\lambda) = \hat{p} + m, \qquad \sum_\lambda v(p,\lambda)\bar{v}(p,\lambda) = \hat{p} - m. \qquad \text{(E.14)}$$

In the case of photon (massless vector boson), the polarization is only transverse:

$$\epsilon^\mu = (0, \vec{\epsilon}) \quad \text{with} \quad \vec{p}\cdot\vec{\epsilon} = 0. \qquad \text{(E.15)}$$

Therefore, for an unpolarized cross-section involving (massless) photons, one has to sum over the photon polarizations:

$$\sum_{polar.} \epsilon^*_\mu \epsilon^\mu = -\left(g_{\mu\nu} - \frac{q^\mu q^\nu}{q^2}\right), \qquad \text{(E.16)}$$

where q is the photon four-momentum.

◇ Polarized Cross-Section

For polarized cross-section, one inserts the projection matrices:

$$u\left(p,\lambda = \pm\frac{1}{2}\right)\bar{u}\left(p,\lambda = \pm\frac{1}{2}\right) = \frac{1}{2}(\hat{p} + m)\left(\frac{1 \pm \gamma_5\hat{s}}{2}\right),$$

$$v\left(p,\lambda = \pm\frac{1}{2}\right)\bar{v}\left(p,\lambda = \pm\frac{1}{2}\right) = \frac{1}{2}(\hat{p} - m)\left(\frac{1 \pm \gamma_5\hat{s}}{2}\right), \qquad \text{(E.17)}$$

where: s is the polarization four-vector of the (anti-)particle with energy-momentum p:

$$s\cdot p = 0 \quad \text{and} \quad s^2 = -1. \qquad \text{(E.18)}$$

E.8 Feynman Momentum Integrals

The Feynman parametrization is needed to recombine the product of denominators appearing in the momentum integral. We shall discuss the most usual ways of parametrization.

♣ Schwinger Representation

The first one consists of an exponentiation of the propagator denominators and leads to:

$$\frac{1}{a_1 \ldots a_n} = \int_0^\infty dz_1 \ldots \int_0^\infty dz_n \exp\left(-\sum_{i=1}^n a_i z_i\right). \tag{E.19}$$

In connection to this parametrization, the following Gaussian integral is useful:

$$\int \frac{d^n k}{(2\pi)^n} e^{-\alpha k^2} = \frac{1}{(4\pi\alpha)^{n/2}}. \tag{E.20}$$

◇ Original Feynman Parametrization

The second alternative is obtained from the original Feynman parametrization:

$$\frac{1}{a_1 \ldots a_n} = (n-1)! \int_0^1 dz_1 \ldots \int_0^1 dz_n \; \delta\left(1 - \sum_i z_i\right) \frac{1}{\left(\sum_i a_i z_i\right)^n}. \tag{E.21}$$

After a suitable change of variables, one can eliminate the δ-function and one obtains:

$$\frac{1}{a_1 \ldots a_n} = (n-1)! \int_0^1 u_1^{n-2} du_1 \int_0^1 u_2^{n-3} du_2 \ldots \int_0^1 du_{n-1}$$

$$\times \left[(a_1 - a_2)u_1 \ldots u_{n-1} + (a_2 - a_3)u_1 \ldots u_{n-2} + \ldots + a_n\right]^{-n}. \tag{E.22}$$

This parametrization is quite convenient as it allows possible cancellations among terms of two propagators, and has the advantage to provide finite bounds of integration, which is convenient in various numerical integration calculations encountered, e.g. in QED calculations ($g - 2, \ldots$). Particularly useful cases of Eq. (E.22) are:

$$\frac{1}{a^\alpha b^\beta} = \frac{\Gamma(\alpha + \beta)}{\Gamma(\alpha)\Gamma(\beta)} \int_0^1 dx \frac{x^{\alpha-1}(1-x)^{\beta-1}}{\left[(a - b)x + b\right]^{\alpha+\beta}}, \tag{E.23}$$

and:

$$\frac{1}{a^n b^m c^r} = \frac{\Gamma(n+m+r)}{\Gamma(n)\Gamma(m)\Gamma(r)} \int_0^1 dx\, x^{m+n-1}(1-x)^{r-1}$$

$$\times \int_0^1 dy \frac{(1-y)^{m-1} y^{n-1}}{\left[(a-b)xy + (b-c)x + c\right]^{m+n+r}}, \quad \text{(E.24)}$$

entering in a one-loop calculation respectively for a two-point and three-point functions.

E.9 Dimensional Regularization

Dimensional regularization instead of Pauli–Villars regularization (PVR) (successful for QED) is used for non-abelian gauge theories as it preserves gauge invariance. In the following, we shall evaluate momentum integrals in $n = 4 - \epsilon$-dimensions where $\epsilon \to 0$ at the end of calculations. In this way, ultra-violet and/or infrared divergences encountered in the calculation are transformed into $1/\epsilon^p$ poles which replace the cut-off introduced in QED for controlling infinities. After a suitable choice of integration variable and the previous Feynmann parametrization of the propagators, the momentum integrals take the generic form:

$$I(m,r) \equiv \int \frac{d^n k}{(2\pi)^n} \frac{(k^2)^r}{\left[k^2 - R^2\right]^m}$$

$$= \frac{i(-1)^{r-m}}{(16\pi^2)^{n/4}} (\mathbf{R}^2)^{r-m+n/2} \frac{\Gamma(r+n/2)\Gamma(m-r-n/2)}{\Gamma(n/2)\Gamma(m)}.$$

$$\text{(E.25)}$$

Using the symmetry of the integration, it is easy to show that:

$$\int \frac{d^n k}{(2\pi)^n} \frac{k_\mu k_\nu}{\left[k^2 - R^2\right]^m} = \frac{1}{n} g_{\mu\nu} \int \frac{d^n k}{(2\pi)^n} \frac{k^2}{\left[k^2 - \mathbf{R}^2\right]^m}. \quad \text{(E.26)}$$

In the case where r is odd:

$$\int \frac{d^n k}{(2\pi)^n} \frac{k_{\mu_1} \dots k_{\mu_r}}{\left[k^2 - \mathbf{R}^2\right]^m} = 0. \quad \text{(E.27)}$$

Finally, it is important to notice that *tadpole type integral vanishes identically in dimensional regularization*:

$$\int \frac{d^n k}{(2\pi)^n} (k^2)^{\beta-1} = 0 \quad \text{for} \quad \beta = 0, 1, 2, \dots \quad \text{(E.28)}$$

More momentum integrals are tabulated in Refs. [33–35] and will not be repeated here.

E.10 The Γ and Beta Function $B(x, y)$

These functions appear during the evaluation of momentum integrals and it is useful to know their properties.

♣ The Γ Function

It is defined for complex z by the *Euler integral*:

$$\Gamma(z) = \int_0^\infty dt \ t^{z-1} e^{-t}, \tag{E.29}$$

If the previous integral does not exist, it can be defined, using an analytic continuation, by:

$$\Gamma(z) = \sum_{n=0}^\infty \frac{(-1)^n}{n!(z+n)} + \int_1^\infty dt \ t^{z-1} e^{-t}, \tag{E.30}$$

which expresses that $\Gamma(z)$ is analytic in the entire z-plane but contains simple poles at $z = 0, -1, -2, \ldots$. It has the properties:

$$\Gamma(n) = (n-1)! \,, \tag{E.31}$$

for integer n and:

$$\Gamma(1+z) = z\Gamma(z)$$

$$= \exp\left\{ -z\gamma_E + \sum_{n=2}^\infty (-1)^n \frac{z^n}{n} \zeta(n) \right\}, \tag{E.32}$$

with:

$$\gamma_E \equiv \gamma = \lim_{n \to \infty} \left\{ 1 + \frac{1}{2} + \cdots + \frac{1}{n} \right\} = 0.5772156649\ldots \tag{E.33}$$

is the *Euler constant* and:

$$\zeta(n) = \sum_{k=1}^\infty \frac{1}{k^n}, \tag{E.34}$$

is the *Riemann function*. For $\epsilon \to 0$, it has the expansion:

$$\lim_{\epsilon \to 0} \Gamma(1+\epsilon) = 1 - \epsilon\gamma_E + \frac{\epsilon^2}{2} \left(\gamma_E^2 + \zeta(2) \right) - \frac{\epsilon^3}{3} \left(\frac{\gamma_E^3}{2} + \frac{\pi^2}{4}\gamma_E + \zeta(3) \right) + \mathcal{O}(\epsilon^4), \tag{E.35}$$

from which one can deduce $\Gamma(\epsilon)$ with the help of Eq. (E.32) which is very useful for dimensional regularization where the space–time is taken to be $n = 4 - \epsilon$; $\zeta(2) = \pi^2/6$, $\zeta(3) = 1.2020569031\ldots$ are the Riemann functions.

◇ The Beta Function $B(x, y)$

It is defined as:

$$B(x, y) = \int_0^1 dt \; t^{x-1}(1-t)^{y-1} = \frac{\Gamma(x)\Gamma(y)}{\Gamma(x+y)}, \qquad \text{(E.36)}$$

and has the useful properties:

$$B(x+1, y) = \left(\frac{x}{x+y}\right) B(x, y), \qquad B(x, 1+y) = \left(\frac{y}{x+y}\right) B(x, y).$$

$$\text{(E.37)}$$

Therefore, one can deduce:

$$B(1+az, 1+bz) = \frac{1}{1+(a+b)z} \exp\left\{ -z\gamma_E \right.$$

$$\left. + \sum_{n=2}^{\infty} (-1)^n \frac{z^n}{n} \zeta(n) \left[a^n + b^n - (a+b)^n \right] \right\}. \qquad \text{(E.38)}$$

In the limit $\epsilon \to 0$, it has the Taylor expansion:

$$\lim_{\epsilon \to 0} B\left(1 - a\epsilon, 1 - b\epsilon\right) = 1 - \epsilon(a+b) + \epsilon^2 \left[(a+b)^2 - ab\frac{\pi^2}{6} \right]$$

$$+ \epsilon^3 (a+b) \left[-(a+b)^2 + ab\zeta(2) + ab\zeta(3) \right] + \cdots,$$

$$\lim_{\epsilon \to 0} B\left(n - \frac{\epsilon}{2}, 1 - \frac{\epsilon}{2}\right) = \frac{1}{n} \left\{ 1 + \frac{\epsilon}{2} \left[\frac{2}{n} + \sum_{j=1}^{n-1} \frac{1}{j} \right] \right\} + \mathcal{O}\left(\epsilon^2\right),$$

$$\lim_{\epsilon \to 0} B\left(n - \frac{\epsilon}{2}, 2 - \frac{\epsilon}{2}\right) = \frac{1}{n(n+1)} \left\{ 1 - \frac{\epsilon}{2} \left[1 - \frac{2}{n} - \frac{2}{n+1} - \sum_{j=1}^{n-1} \frac{1}{j} \right] \right\}$$

$$+ \mathcal{O}\left(\epsilon^2\right), \qquad \text{(E.39)}$$

E.11 The Pseudoscalar Two-Point Correlator

In order to illustrate the previous discussions, let us consider the pseudoscalar correlator:

$$\Psi_5(q^2) \equiv i \int d^4x \; e^{iqx} \langle 0 | \mathcal{T} J_P(x) \left(J_P(0) \right)^\dagger | 0 \rangle, \qquad \text{(E.40)}$$

where:

$$J_P = (m_i + m_j) : \bar{\psi}_i (i\gamma_5) \psi_j :, \qquad \text{(E.41)}$$

is the light quark pseudoscalar current; m_i is the mass of the quark ψ_i. In order to simplify the discussion, we shall work to the lowest order of perturbative QCD and work with massless quarks in the fermion loop given in the following diagram

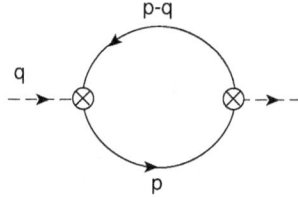

$$(E.42)$$

Using Feynman rules given previously, it reads:

$$i\Psi_5(q^2) = (m_i + m_j)^2(-1)N \int \frac{d^4p}{(2\pi)^4} \mathbf{Tr}\left\{(i\gamma_5)\frac{i}{\hat{p} + i\epsilon'}(i\gamma_5)\frac{i}{\hat{p} - \hat{q} + i\epsilon'}\right\},$$
$$(E.43)$$

where one can notice that for large k^2, one has a divergent integral:

$$I = \int \frac{d^4k}{k^2} = \infty. \qquad (E.44)$$

One can also use Pauli–Villars regularization (PVR), but it is more convenient to use dimensional regularization. In so doing, one works in $n \equiv 4 - \epsilon$ space–time dimensions, such that the previous expression becomes:

$$\nu^\epsilon \Psi_5(q^2) = (m_i + m_j)^2(-i)N \int \frac{d^np}{(2\pi)^n} Tr\left\{(i\gamma_5)\frac{1}{\hat{p} + i\epsilon'}(i\gamma_5)\frac{1}{\hat{p} - \hat{q} + i\epsilon'}\right\}.$$
$$(E.45)$$

The arbitrary scale ν has been introduced for dealing with dimensionless quantities in $4 - \epsilon$ dimensions. One can parametrize the quark propagators à la Feynman (see Eq. (E.23)):

$$\frac{1}{ab} = \int_0^1 \frac{dx}{\left[(a-b)x + b\right]^2} \equiv \int_0^1 \frac{dx}{\left[(p-l)^2 - \mathbf{R}^2\right]^2}, \qquad (E.46)$$

where:

$$a = (p-q)^2 + i\epsilon',$$
$$b = p^2 + i\epsilon',$$
$$l = qx,$$
$$\mathbf{R}^2 = -q^2x(1-x) - i\epsilon'. \qquad (E.47)$$

One uses the properties of the Dirac matrices in n-dimensions given previously:

$$Tr\ \hat{p}\gamma_5(\hat{p} - \hat{q})\gamma_5 = -4p(p - q),\qquad (E.48)$$

and does the shift:

$$p \to \tilde{p} + l.\qquad (E.49)$$

Therefore, one arrives at the momentum integration of the type:

$$\int \frac{d^n\ \tilde{p}}{(2\pi)^n}\ \frac{\tilde{p}^k}{\left[\tilde{p}^2 - \mathbf{R}^2\right]^2},\qquad (E.50)$$

which one can evaluate using the formula given in the previous section. It is easy to obtain:

$$\nu^\epsilon \Psi_5^B(q^2) = (m_i + m_j)^2 \frac{N}{4\pi^2} \int_0^1 dx\ \Gamma\left(\frac{\epsilon}{2}\right) \left(\frac{\mathbf{R}^2 - i\epsilon'}{\nu^2}\right)^{-\epsilon/2}$$

$$\times \left(3 + \frac{\epsilon}{2}\right) q^2 x(1 - x).\qquad (E.51)$$

where $\gamma_E = 0.5772\dots$ is the Euler constant. The loop ultraviolet divergence appears as a pole at $\epsilon = 0$ of the Γ-function:

$$\lim_{\epsilon \to 0} \Gamma\left(\frac{\epsilon}{2}\right) \simeq \frac{2}{\epsilon} + \ln 4\pi - \gamma_E + \mathcal{O}(\epsilon),\qquad (E.52)$$

which, as you may have noticed simplifies the calculation, which is remarkable when one does a higher order calculation. For large value of q^2, one then obtains the leading order result:

$$\nu^\epsilon \Psi_5^B\left(q^2\right) = (m_i + m_j)^2 q^2 \frac{N}{8\pi^2} \left\{\frac{2}{\epsilon} + \ln 4\pi - \gamma_E - \ln\left(\frac{-q^2}{\nu^2}.\right)\right\}.$$

$$(E.53)$$

More involved calculations can be done following the previous procedure used at lowest order for this simple example.

Part IX
Annex

Some Significant Dates
in Particle Physics

This list is not exhaustive but mentions works in direct relation with the subject we are dealing with. Due to the multitude of recent publications, only contributions deemed important by the author and his collaborators are cited.

- 610–547 BC: Thinking of Anaximander of Miletus (greek philosopher and founder of the Ionian school) on infinity and on the formation of the Universe.
- 383–322 BC: Aristotle (inventor of the word "Lycée" meaning high school) refutes the hypotheses of Anaximander of Miletus and based his theory on the movement (infinity is a circular motion), that the Universe is divided into sublunary (corruption) and supralunary, and on the hypothesis on the fall of bodies questioned by Galileo.
- 1633: R. Descartes assumes that the Universe is full of animated substances and swirls.
- 1686: I. Newton introduced the mechanics which contradicts the thought of Descartes.
- 1865: J.C. Maxwell introduced the law on the electromagnetic waves which unify electric and magnetic forces.
- 1869: Dimitri Mendeleev introduces the periodic table of chemical elements.
- 1897: J. Thompson discovers the electron.

- 1900: M. Planck (Nobel Prize of 1918) introduced non-relativistic quantum mechanics.
- 1902: A.H. Lorentz (Nobel Prize of 1902) introduces Lorentz transformation.
- 1905: A. Einstein (Nobel Prize of 1921) discovers the photoelectric effect from the particle nature of light (photon) and introduced special relativity.
- 1911: E. Rutherford discovers the nucleus.
- 1913: N. Bohr emits the notion of atoms.
- 1927: Heisenberg (Nobel Prize of 1932) discovers the uncertainty principle in quantum mechanics.
- 1927: P. Dirac introduced the equation of a system containing charge and electromagnetic fields. It is the beginning of the relativistic quantum field theory unifying quantum mechanics and relativity.
- 1930: W. Pauli assumes the existence of the neutrino to explain the electron energy spectrum in β radioactivity.
- 1930: E. Fermi proposed the theory of weak interactions in analogy to the electromagnetic interaction.
- 1932: Discovery of the anti-electron called positron e^+.
- 1937: Discovery of the muon μ (brother of the electron and 200 times heavier).
- 1930–1940: Development of the program of renormalization by S.-I. Tomonaga, J. Schwinger, and R.P. Feynman (Nobel Prize of 1965) to remove the infinities in the theoretical QED calculations.
- 1947: B. Pontecorvo suggests that the electron and the muon have the same charge by the weak interaction.
- 1947: Experimental evidence of the $\theta^+ \to \pi^+\pi^0$ with parity $P = 1$ and a mass about 500 MeV.
- 1949: Observation of the $\tau^+ \to \pi^+\pi^+\pi^-$ with parity $P = -1$ which have the same mass and lifetime as the θ particle: the $\theta-\tau$ puzzle.
- 1954: Lagrangian formulation by C.N. Yang and R. Mills.
- 1954: 12 European countries come together to create the European Centre for Nuclear Research (CERN) in Geneva.
- 1956: Proposal of tests by T.D. Lee and C.N. Yang (Nobel Prize of 1957) to verify the violation of parity. The θ and τ^+ are identified as the same particle: The kaon (K).
- 1956: Detection of the neutrino of the electron (ν_e) suggested by W. Pauli in a U.S. nuclear power station.
- 1957: Observation of the parity violation by C.S. Wu and R. Garwin, L. Lederman and M. Weinrich.

- 1962: Discovery of the neutrino of the muon (ν_μ) at Brookhaven (USA).
- 1964: Observation at Brookhaven of the different behaviors of the K^0 and its antiparticle $\bar{K}^0$ indicating a violation of parity.
- 1964: P. Higgs, F. Englert and R. Brout propose a mechanism to give masses to matter constituents and gauge bosons.
- 1964: M. Gell-Mann (Nobel Prize of 1969) and H. Fritzsch introduce the Quark Model and the notion of hadrons which are bound states of quark and antiquark (mesons) and three quarks (baryons).
- 1967: Observation at SLAC, Stanford that proton contains point-like basic grains "partons" which were subsequently identified from the jets of quarks.
- 1967: Presentation of the Standard Model (SM) by S.L. Glashow, A. Salam and S. Weinberg for unifying electromagnetic and weak interactions. The model predicts the existence of neutral and charged currents. It is a $SU(2)_L \otimes U(1)$ gauge theory.
- 1967: G. Veneziano presents the dual model to explain the hadronic spectrum. His model is the origin of string theory.
- 1967: S. Weinberg as well as T. Das, S. Gurlanik, V. Mathur and S. Okubo introduce the spectral sum rules for controlling chiral symmetry breakings.
- 1968: G. Charpak designs a multi-wire chamber detector: he received the Nobel Prize in 1992.
- 1968: Solar neutrino deficit. Sign of the neutrino oscillation?
- 1970: GIM mechanism proposed by S.L. Glashow, J. Illiopoulos and L. Maiani for explaining the $K_L - K_S$ mass difference.
- 1971: G. 't Hooft and M. Veltman show that one can make complete calculations in Standard Model like in QED. It is said that the Standard Model is renormalizable. They receive the Nobel Prize in 1999.
- 1972: M. Kobayashi and T. Maskawa show that CP violation is natural if there are three families of quarks, which allowed them to win the Nobel Prize in 2008 with Y. Nambu.
- 1973: Discovery of neutral currents in the Gargamelle experiment at CERN, which permitted L. Lederman, M. Schwartz, and J. Steinberger to get the Nobel Prize in 1988.
- 1973: Evidence by G. 't Hooft and D. Gross, D. Politzer and F. Wilczek (Nobel Prize of 2004) on the asymptotic freedom of the $SU(3)$ gauge theory of Quantum Chromodynamics (QCD) for explaining the strong interaction mediated by eight massless gluons. Quarks that have three different colors as well as gluons may be released at high energy.
- 1974: Discovery of the fourth quark: charm (c) by B. Richter and S. Ting (Nobel Prize of 1976).

• 1974: Discovery of the lepton heavy tau (τ) at SLAC by M. Perl, Nobel Prize, 1995 (shared with F. Reines on his discovery of neutrinos).

• 1975: H. Fritzsch and P. Minkowski make a systematic study of the bound states of gluons (glueballs or gluonia).

• 1977: Discovery at Fermilab (USA) by the group of L. Lederman of the Υ (upsilon) meson bound state of the fifth b quark and of its antiquark.

• 1979: Observation of the gluons jets at DESY, Hamburg.

• 1979: E. Floratos, S. Narison and E. de Rafael control the breaking of S. Weinberg sum rules (1967) by the masses of the quarks and propose a definition of these perturbative masses in QCD.

• 1979: Introduction of the semi-non-perturbative method for the spectral sum rules in QCD by M. Shifman, A. Vainshtein and V.I. Zakharov (Sakurai Prize of 1999).

• 1981: C. Becchi, S. Narison, E. of Rafael and F.J. Yndurain estimate for the first time the absolute values of "running masses" of light quarks u, d and s using the QCD spectral sum rules.

• 1983: Discovery of the $W^{\pm}$ and Z bosons by the UA1 and UA2 groups at CERN, Geneva which allowed C. Rubbia and Van der Meer to receive the Nobel Prize in 1984.

• 1987: S. Narison discovers that the symmetry of the heavy quarks is largely violated at the b-quark scale because of the large mass corrections for the D and B mesons decay constants. This result is confirmed later by lattice simulations and experiments.

• 1987: S. Narison estimates the running masses of the quarks c and b with QCD spectral sum rules.

• 1989: Start-up of LEP after 10 years of design and construction.

• 1989: S. Narison and G. Veneziano show that the scalar meson σ or $f^0(600)$ may contain a large component of gluons in its wave function which was confirmed later by R. Kaminski, G. Mennessier, S. Narison, W. Ochs and G. Wang (2010–2012) from the analysis of data from the $\pi\pi$ and $\gamma\gamma$ scattering experiments.

• 1990: Invention at CERN of the Web pages: World Wide Web (www) by T. Berners-Lee and R. Cailliau for instantaneously communicating the results of the experiment at CERN to other members of the international collaboration.

• 1990: E. Braaten, A. Pich and S. Narison extract, for the first time, at relatively low energy, the QCD coupling α_s from the semi-hadronic decay of the heavy lepton τ.

- 1992: Confirmation by GALLEX (Italy) and SAGE (Russia) of the deficit solar neutrinos observed in 1968.
- 1994: Discovery of the sixth quark top (t) at Fermilab.
- 1995: Start-up of the LEP200 program of 208 GEV total energy where a W^+W^- pair production threshold is crossed. The experiment is stopped in 2002 for the preparation of the LHC. The Higgs boson is still missing but a specific area for its mass is bounded between 114 and 200 GeV.
- 1995: First observation of an anti-hydrogen at CERN.
- 1998: Observation of the oscillation of the muon neutrino (ν_μ) in the KAMIOKANDE (Japan) experiment proving that it is massive.
- 1999: Starting the B-factory (BELLE) at KEK in Japan and BaBar at SLAC Stanford in USA for measuring CP violation and for studying the properties of the B (beautiful) mesons.
- 2000: Proof of the existence of the tau neutrino (ν_τ) at Fermilab.
- 2001: SNO (Canada) experiment shows that the sun produces the number of neutrinos expected but some parts of them are transformed into muon neutrino (ν_μ) and tau neutrino (ν_τ) by oscillations.
- 2001: Indication that the plasma of quarks and gluons would be produced in collisions of heavy ions at CERN.
- 2003: Discovery of the exotic X(3872) meson ($J^P = 1+$) by BELLE through its decay $J/\psi\pi^+\pi^-$ and confirmed later by different groups. A four-quark or meson–antimeson molecule candidate.
- 2005: Discovery of the exotic Y(4260) meson ($J^P = 1-$) by BaBar and confirmed later by different groups.
- 2008: Shutdown of the beauty factory BaBar of SLAC. At BELLE, KEK, Japan, they showed that the standard model CP violation is too weak to explain the asymmetry of matter and antimatter in the universe. Another track perhaps on the side of neutrinos.
- 2008: Start-up of the LHC at CERN.
- 2012: Closing of the $\bar{p}p$ Tevatron experiment of 1 TeV per beam at Fermilab.
- 2012: Discovery by ATLAS and CMS experiments at LHC, CERN (Geneva) of a boson having a mass of 125 GeV which looks like the Higgs boson of the Standard Model.
- 2013: Discovery of the Z_c(3900) exotic meson ($J^P = 1+$) by BELLE and BESIII through its $J/\psi\pi^+$ decay. A four-quark or meson–antimeson candidate.
- 2015: Discovery of the P_c(4380) and P_c(4450) by LHCb from $\Lambda_b \rightarrow J/\psi_p K^+$ decays. A pentaquark or baryon–meson molecule candidate.

Nobel Prize in Physics

Search for new particles and their interactions led to different Nobel Prizes in Physics

- 1901: W.C. Rontengen (Munich) for the discovery of X-rays.
- 1903: H.A. Becquerel (Paris), P. Curie and M. Curie for the discovery of radioactivity and its properties.
- 1906: J.J. Thomson (Cambridge) for the discovery of electron and its properties.
- 1921: A. Einstein (Berlin) for the discovery of the photoelectric effect (photon).
- 1923: R. Millikan (Pasadena) for the photoelectric and electron charge experiments.
- 1927: A.H. Compton (Chicago) for the discovery of the Compton effect, C.T.R. Wilson (Cambridge) for the visualization of charged particles trajectory in a fog chamber.
- 1929: L. de Broglie for his discovery of the wave nature of electrons.
- 1935: J. Chadwick (Liverpool) for the discovery of the neutron.
- 1936: V.F. Hess (Innsbruck) for the discovery of the cosmic rays, C.D. Anderson (Pasadena) for the discovery of the positron (anti-electron).
- 1939: E.O. Lawrence (Berkeley) for the invention of the cyclotron.
- 1943: Otto Stern (the award citation omitted mention of the SternGerlach experiment, as Gerlach had remained active in Nazi-led Germany).
- 1944: I.I. Rabi (Columbia, MIT) for the discovery of nuclear magnetic resonance to discern the magnetic moment and nuclear spin of atoms.

- 1948: P.S.M. Blackett (Manchester) for developing the fog chamber and for studying cosmic rays.
- 1949: H. Yukawa (Kyoto) predicts the existence of pions as a mediator of nuclear forces.
- 1950: C.F. Powell (Bristol) discovers the pion using the photograph method for studying nuclear forces.
- 1955: W. Lamb (Arizona, Oxford) for the discoveries concerning the fine structure of the hydrogen spectrum. P. Kusch (Columbia) for his accurate determination of the magnetic moment of the electron.
- 1957: T.D. Lee (New York) and C.N. Yang (Princeton) for the parity violation.
- 1958: P.A. Tcherenkov, I.M. Franck and I.E. Tamm (Moscow) for the discovery of the Tcherenkov effect.
- 1959: E. Segré, O. Chamberlain (Berkeley) for the discovery of the anti-proton.
- 1960: D.A. Glaser (Berkeley) for the invention of a bubble chamber.
- 1961: R. Hofstadter (Stanford) for the diffusion of electrons and for the study of the structure of the nucleons.
- 1962: L. Landau (Moscow) for his theories on condensed matter, especially for liquid helium.
- 1965: R.P. Feynman (Pasadena), J. Schwinger (Cambridge, Massachussets) and S. Tomonoga (Tokyo) for the renormalization program in quantum electrodynamics (QED).
- 1967: H. Bethe (Cornell) on the theory of stellar nucleosynthesis.
- 1968: L.W. Alvarez (Berkeley) for the discovery of hadronic resonances and for the development of bubble chambers.
- 1969: M. Gell-Mann (Pasadena) for the Quark Model (classification of particles).
- 1976: B. Richter (Stanford) and S.C. Ting (Cambridge, Massachussets) for the discovery of the J/ψ meson bound state of the charm–anticharm quarks.
- 1979: S.L. Glashow (Harvard), A. Salam (Trieste and London) and S. Weinberg (Harvard) for the electroweak Standard Model.
- 1980: J.W. Cronin (Chicago) and V.L. Fitch (Princeton) for the discovery of CP violation in the K-decays.
- 1984: C. Rubbia and S. Van der Meer (CERN, Geneva) for the discovery of the $W^{\pm}$ and Z^0 bosons.

- 1988: L. Lederman (Fermilab), M. Schwartz (Mountain View) and J. Steinberger (CERN, Geneva) for the discovery of neutral current and of the neutrino ν_μ.
- 1990: J.I. Friedman, H.W Kendall (MIT, Massachussets) and R.E Taylor (Stanford) for their pionniering work on deep inelastic scattering for quarks and parton models.
- 1992: G. Charpak (Paris and CERN, Geneva) for the invention of multi-wire detectors.
- 1995: M.L. Perl (Stanford) for the discovery of the τ lepton, F. Reines (University of California, Irvine) for the discovery of the neutrino ν_e.
- 1999: G. 't Hooft and M. Veltman (Utrecht) for the quantum structure of the electroweak interactions.
- 2004: D. Gross, D. Politzer and F. Wilczek for the discovery of asymptotic freedom in QCD.
- 2008: H. Nambu, T. Kobayashi and S. Maskawa for the discovery of the mechanism of spontaneous broken symmetry and CP-violation.
- 2013: F. Englert and P. Higgs for the so-called Higgs mechanism after the discovery of the Higgs boson of mass 125 GeV at CERN in 2012.

Part X
Bibliography and Index

Bibliography

[1] *Some Inspiring Popular and Introductory Books*

- Courier-CERN (CERN monthly) where I also took some pictures;
- Elementary: Biannual Journal Published by the CNRS's IN2P3.
- Some Wikipedia articles on the internet.
- Series of the Encyclopedia Universalis Books.
- M. Veltman, Facts and Mysteries in Elementary Particle Physics, World Scientific, Singapore (2003).
- H. Fritzsch, Quarks, the stuff of Matter, Penguin books Ltd., London (1983).
- E. Augé *et al.*, *Voyage au coeur de la matière*, Edition Belin-CNRS, France, 2002.
- M. Davier, *LHC: The Higgs Boson*, Edition Pommier, France (2013).
- G. Cohen-Tannoudji and M. Spiro, *The Boson and the Mexican Hat*, Edition Folio Essais Inedit, Gallimard, France (2013).
- J. Baggott, *The God Particle: The Discovery of the Higgs Boson*, Edition Dunod, France (2013).
- J.-C. Boudenot, Histoire de la physique des particules, de Thalès au boson de Higgs, 2nd edition, Edition Ellipses, France (2013).

[2] *Some Books in Quantum Mechanics*

- A. Messiah, *Quantum Mechanics*, Dunod, Paris (1960).
- Raoelina Andriambololona, *Mecanique Quantique*, Collection Lira — Madagascar (1990).

[3] *Some Classic Books in Quantum Field and Gauge Theories*

- N. Bogolioubov and D.V. Chirkov, *Introduction to Quantum Field Theory*, Dunod, Paris (1960).
- L. Landau and E. Lifchitz, *Relativistic Quantum Field Theory*, Mir Editions, Moscow (1972).
- J.D. Bjorken and S.D. Drell, *Relativistic Quantum Fields*, McGraw-Hill Book co, New York (1965).
- R.P. Feynmann, The Theory of Fundamental Processes, *Lecture Note Series, Frontiers in Physics* (1962).
- R.P. Feynmann, Quantum Electrodynamics, *Lecture Note Series, Frontiers in Physics* (1962).
- E. de Rafael, *Lectures on Quantum Electrodynamics*, Univ. Autonoma, Barcelona (1976).

[4] *Some Books in Particle Physics*

- V. De Alfaro, S. Fubini, G. Furlan and C. Rossetti, *Current in Hadron Physics*, Elsevier (1973).
- R.P. Feynmann, Photon–Hadron Interactions, *Lecture Note Series, Frontiers in Physics* (1972).
- S. Donnachie, H.G. Dosch, O. Nachtmann, P. Landshoff, *Camb. Monogr. Part. Phys. Nucl. Phys. Cosmol.* **19** (2002) 1.
- F.J. Ynduráin, *The Theory of Quark and Gluon Interactions*, 3rd edition, Springer-Verlag (1999).
- P. Pascual and R. Tarrach, QCD: Renormalization for the Practitioner, *Lecture Notes in Physics*, **194**, Springer-Verlag (1984).
- R.E. Marshak, Riazuddin and C.P. Ryan, *Theory of Weak Interactions in Particle Physics, Monographs and texts in physics and astronomy*, **24**, John Wiley and sons, New York (1969).

Some Reviews and Original Papers

[5] Review of Particle Physics, K.A. Olive *et al.*, (Particle Data Group: PDG), *Chin. Phys.* **C38** (9) (2014) 090001 and references therein (online at: http://pdg.lbl.gov/).

[6] E. of Rafael, arXiv: 1210.4705 [hep-ph];
A. Hoecker and W. Marciano in PDG[5].

[7] T. Blum *et al.*, arXiv: 1311.2198 [hep-ph].

[8] BaBar and BESIII collaborations, talks paresented respectively by P. Lukin and H. Hu at the 18th international conference QCD15-Montpellier, 29 June–3 July 2015 (30th anniversary).

[9] M. Davier *et al.*, arXiv: 0906.5443 (2009).

[10] G. Amelino-Camelia *et al.* (KLOE-2 Collaboration), *Euro. Phys. J.* **C68** (2010) 619.

[11] L. Lukaszuk and B. Nicolescu, *Nuovo Cim. Lett.* **8** (1973) 405.

[12] A. Pich, *Prog. Part. Nucl. Phys.* **75** (2014) 41–85; D. Boito *et al.*, *Phys. Rev.* **D85** (2012) 093015.

[13] S. Bethke in PDG [5].

[14] For a review, see e.g. J.M. Richard, *An introduction to the quark model*, arXiv: 1205.4326 [hep-ph].

[15] For a review, see e.g. M. Neubert, Heavy quark symmetry, *Phys. Rep.* **245** (1994) 249.

[16] J. Gasser and H. Leutwyler, *Ann. Phys.* **158** (1984) 142; *Nucl. Phys.* **B250** (1985) 465.

[17] For a review, see e.g.: E. de Rafael, *Chiral Lagrangians and Kaon CP-violation*, arXiv: hep-ph/9502254;
A. Pich, *Chiral Perturbation, Rept. Prog. Phys.* **58** (1995) 563.

[18] M.A. Shifman, A.I. Vainshtein and V.I. Zakharov, *Nucl. Phys.* **B147** (1979) 385, 448.

[19] B.L. Ioffe, *Nucl. Phys.* **B188** (1981) 317; B.L. Ioffe, **B191** (1981) 591; Y. Chung *et al.*, *Z. Phys.* **C25** (1984) 151; H.G. Dosch, Non-Perturbative Methods (Montpellier 1985); A.A. Ovchinnikov and A.A. Pivovarov, *Yad. Fiz.* **48** (1988) 1135.

[20] R. Tarrach, *Nucl. Phys.* **B183** (1981) 384.

[21] H. Fritzsch and P. Minkowski, *Nuov. Cim.* **30A** (1975) 393.

[22] G. Mennessier, *Z. Phys.* **C16** (1983) 241.

[23] J. Govaerts *et al.*, *Nucl. Phys.* **B258** (1985) 215; *Nucl. Phys.* **B262** (1985) 575;

[24] R.T. Kleiv *et al.*, *Nucl. Phys.* (Proc. Suppl.), 234 (2013) 150; *ibid. Nucl. Phys.* (Proc. Suppl.) **B234** (2013) 154.

[25] F.S. Navarra, M. Nielsen, S.H. Lee, *Phys. Rep.* **497** (2010) 4; S.L. Zhu, *Int. J. Mod. Phys.* **E17** (2008); R. Albuquerque, Montpellier–Sao Paulo thesis 2012.

[26] P. Kovtun, Dan T. Son and Andrei O. Starinets, *Phys. Rev. Lett.* **94** (2005) 111601.

[27] S.J. Brodsky, G.F. de Teramond and H.G. Dosch, [arXiv:1301.1651] (2013).

[28] P. Colangelo *et al.*, *Phys. Rev.* **D85** (2012) 035013; For a review of works prior 2009, see e.g.: F. Jugeau [hep-ph/0902.3864].

[29] O. Andreev, *Phys. Rev.* **D73** (2006) 107901; O. Andreev and V.I. Zakharov, *Phys. Rev.* **D76** (2007) 047705;

[30] J. Ellis, arXiv:1501.05418 [hep-ph] (2015).

[31] G. Aad *et al.*, [ATLAS Collaboration], *Phys. Rev.* **D90** (2014) 052004 [arXiv:1406.3827 [hep-ex]].

[32] V. Khachatryan *et al.*, [CMS Collaboration], arXiv:1412.8662 [hep-ex].

Author's Books and Some Review Articles

[33] S. Narison, *QCD as a theory of hadrons: From partons to confinement*, Cambridge University Monograph Series 2002.

[34] S. Narison, *QCD spectral sum rules, Lecture Series in Physics*, Vol. 26, World Scientific, Singapore (1987).

[35] S. Narison, Techniques of dimensional regularization, *Phys. Rept.* **84** (1982) 263; S. Narison, *Acta Phys. Pol.* **B26** (1995) 687; S. Narison, arXiv:1409.8148 (to appear in *Nucl. Phys. Proc. Suppl.*).

Some Selected Author's Original Contributions

[36] J. Calmet, M. Perrottet, S. Narison and E. de Rafael, *Rev. Mod. Phys.* **49** (1977) 21; S. Narison, *Phys. Lett.* **B513** (2001) 53; *Erratum-ibid.* **B526** (2002) 414.

[37] E. Braaten, S. Narison and A. Pich, *Nucl. Phys.* **B373** (1992) 581.

[38] S. Narison, *Phys. Lett.* **B673** (2009) 30.

[39] S. Narison, G.M. Shore, G. Veneziano, *Nucl. Phys.* **B546** (1999) 235. *Nucl. Phys.* **B391** (1993) 69–99; *Nucl. Phys.* **B433** (1995) 209–233.

[40] E.G. Floratos, S. Narison and E. de Rafael, *Nucl. Phys.* **B155** (1979) 155.

[41] K. Chetyrkin, S. Narison and V.I. Zakharov, *Nucl. Phys.* **B550**, 353 (1999); S. Narison and V.I. Zakharov, *Phys. Lett.* **B522** (2001) 266; S. Narison and V.I. Zakharov, *Phys. Lett.* **B679** (2009) 355.

[42] H.G. Dosch, M. Jamin and S. Narison, *Phys. Lett.* **B220** (1989) 251.

[43] C. Becchi, S. Narison, E. de Rafael and F.J. Yndurài
n, *Z. Phys.* **C8** (1981) 335; S. Narison and E. de Rafael, *Phys. Lett.* **B103** (1981) 57.

[44] S. Narison, *Phys.Rev.* **D74** (2006) 034013; S. Narison, *Phys. Lett.* **B738** (2014) 346.

[45] S. Narison, *Phys. Lett.* **B197** (1987) 405.

[46] S. Narison, *Phys. Lett.* **B693** (2010) 559; *Erratum ibid.* 705 (2011) 544; S. Narison, *Phys. Lett.* **B706** (2012) 412; S. Narison, *Phys. Lett.* **B707** (2012) 259.

[47] S. Narison, *Phys. Lett.* **B104** (1981) 485.

[48] R.M. Albuquerque, S. Narison, *Phys. Lett.* **B694** (2010) 217; R.M. Albuquerque, S. Narison and M. Nielsen, *Phys. Lett.* **B684** (2010) 236.

[49] S. Narison, *Phys. Lett.* **B626** (2005) 101; S. Narison, *Phys. Lett.* **B300** (1993) 293.

[50] S. Narison, *Phys. Lett.* **B605** (2005) 319.

[51] G. Launer, S. Narison and R. Tarrach, *Z. Phys.* **C26** (1984) 433.

[52] K. Hagiwara, S. Narison and D. Nomura, *Phys. Lett.* **B540** (2002) 233; S. Narison and A.A. Pivovarov, *Phys. Lett.* **B327** (1994) 341.

[53] S. Narison, *Phys. Lett.* **B198** (1987) 104.

[54] S. Narison, *Phys. Lett.* **B718** (2013) 1321; S. Narison, *Phys. Lett.* **B721** (2013) 269.

[55] S. Narison, *Int. J. Mod. Phys.* **A30** (2015) 1550116.

[56] S. Narison, *Phys. Lett.* **B605** (2005) 319.

[57] S. Narison and G. Veneziano, *Int. J. Mod. Phys.* **A4** (1988) 2751; S. Narison, *Nucl. Phys.* **B509** (1998) 312; S. Narison, *Phys. Rev.* **D73** (2006) 114024.

[58] G. Mennessier, S. Narison, X.-G. Wang, *Phys. Lett.* **B696** (2011) 40; *Phys.Lett.* **B688** (2010) 59; R. Kaminski, G. Mennessier, S. Narison, *Phys. Lett.* **B680** (2009) 148; G. Mennessier, S. Narison and W. Ochs, *Phys. Lett.* **B665** (2008) 205.

[59] J.I. Latorre, S. Narison and P. Pascual, *Z. Phys.* **C34** (1987) 347 and references therein; K.G. Cheyrkin and S. Narison, *Phys. Lett.* **B485** (2000) 145.

[60] S. Narison, *Phys. Lett.* **B675** (2009) 319.

[61] F. Fenosoa, S. Narison and A. Rabemananjara, *Nucl. Phys. Proc. Suppl.* **258–259** (2015) 152, *arXiv:* 1409.8591; *Nucl. Phys. Proc. Suppl.* **258–259** (2015) 156, *arXiv:* 1410.0137, and references therein.

[62] R.D. Matheus, S. Narison, M. Nielsen and J.M. Richard, *Phys. Rev.* **D75** (2007) 014005; S. Narison, F.S. Navarra and M. Nielsen, *Phys. Rev.* **D83** (2011) 016004; R.D. Matheus and S. Narison, *Nucl. Phys. Proc. Suppl.* **152** (2006) 236.

[63] F. Jugeau, S. Narison and H. Ratsimbarison, *Phys. Lett.* **B722** (2013) 111.

Index

Part XI
About the Author

Stephan Narison

Directeur de Recherche 1ère classe au Laboratoire Univers & Particules de Montpellier
Founder of the QCD, Montpellier, FR and HEPMAD, Antananarivo, MG Conferences
Founder of the Research Institute iHEPMAD (Antananarivo, Madagascar)
President of the Association for Developing Science (AGMM, Madagascar)
Associated Permanent Member of the "Académie des Sciences Malgache"
Grand Officer of the "Ordre National Malgache"

- Born on 17 September, 1951 at Antsakabary, Majunga (Madagascar)
- Married, 2 children, 1 grandchild
- Nationalities: Malagasy and French
- Emails: snarison at yahoo.fr/snarison at gmail.com
- Phone/Fax: (00-33) 4 67 14 35 68
- Mobile phone: (00-33) 6 08 07 86 91
- Conference and personal site: http://www.lupm.univ-montp2.fr/users/qcd/

Summary

♣ Stephan Narison is, presently, Director of Research 1st class at the Laboratory Universe and Particles of Montpellier (LUPM) attached to the CNRS-IN2P3 and the University of Montpellier 2, France.[1]

---

[1]A parallel presentation can be found in the online wikipedia site *http://en.wikipedia.org/wiki/Stephan_Narison* and his personal site.

◇ He is a theoretical high-energy physicist specialized in the non-perturbative spectral sum rules approach of quantum chromodynamics (QCD), which has been introduced by M.A. Shifman, A.I. Vainshtein and V.I. Zakharov (SVZ) [18] in 1979 for studying the hadron properties (masses, decay constants and widths) in terms of the QCD fundamental parameters (coupling constant, quark masses, quark and gluon condensates) via some improved forms of the well-known dispersion relations.

♡ He completed his primary, secondary and university studies in Madagascar and prepared his "Doctorat 3ème cycle" and "Doctorat d'Etat" at the University of Marseille, Luminy (France).

♠ He has been invited by different world laboratories and has participated in various international workshops and conferences during his career as a physicist.

1. Biography

♣ Stephan Narison is a Malagasy born on 17 September, 1951 (Antsakabary, Majunga, Madagascar). He has both malagasy and french citizenship since 1980. He is the oldest among three brothers. He is married to Lydia and has two sons Larry and Rindra. The older son is an engineer in industrial mechanics and the younger is a neuro-psychologist. Stephan's father was a teacher in various primary schools and the family moved to different villages in Madagascar where Stephan came in contact with villagers and learnt about their ways of living and the fundamental Malagasy traditions.

◇ After 3 years (normally 5 years) in primary school, he succeeded to pass a competitive exam in 1962, enter to the best high school of Madagascar, the Lycee Gallieni of Antananarivo belonging at that time to the academy of Aix-Marseille in France. In 1969, he entered the University of Madagascar and after his Master's degree in 1973, he became a teacher in different high schools in Antananarivo.

♡ In 1974, he got a fellowship from the European commission in Brussels to study engineering. Based on his qualifications he was admitted to the "grandes écoles" Supelec and Ecole Centrale in Paris. However, his quantum mechanics teacher had convinced him to continue in theoretical physics where he appreciated the researcher freedom and the nobility of science. Transforming engineering into a research fellowship, he was the admitted to the famous "Diplome d'etudes approfondies (DEA)" of Brossel in Paris but he had chosen to go to the Center of Theoretical Physics (CPT) of Marseille (FR) (where his quantum mechanics teacher has studied) to pursue his PhD in theoretical High-Energy Physics (HEP) under the supervision of Prof. Eduardo de Rafael.

2. Careers

During the preparation of his Doctorate thesis, Stephan was a teacher at the Universities of Marseille, Provence and Marseille, Luminy from 1977–1979 and was offered by Prof. Abdus Salam a post-doc position at the International Center for Theoretical Physics (ICTP) in Trieste (IT) from 1979–1981. Later on and jobless, he was invited as a visiting researcher at LAPP, Annecy in 1981–1982 by Prof. M.K. Gaillard. He obtained a fellowship at the theory division of CERN, Geneva (CH) from 1982–1984 and a permanent position at the Laboratory of Physics and Mathematics (LPM) of Montpellier (FR) in October 1984. There, he was promoted as the Directeur de Recherche, 2nd class in 1995 and 1st class in 2011.

3. Visits and Collaborations

♣ Stephan has been invited as an appointed visitor for collaborations in different HEP laboratories: SLAC, Stanford (USA) in 1986; *France–Spain cooperation*: University Diagonal and Autonoma, Barcelona in 1986–1991; CERN, Geneva (Switzerland) scientific associate in 1984–1996; *France–Germany cooperation:* Institute of Theoretical Physics, Heidelberg as a DFG visitor in 1986 and Von-Humboldt fellow in 1992–1993; *France–Austria cooperation:* Institute of Theoretical Physics, Vienna in 1993 and 1995; KEK, Tsukuba (Japan) in 2000; NCTS, Hsinchu and NTU, Taipei (Taiwan) in 2000–2001; University Ann-Arbor, Michigan (USA) in 2001; Max-Planck Institute, Munich (Germany) in 2003; *France–China cooperation:* Pekin University and Academy of Sciences, Beijing in 2010; *France–Brazil cooperation:* University Sao Paulo in 2011 and 2013.

◇ Besides different international conferences which he has attended (ICHEP, EPS, Hadron conferences,...), he has also briefly visited some other laboratories for delivering seminars: LBL, Berkeley, San Francisco (USA) in 1986; BNL, New York (USA) in 1988; Cracovia (Poland) in 1995, Varsovia (Poland) in 1996; Napoli, Milano, Padova, Bologna, Pisa (Italy) in the period of 1984 to 1996; Kias (Seoul) in 2000; Hiroshima, Kanazawa, Kyoto, Tokyo, Osaka and Sendai in 2000 (Japan); Beijing, Shanghai (China) in 2010.

4. Conference Organisations

♣ Besides his active research, Stephan Narison is the founder and chairman of the Series of QCD–Montpellier International Conference which he started in 1985 with the title "Non perturbative methods" and modified since 1990 as QCD international conference held every 2 years in Montpellier. The QCD–Montpellier conference involves about 100–120 participants and has become a classic where experts in this field including Nobel Prize winners Profs J. Steinberger (QCD97), D. Gross (QCD98) and G. 't Hooft (QCD98 and QCD02) have participated. The novelty of these meetings is the balance between experimental and theorists participants and the fact that many PhD students and post-docs present, for the first time, their international contributions in the presence of experts in the field. The QCD, Montpellier conference was among the first Euroconferences (1994–2000) supported by the European commission in Brussels and is regularly supported by the CNRS–IN2P3, the University of Montpellier 2, the Region of Languedoc-Roussillon, the Mayor of Montpellier and in 2014 and 2015 by the OCEVU-LABEX of Marseille. The QCD conference proceedings are published by Elsevier in *Nuclear Physics Proceedings Supplement.*

◇ He started, in 2001, the Series of High-Energy Physics (HEP) international conference in Madagascar (HEPMAD) in alternance with the QCD, Montpellier conference for promoting high-energy physics there. While having the same spirit as that of the QCD, Montpellier conference, the subject of the HEPMAD conference is wider and the speakers are asked to give more pedagogical introductory talks. This conference is published online by SLAC, Stanford in the eCONF proceedings site: *www.slac.stanford.edu/econf/*

5. Promoting HEP in Madagascar

♣ Since the organization of the HEPMAD conference in 2001, Stephan is involved in promoting HEP in Madagascar. In 2004, he founded the iHEP-MAD research institute at the University of Antananarivo where Master degrees and PhD students are admitted. Most of them attend, each year, the CERN, Geneva, ICTP, Trieste and African summer schools. Some of them get fellowships to continue their studies abroad. To create more impact about this action among the community, he delivers seminars for large public, institutions, high school and university students in different cities of Madagascar.

◇ He has founded respectively in 1993 and 2009 the Association Culturelle Malgache of Montpellier (ACMM) and Association Gasy–Miara–Mandroso (AGMM) which serve as a platform in the realization of these different actions for developing science in this country.[2]

♡ This book is, in some sense, the written version of these different seminars and lectures.

6. Appartenances and Honours

- 2013: Award by the University of Sao Paulo as the co-supervisor of Dr R.M. de Albuquerque best thesis in science for the University.
- 2012: Associated Member of the Malagasy Academy of Sciences.
- 2012: "Grand Officier" of the Malagasy National Order.
- 2010: President of the "Association Gasy–Miara Mandroso" (AGMM).
- 2006: "Commandeur" of the Malagasy National Order.
- 2000: Outstanding Man of the 20th century by the American Biographical Institute (ABI).
- 2000: Japanese Promotion of Science, senior fellow.
- 1998: Outstanding People of the 20th century by the International Biographical Center (Cambridge).
- 1998: Included in the Marquise Who's Who in Science and Engineering.
- 1996: Correspondant Member of the Malagasy Academy of Sciences.
- 1996: Included in the International Directory of Distinguished Leadership by ABI.
- 1995: Invited member of the New York Academy of Science.
- 1993: Invited member of the American Association for the Advancement of Science.
- 1993: President of the "Association Culturelle Malgache" of Montpellier.
- 1992: Alexander Von Humboldt senior fellow.
- 1983: Member of the French and European Physical Societies.

---

[2]If you plan to help us in this long-term project for developing science in Madagascar, your help and/or donations are welcome. The association site is: *http://www.lupm.univ-montp2.fr/users/qcd/agmm.html/*

7. Scientific Publications and Selected Works

♣ At the time of writing (July 2015), according to the SLAC-inspire site:
inspirehep.net/search?p=FIND+AUTHOR+NARISON
Stephan has 238 publications where 187 of them are published or/and in arXiv and have mainly contributed to the development of hadron physics within the QCD Spectral Sum Rules (QSSR) approach introduced by M.A. Shifman, A.I. Vainshtein and V.I. Zakharov (SVZ) [18] for studying the hadron properties in terms of the QCD fundamental parameters.
◇ His different works prior to 2001 are reviewed in textbooks [33, 34] and in review articles [35].
♡ His most recognized results are compiled in the Particle Data Group (PDG) booklet [5].
♠ Among his different contributions, one can mention the following:

- Estimate of the hadronic contributions to the muon anomalous magnetic moment (content of his "3ème cycle" thesis [36]).
- Introduction of the renormalization group invariant and running quark masses in QCD [40].
- Estimate of the running light quark masses [43] which has been improved recently in [44].
- Definition of the heavy quark running and pole masses in the dimensional $\overline{MS}$ renormalization scheme [45].
- Estimate of the charm and bottom quark running masses and of the gluon condensates [46].

- SU(3) breaking of the light quark condensate [47, 48].
- Estimate of the 4-quark condensates and of its deviation from the vacuum saturation estimate [42, 51].
- Estimate of the $B^0 - \bar{B}^0$ matrix elements [52].
- Estimate of the open charm and bottom decay constants [53, 54], [55, 56].
- Analysis of the glueball masses and widths [57].
- Glueball test on the nature of the scalar meson $\sigma(500)$ using $\pi\pi$ and $\gamma\gamma$ scatterings data [58].
- Study of the spectra of light hybrid mesons [59, 60]
- Analysis of the spectra of charmonium and bottomium 4-quark and molecule states [61, 62].

- Extraction of the QCD coupling α_s from inclusive hadronic tau decays which leads to a very precise determination of this coupling when evaluated at the Z-boson mass [37, 38] and demonstrates its $1/\log Q$-behavior versus the energy Q as predicted by asymptotic freedom.
- Introduction of the tachyonic gluon mass for parametrizing the uncalculated higher order terms of the perturbative QCD series and their equivalence [41, 49] and proof of its existence using some AdS/QCD models [63].